# Geophysics and Environmental Physics

The series *Geophysics and Environmental Physics* (GEP), reports new developments in geophysics research and teaching with a high quality and the explicit aim to summarize and communicate current knowledge in an accessible way. Books published in this series are conceived as bridging material between graduate textbooks and the forefront of research.

The research fields covered by the series are:

- Atmospheric Physics, Environmental Physics, Biogeophysics and Physics of Climate;
- Earth Magnetism and Rock Physics, Seismology and Physics of Volcanoes;
- Geodynamics and Tectonics;
- Physical Oceanography and Hydrology;
- Surface Processes;
- Geodesy;
- Instrumentation and Data Systems;
- Nonlinear Processes in Geophysics;
- Planetary and Solar System Physics.

Proposal should be sent to a member of the Editorial Board, or directly to the Responsible Editor at Springer:

Dr Lisa Scalone - lisa.scalone@springernature.com

Vincenzo Levizzani • Christopher Kidd

# Precipitation

## From Cloud Physics to Satellite Observations

Springer

Vincenzo Levizzani
Castelvetro di Modena, Italy

Christopher Kidd
Earth Syst. Sci. Interdiscipl. Center
University of Maryland, College Park
College Park, USA

ISSN 2948-2186 ISSN 2948-2194 (electronic)
Geophysics and Environmental Physics
ISBN 978-3-031-97095-5 ISBN 978-3-031-97096-2 (eBook)
https://doi.org/10.1007/978-3-031-97096-2



This Springer imprint is published by the registered company Springer Nature Switzerland AG
The registered company address is: Gewerbestrasse 11, 6330 Cham, Switzerland

*Dedicated to Angela and my family whose love makes everything possible, including this book.*
*Vincenzo*
*Dedicated to Judi, Sarah, and Rachel with whom all is complete.*
*Chris*

# Preface

The Earth system water balance general equation is roughly expressed as

$$P = Q + ET + \Delta S$$

where $P$ is the precipitation, $Q$ the streamflow, $ET$ the evapotranspiration, and $\Delta S$ the change in storage. Such an equation and its modified versions are used to describe the flow of water in and out of a system, by balancing the water cycle, model the water balance in basins, analyse water resources, and forecast floods or droughts. Statistical models are then used by hydrologists to describe relationships between variables.

Precipitation is an essential component in the form of rain, snow, and hail while describing the Earth's water cycle and the movement of water on land. It is the downward section of the water movement on our planet, the one that brings water back from the atmosphere over land and ocean.

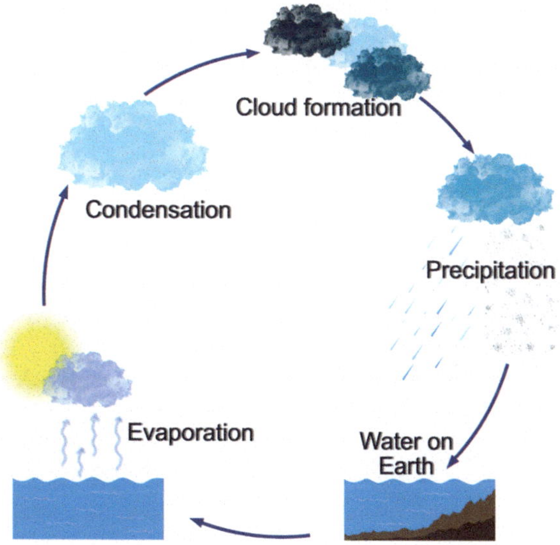


Precipitation affects the water cycle in many ways:

- Infiltration: Precipitation that soaks into the soil.
- Runoff: Precipitation that flows on the Earth's surface towards streams.
- Evaporation: Water that changes from liquid to gas and returns to the atmosphere.
- Evapotranspiration: Water that is pulled out of the soil and released into the atmosphere by plants.

It is thus of paramount importance to understand precipitation formation mechanisms and the way that water condenses in the clouds. Without a clear understanding of such mechanisms, it would be impossible to make weather forecasts and explore the influence of global change on the water cycle.

After a lifetime spent on studying cloud physics, precipitation formation, and satellite meteorology, we decided it was time to write a book that contains the state of the art of precipitation science and make it available to students and scholars who need to approach precipitation data for a variety of uses. We start with a short look at the clouds in the Earth system. The cloud observation technologies are described in Chap. 2 to make the reader aware that cloud observations are most of the time indirect in nature and thus need attention when using data. Chapter 3 contains a somewhat in-depth outline of the necessary cloud physics notions that lay behind precipitation formation. Precipitation systems are the subject of Chap. 4 where many different weather structures are presented as well as their changes in a warming climate. The observations of precipitation examined in Chap. 5 conclude the book, setting the foundations for the use of precipitation datasets.

Precipitation science is in continuous evolution, and we realise that the book content will soon be outdated, but the basic principles contained in the following pages lay the foundations to understand how precipitation falls from clouds. We hope this will be useful to our readers.

Castelvetro di Modena, Italy and Annapolis, MD, USA
15 April, 2025

Vincenzo Levizzani
Chris Kidd

# Acknowledgements

This book was made possible by a nice discussion we had with our Editor Dr. Lisa Scalone who took the time to ask us whether we could write a book on what we know in science. Springer Nature has supported us for several books and they deserve gratitude for their commitment to produce fine scientific literature and for the quality of their publishing.

VL wishes to acknowledge its former institution, the Italian National Council of Research (CNR) for supporting me throughout my long career, which was concluded on 1 July 2024. My former research group in Bologna gave me an invaluable support.

CK would like to acknowledge the many that I have had the good fortune to know along my career path at the Universities of Nottingham, Bristol, and Birmingham, together with my current colleagues at the University of Maryland and NASA's Goddard Space Flight Center.

We are in debt to the International Precipitation Working Group (IPWG, https://www.eorc.jaxa.jp/IPWG/) for providing a fertile environment for discussion and collaboration among all scientists around the world. IPWG is composed of fine scientists and individuals committed to working together and disseminating knowledge on precipitation.

Our families always support us and patiently bear the burden of our relentless work in science. Angela and Judi deserve more than a few words for all they do and for encouraging us to do what we like most in life.

**Competing Interests** The authors have no competing interests to declare that are relevant to the content of this manuscript.

# Contents

# Abbreviations and Acronyms

| | |
|---|---|
| ABI | Advanced Baseline Imager |
| AE | Auto Estimator |
| AGRI | Advanced Geostationary Radiation Imager |
| AHI | Advanced Himawari Imager |
| AI | Artificial intelligence |
| AIP | Algorithm Intercomparison Programme |
| AMeDAS | Radar-Automated Meteorological Data Acquisition System |
| AMS | American Meteorological Society |
| AMSR | Advanced Microwave Scanning Radiometer |
| AMSU | Advanced Microwave Sounding Unit |
| AMW | Active microwave |
| ANN | Artificial neural networks |
| AOD | Aerosol optical depth |
| APHRODITE | Asian Precipitation—Highly Resolved Observational Data Integration Towards Evaluation of Water Resources |
| ATMS | Advanced Technology Microwave Sounder |
| ATN | Advanced TIROS-N |
| AVHRR | Advanced Very-High-Resolution Radiometer |
| BMP | Bulk microphysics parameterisation |
| BR | Bulk Richardson number |
| CALIPSO | Cloud-Aerosol Lidar and Infrared Pathfinder Satellite Observation |
| CAPE | Convective available potential energy |
| CAPSAT | Clouds-Aerosols-Precipitation Satellite Analysis Tool |
| CARE | Centre for Atmospheric Research Experiments |
| CAS | Cloud Appreciation Society |
| CATS | Cloud-Aerosol Transport System |
| CA | Cloud to air lightning |
| CCN | Cloud condensation nucleus |
| CCS | Cloud classification system |
| CDA | Command and Data Acquisition station |
| CDR | Climate data record |
| CEDA | Centre for Environmental Data Analysis |
| CERES | Clouds and the Earth's Radiant Energy Sensor |
| CG | Cloud to ground lightning |

| | |
|---|---|
| CHRS | Center for Hydrometeorology and Remote Sensing |
| CIP | Cloud Imaging Probe |
| CLASS | Comprehensive Large Array-data Stewardship System |
| CMA | China Meteorological Agency |
| CMAP | Climate Prediction Center Merged Analysis of Precipitation |
| CMORPH | Climate Prediction Center Morphing Technique |
| CM SAF | EUMETSAT Climate Monitoring Satellite Application Facility |
| CNES | Centre National d'Études Spatiales |
| CNN | Convolutional neural network |
| CoCoRaHS | Community Collaborative Rain, Hail and Snow Network |
| COADS | Comprehensive Ocean–Atmosphere Data Set |
| CPC | Condensation particle counter |
| CPC | Climate Prediction Center |
| CPI | Cloud Particle Imager |
| CPL | Cloud Physics Lidar |
| CRM | Cloud resolving model |
| CSPS | Convective/stratiform precipitation separation |
| CWP | Cloud water path |
| DLR | Deutsches Zentrum für Luft- und Raumfahrt |
| DMSP | Defense Meteorological Satellite Program |
| DNS | Direct numerical simulation |
| DOW | Doppler on Wheels |
| DPC | Drop-to-particle conversion |
| DPR | Dual-frequency Precipitation Radar |
| DWR | Dual-wavelength ratio |
| EarthCARE | Earth Cloud Aerosol and Radiation Explorer |
| ECT | Equatorial crossing time |
| EIA | Earth incidence angle |
| ELVES | Emissions of light and Very low frequency perturbations due to electromagnetic pulse sources |
| ENSO | El Niño Southern Oscillation |
| EOL | Earth Observing Laboratory |
| EPS-SG | European Polar Satellite—Second Generation |
| ESA | European Space Agency |
| ESM | Earth system model |
| ESMR | Electrically Scanning Microwave Radiometer |
| eTRaP | Ensemble Tropical Rainfall Potential |
| EUMETSAT | European Organisation for the Exploitation of Meteorological Satellites |
| FAAM | Facility for Airborne Atmospheric Measurements |
| FEWS Net | Famine Early Warning System Network |
| FFMP | Flash Flood Monitoring and Prediction System |
| FP | Fine particle |
| FY-2 | Feng-Yun-2 |
| GEWEX | Global Energy and Water Exchanges |

| | |
|---|---|
| GDAS | Global Data Assimilation System |
| GEO | Geosynchronous Earth Orbit satellite |
| GEO | Group on Earth Observations |
| GeoXO | Geostationary Extended Observations satellite |
| GES DISC | Goddard Earth Sciences Data and Information Services Center |
| GFMS | Global Flood Monitoring System |
| GFS | Global Forecast System |
| GFWED | Global Fire Weather Database |
| GHCN | Global Historical Climatology Network |
| GHOST | Greenish optical emission from sprite tops |
| GISS | Goddard Institute for Space Studies |
| GMAO | Global Modeling and Assimilation Office |
| GMI | GPM Microwave Imager |
| GMS | Geostationary Meteorological Satellite |
| GMSRA | GOES Multispectral Rainfall Algorithm |
| GOES | Geostationary Operational Environmental Satellite |
| GPC | Gas-to-particle conversion |
| GPCC | Global Precipitation Climatology Centre |
| GPCP | Global Precipitation Climatology Project |
| GPI | Global Precipitation Index |
| GPM | Global Precipitation Measurement mission |
| GPROF | Goddard PROFiling algorithm |
| GPS | Global Positioning System |
| GridSat | Gridded Satellite |
| GSFC | Goddard Space Flight Center |
| GSMaP | Global Satellite Mapping of Precipitation |
| HDF | Hierarchical Data Format |
| HALO | High Altitude and long Range Aircraft |
| HIAPER | High-performance Instrumented Airborne Platform for Environmental Research |
| HOAPS | Hamburg Ocean Atmosphere Parameters and Fluxes from Satellite Data |
| HPC | High-performance computer |
| H-SAF | EUMETSAT Satellite Application Facility in Support to Operational Hydrology and Water Management |
| HSRL | High Spectral Resolution Lidar |
| IAMAS | International Association of Meteorology and Atmospheric Sciences |
| IC | Intracloud lightning |
| ICAO | International Civil Aviation Organization |
| ICHARM | International Centre for Water Hazard and Risk Management |
| IFAS | Integrated Flood Analysis System |
| IMERG | Integrated Multi-satellitE Retrievals for GPM |
| IN | Ice nucleus |
| INA | Ice nucleation active protein |

| | |
|---|---|
| INCUS | INvestigation of Convective UpdraftS |
| INSAT | Indian National Satellite |
| IPWG | International Precipitation Working Group |
| ISCCP | International Satellite Cloud Climatology Project |
| ISS | International Space Station |
| ITCZ | Intertropical Convergence Zone |
| IWC | Ice water content |
| JAXA | Japan Aerospace Exploration Agency |
| JPSS | Joint Polar Satellite System |
| LDAS | Land Data Assimilation System |
| LDR | Linear depolarisation ratio |
| LEO | Low Earth Orbit satellite |
| LES | Large eddy simulation |
| LHASA | Landslide Hazard Assessment for Situational Awareness |
| LIDAR | Light detection and ranging |
| LIS | Lightning Imaging Sensor |
| LPVEx | Light Precipitation Validation Experiment |
| LWC | Liquid water content |
| LWP | Liquid water path |
| MAPLE | McGill Algorithm for Precipitation forecasting using Lagrangian Extrapolation |
| MASC | Multi-Angle Snowflake Camera |
| MCC | Mesoscale convective complex |
| MCS | Mesoscale convective system |
| MEaSUREs | Making Earth System Data Records for Use in Research Environments |
| METAR | Meteorological Aerodrome Report |
| MetOp | Meteorological Operational satellite |
| MHS | Microwave Humidity Sounder |
| MICRA | Microwave-Infrared Combined Rainfall Algorithm |
| MiCRO | Micro insurance Catastrophe Risk Organisation |
| MIRA | Microwave-adjusted IR Algorithm |
| MISR | Multi-angle Imaging Spectroradiometer |
| MIT | Massachusetts Institute of Technology |
| ML | Machine learning |
| MODIS | Moderate-Resolution Imaging Spectroradiometer |
| MRMS | Multi-Radar/Multi-sensor System |
| MRR | Micro rain radar |
| MSG | Meteosat Second Generation |
| MSU | Microwave Sounding Unit |
| MSWEP | Multi-Source Weighted-Ensemble Precipitation |
| MTG | Meteosat Third Generation |
| MT-SAT | Multi-Functional Transport Satellite |
| MWHS | Microwave Humidity Sounder |
| MWI | MicroWave Imager |

| | |
|---|---|
| MWS | MicroWave Sounder |
| NASA | National Aeronautics and Space Administration |
| NCAR | National Center for Atmospheric Research |
| NCEP | National Centers for Environmental Prediction |
| NCFR | Narrow cold-frontal rainband |
| NERC | Natural Environmental Research Council |
| NESDIS | National Environmental Satellite, Data, and Information Service |
| netCDF | Network Common Data Form |
| NEXRAD | Next-Generation Radar |
| NGO | Non-governmental organisation |
| NICT | National Institute of Information and Communications Technology |
| NN | Neural networks |
| NOAA | National Oceanic and Atmospheric Administration |
| NPF | New particle formation |
| NSF | National Science Foundation |
| NSSL | National Severe Storms Laboratory |
| NWS | National Weather Service |
| OPERA | Operational Programme for the Exchange of Weather Radar Information |
| OTD | Optical Transient Detector |
| PACRAIN | Comprehensive Pacific Rainfall Database |
| PANI | Provision of Advisory for Necessary Irrigation |
| PCF | Pore condensation and freezing |
| PDIR-now | PERSIANN Dynamic Infrared Rain Rate near real-time |
| PERSIANN | Precipitation Estimation from Remotely Sensed Information using Artificial Neural Networks |
| PHIVOLCS | Philippine Institute of Volcanology and Seismology |
| PIA | Path integrated attenuation |
| PIP | Precipitation Intercomparison Projects |
| PMW | Passive microwave |
| PPS | Precipitation Processing System |
| PR | Precipitation Radar |
| PSC | Polar stratospheric cloud |
| QLL | Quasi-liquid layer |
| RADAR | Radio detection and ranging |
| RainCube | Radar in a CubeSat |
| RF | Random forest |
| RFI | Radio frequency interference |
| SAP | Secondary aerosol particle |
| SAPHIR | Sondeur Atmospherique du Profil d'Humidité Intertropicale par Radiométrie |
| SBM | Spectral bin microphysics |
| SCaMPR | Self-Calibrating Multivariate Precipitation Retrieval |
| SDG | Sustainable development goals |
| SDM | Super droplet method |

| | |
|---|---|
| SDSMT | South Dakota School of Mines and Technology |
| SeaWiFS | Sea-viewing Wide Field-of-view Sensor |
| SEVIRI | Spinning Enhanced Visible Infrared Imager |
| SIP | Secondary ice particle |
| SMAP | Soil Moisture Active Passive mission |
| SMMR | Scanning Multichannel Microwave Radiometer |
| SOC | Self-organised criticality |
| SOFM | Self-organising feature map |
| SPICE | Solid Precipitation Intercomparison Experiment |
| SSM/I | Special Sensor Microwave/Imager |
| SSMIS | Special Sensor Microwave/Imager Sounder |
| SSM/T | Special Sensor Microwave/Temperature |
| STAR | Center for Satellite Applications and Research |
| STEPS | Short-Term Ensemble System |
| SWOT | Surface Water and Ocean Topography mission |
| TAMSAT | Tropical Applications of Meteorology using SATellite data |
| TBR | Tipping bucket raingauge |
| TEMPEST | Temporal Experiment for Storms and Tropical Systems |
| TGF | Terrestrial gamma-ray flash |
| TIROS | Television Infrared Observation Satellite |
| TLE | Transient luminous event |
| TMI | TRMM Microwave Imager |
| TMPA | TRMM Multi-Satellite Precipitation Algorithm |
| TRMM | Tropical Rainfall Measuring Mission |
| TROLLs | Transient red optical luminous lineaments |
| TROPICS | Time-Resolved Observations of Precipitation structure and storm Intensity with a Constellation of Smallsats |
| UAV | Unmanned aerial vehicle |
| UFP | Ultrafine particle |
| USGS | United States Geological Survey |
| UW CEE | University of Washington Civil and Environmental Engineering |
| VCS | Vidicon Camera System |
| VHRR | Very High Resolution Radiometer |
| VIRS | Visible and InfraRed Scanner |
| WBF | Wegener-Bergeron-Findeisen process |
| WCRP | World Climate Research Program |
| WFF | Wallops Flight Facility |
| WIVERN | Wind Velocity Radar Nephoscope |
| WMO | World Meteorological Organization |
| WRSI | Water Requirement Satisfaction Index |

# List of Figures

# List of Tables

# 1 Clouds and Precipitation in the Earth System

Clouds are not just those very variable elements of our skies, changing from white fair-weather cumulus to stormy black cumulonimbus. We all have experience of clouds and maybe don't even pay attention to their visual appearance, not to mention their role in the weather and climate system of our planet. However, clouds play a fundamental role in the water cycle that sustains life on Earth. Without them we would not have water available, and we would disappear together with the rest of living species. It is with Leonardo da Vinci in the 1500 that all the pieces started to fall in place. In the Leicester Codex he stated that he wanted to write the *Treatise on water* (*Trattato dell'acqua*) that he eventually never completed. He had in mind to unveil the elemental structure of water, to illustrate the reasons behind natural disasters, to investigate turbulent motion of water (Fig. 1.1), to conceive solutions to prevent damages to dykes and dams, to underline analogies between water and air and movements of fishes and birds.

In his *Manuscript C*, Leonardo stated that "*Acqua è fra i quattro elementi il secondo men greve e di seconda volubilità. Questa non ha mai requie insino che si congiunge al suo marittimo elemento [...]. Volentieri si leva per lo caldo in sottile vapore per l'aria. Il freddo la congela, stabilità la corrompe*" ("Water among the four elements is the second for weight and the second for instability. It never rests until it joins its maritime element [...]. It gladly rises because of heat in a thin vapour in the air. The cold freezes it and stability corrupts it"). In summary, Leonardo had clearly in his mind that we cannot understand water if we do not concentrate on the Earth as a whole, especially in its liquid and aerial components. This is what will guide us in this book, although limiting our interest to clouds and precipitation.

The impacts of clouds and precipitation are many: at one extreme they help to moderate the climate through the transport of heat and energy but conversely can exacerbate our weather through the extremes of droughts and floods.



V. Levizzani, C. Kidd, *Precipitation*, Geophysics and Environmental Physics,
https://doi.org/10.1007/978-3-031-97096-2_1

**Fig. 1.1** Leonardo da Vinci, A deluge, c. 1517–1518. Drawing of an apocalyptic deluge in black chalk. (Public domain)

## 1.1 The Water Cycle

Before directing our attention to clouds and precipitation it is of capital importance to remind ourselves of a few notions on how water is present on Earth. This will help in placing clouds and precipitation at their right place. The Earth is the only planet we know of with plentiful usable water, and where the water is present in its liquid, solid and gaseous components at the pressure and temperature conditions that allow life to be sustained.

The total of Earth's water is about $1.39 \times 10^9$ km$^3$, with the bulk of it, about 96.5%, confined to the oceans. Approximately 1.7% is stored in the polar ice caps, glaciers, and permanent snow, and another 1.7% is stored in groundwater, lakes, rivers, streams, and soil. Only a thousandth of 1% of the water on Earth exists as water vapour in the atmosphere. Details are provided by Gleick (1996) and Abbott et al. (2019), among others. It is relevant to note the residency time of water in the various components of the cycle: oceans (4000 years), ice caps/glaciers/permanent snow (10–10,000 years), groundwater (10 days to 10,000 years), soil moisture (14 days–1 year), lakes (10 years), atmosphere (10 days), rivers (14 days), and biological water (7 days).

Despite the small amount, water vapour has a huge influence on the planet. Water vapour is a greenhouse gas contributing about 60% of the Earth's greenhouse effect, but which is vital to life on the planet. It is a major driver of the Earth's weather and climate as it travels around the globe, transporting latent heat, which is obtained by water molecules as they transition from liquid or solid to vapour. Such heat is released when the molecules condense from vapour back to liquid or solid, generating cloud droplets and various forms of precipitation. While the oceans of Earth provide the bulk of the heat (and therefore energy) transfer from the low to high latitudes, they do so over long-time scales. However, as noted above, water has a much shorter residency time in the atmosphere: water evaporated from the surface is transported large distances as water vapour before condensing as clouds and precipitation over a matter of days, thus affecting short-time weather.

All this is integral part of the water cycle (see Fig. 1.2), which is intimately linked with energy exchanges between atmosphere, ocean, and land that determine the Earth's climate and cause much of natural climate variability. The impacts of climate change and variability on the quality of life occur primarily through changes in the water cycle. We may sustain a warmer climate, but certainly we will be heavily affected by lack of water or vice versa the excess of it.

The ocean holds 97% of the total water on the planet; 78% of global precipitation occurs over the ocean, and the ocean is the source of 86% of global evaporation (Fig. 1.3). Besides affecting the amount of atmospheric water vapour and hence rainfall, evaporation from the sea surface is important in the movement of heat. Water evaporates from the surface of the ocean, mostly in warm, cloud-free subtropical seas. Water vapour then condenses as clouds and falls as rain, mostly in the Intertropical Convergence Zone (ITCZ). Condensing water vapour releases latent heat and this drives much of the atmospheric circulation in the tropics. This latent heat release is an important part of the Earth's heat balance.

The physical components of the global water cycle over land and oceans (Trenberth et al. 2007; Trenberth and Fasullo 2013; Launiainen et al. 2014; Rodell et al. 2015) include the evaporation from the ocean and land surfaces, the transport of water vapour by the atmosphere, precipitation onto the ocean and land surfaces, the net atmospheric transport of water from land areas to ocean, and the return flow of fresh water from the land back into the ocean (Fig. 1.4). Precipitation is an irreplaceable step mediated by cloud formation and evolution. Changes in precipitation regimes have a strong influence on all other components of the water cycle and ultimately on life on Earth. Fluxes such as currently estimated (e.g., Douville et al. 2021) provide a general idea of the amounts of water being mobilised, but they are not conclusive since several pieces of the chain are still subject to large uncertainties (e.g., the snowfall component). For this reason, maintaining the observing network and expanding it is very important to meet the urgent need to improve modelling of atmospheric, land surface and human processes and how these components are coupled (Lambert et al. 2025). One more important question regards the discrimination between precipitation trends and the internal variability of the systems/basins as happens in the Mediterranean (e.g., Vicente-Serrano et al. 2025).

Water in the atmosphere not only acts as an agent to transfer large quantities of energy through evaporation/condensation, but the presence of atmospheric moisture directly affects the energy balance. As noted above, water vapour is a potent greenhouse gas, but an essential one. It is responsible to ensuring that the Earth temperature remains tolerable to the wide variety of life on the planet. While it is a greenhouse gas, it is not directly responsible for any global warming, although does respond to warmer climates. Clouds are crucial in the Earth's energy budget: during daylight they strongly reflect incoming solar radiation, although their overall effect is dependent upon many factors such as cloud top height and cloud thickness. At nighttime, clouds act as a thermal "blanket", trapping outgoing longwave (thermal) energy and thus moderating the diurnal temperature fluctuations.

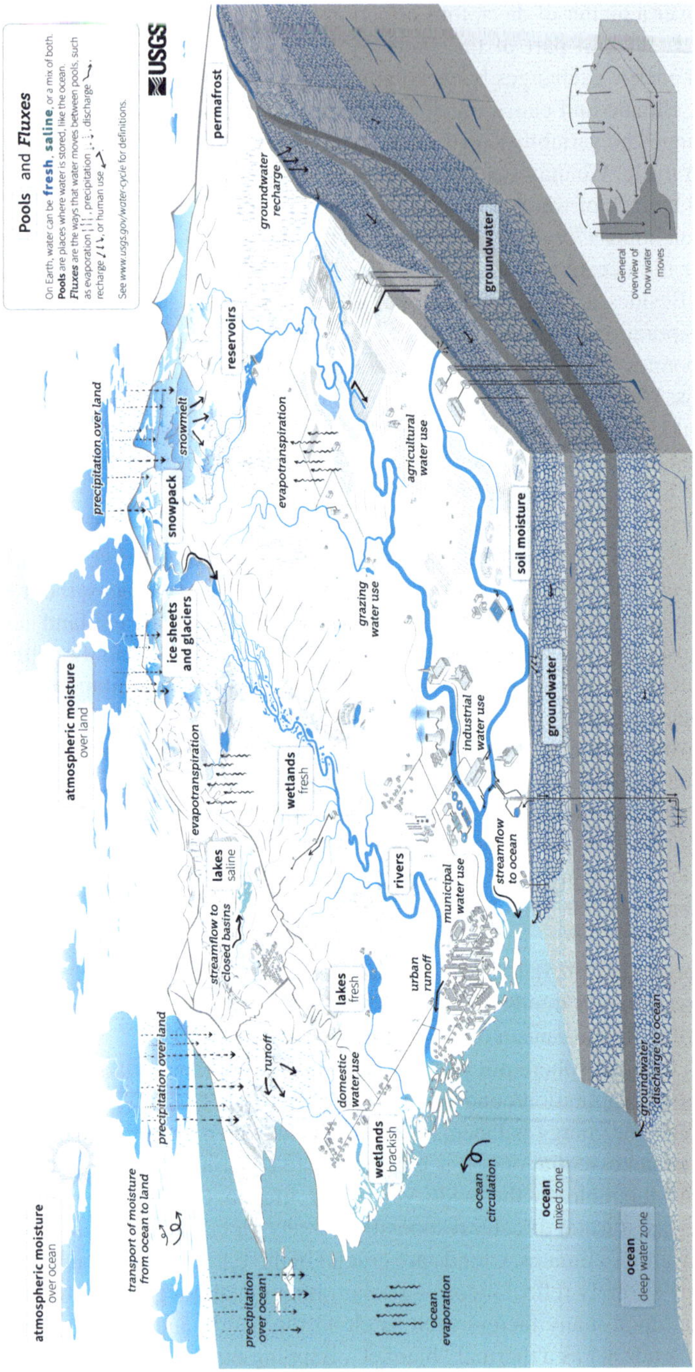


**Fig. 1.2** The Earth's water cycle with its major components. (Original from USGS, public domain)

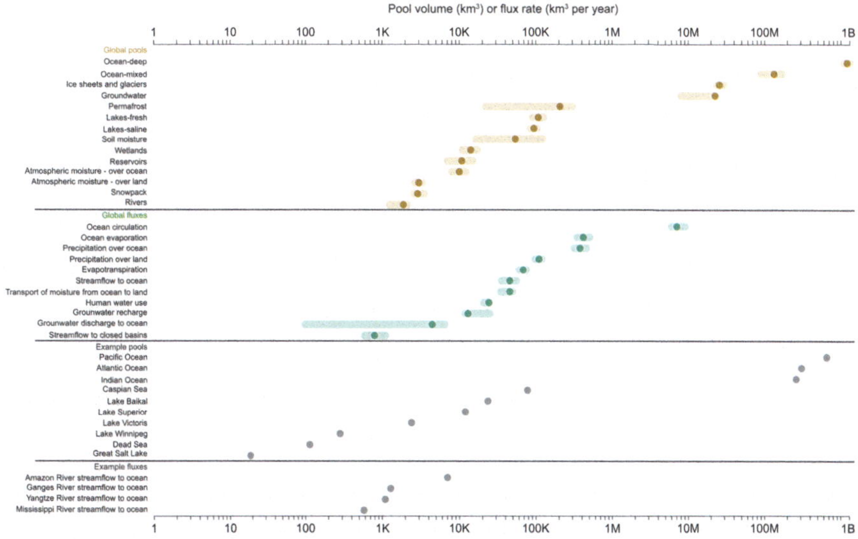


**Fig. 1.3** Size of global pools and fluxes of water with examples of specific pools and fluxes for context. (Original from USGS, public domain. Data from Abbott et al. 2019)

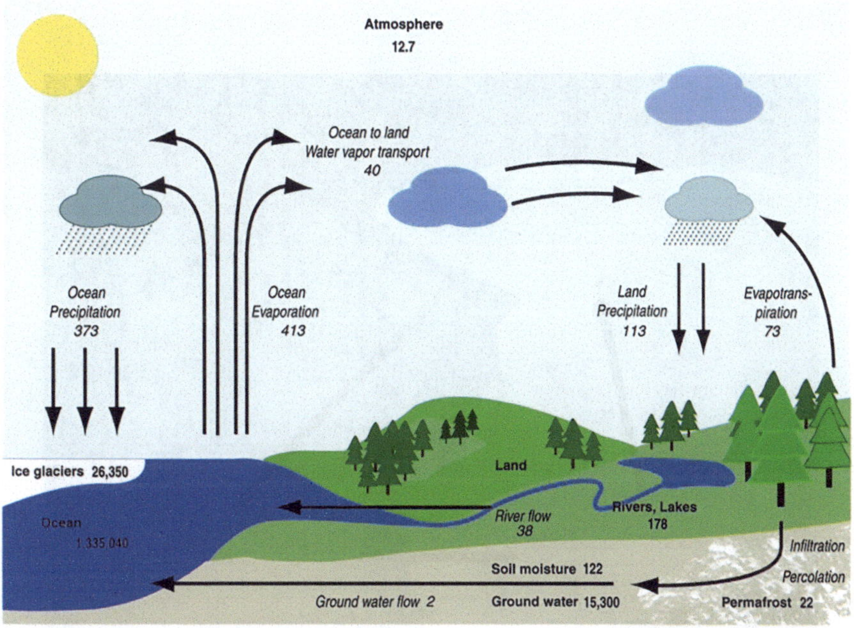


**Fig. 1.4** Schematic of the global hydrologic cycle. The reservoirs (in $10^3$ km$^3$) are shown in **bold** and flows ($10^3$ km$^3$ year$^{-1}$) in *italic*. (Launiainen et al. 2014. © The Authors under CC-BY-4.0 license)

## 1.2 The Cloud Feedback Problem

The most important roles of clouds in climate are to modulate the Earth's basic radiation balance and to produce precipitation. Clouds produce precipitation from water vapour, releasing heat to the atmosphere in the process. Thus, any changes in clouds modifies the radiative energy balance and water exchanges that determine the climate. The trouble is that clouds are produced by the climate, specifically the atmospheric motions (winds) that are produced by the radiative and latent heating influenced by clouds (Stephens 2005). This interconnected loop of relations is called a feedback loop (Fig. 1.5). The ways that clouds respond to changes in the climate are so complex that it is hard to determine their net effect on the energy and water balances and to determine how much climate might change.

Climate sensitivity (the response to a doubling in carbon dioxide in the atmosphere) is highly uncertain, mainly because it remains unclear how clouds will change with warming, either strongly amplifying or dampen global warming, providing climate feedback. Significant uncertainties remain in microphysical mechanisms for cloud feedback, including cloud phase changes, precipitation processes and even aerosol distributions (Gettelman and Sherwood 2016). Observational analyses (e.g., Ceppi and Nowack 2021) suggest that cloud feedback will amplify global warming, making it very unlikely that climate sensitivity is smaller than 2 °C. At the same time, the negative cloud feedback from the change from mixed

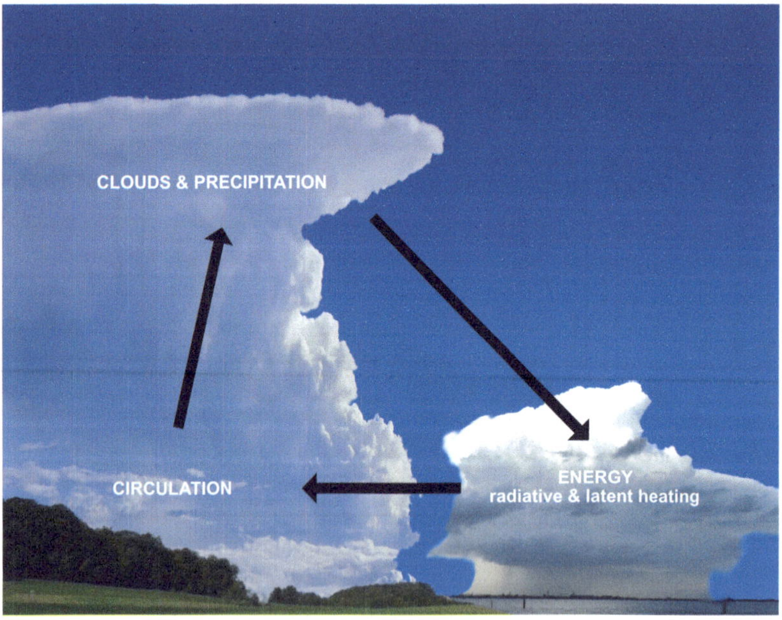


**Fig. 1.5** A schematic depiction of the main elements of the cloud feedback problem

phase to warm clouds because of warming should not be underestimated (Mülmenstädt et al. 2021).

The problem is very difficult in itself. One strategy that has been suggested is the decomposition of the cloud feedback into components due to changes in several gross cloud properties linking to individual processes (Zelinka et al. 2016). Negative feedback mechanisms are rarely investigated. For example, shallow clouds, which contain both ice and supercooled water, respond to warming by transitioning to a state with more liquid water and a greater albedo, resulting in a negative "cloud-phase" climate feedback component. Murray et al. (2021) argue that the magnitude of the negative cloud-phase feedback component depends on the amount and nature of the small fraction of aerosol particles that can nucleate ice crystals.

The problems encountered in diagnosing cloud feedback cannot be solved from observations alone nor proved from simple theoretical argument alone (Stephens 2005). A substantial progress requires a carefully thought and systematic combination of model and observations. As pointed out by Rossow (2024), recent results indicate that the cloud-radiative feedback amplifies the positive cloud-precipitation feedback on the atmospheric circulation from weather-to-annual time scales. Thus, we need to produce consistent long-term datasets allowing for a better understanding of the role of clouds in the Earth system. The International Satellite Cloud Climatology Project (ISCCP) cloud dataset (Schiffer and Rossow 1983; Rossow et al. 2022) has already significantly helped in providing a global picture of the influence of clouds on climate. However, times are mature to avail ourselves of newly conceived combined datasets that include ground and satellite observations together with reanalyses (e.g., Sun and Fu 2021).

## 1.3 Earth System Models

The main problem still is the cloud parameterisations embedded in the global models. This is not only because we haven't got the necessary modelling tools, but also because we do not know enough of certain precipitation forming clouds, their microphysics and internal dynamics, in a word: cloud processes. This is confirmed by several modelling exercises that demonstrate that the simulation of sub seasonal variability by a given model is significantly influenced by the parameterisation of sub grid physical processes (most likely cloud processes) (e.g., Leroux et al. 2016; Hofer et al. 2024). For example, bulk microphysical properties of ice clouds are key since they strongly impact the lifetime and radiative effects of these clouds (e.g., Stubenrauch et al. 2019).

The representation in models of the rapid changes in specific areas is another challenging problem. The Arctic is warming at about twice the rate of the rest of the planet and the representation of Arctic cloud processes in an Earth System Model (ESM) remains difficult (e.g., McIlhattan et al. 2020). Midlatitude clouds seem to contribute to the Arctic amplification via interactions with other climate feedbacks (Bonan et al. 2025). The assumption that low-level clouds over the frozen Antarctic Plateau consist mainly of ice crystals is revealing incorrect. Thus, more datasets are needed to investigate the impact of Antarctic mixed-phase clouds on climate (Lawson and Gettelman 2014; Radenz et al. 2024). Negative sensitivities to global warming appear across warmer regions and they may be attributed to confounding cloud radiative effects, which cool surfaces during rainfall: these effects need to be removed as indicated by Ghausi et al. (2024).

ESMs are now shifting paradigms into km-scale modelling (e.g., Gettelman et al. 2023). For this reason, the regional and global evaluation of ESMs is crucial (Eyring et al. 2019; Reale et al. 2020; Weigel et al. 2021; Stanford et al. 2023; Lee et al. 2024) and should also be focused on evaluating the components of the cloud feedback (e.g., Chao et al. 2024). At the same time, given the high complexity of ESMs codes and their long lifetime, a careful analysis of the software architecture of these models is necessary (e.g., Hasselbring et al. 2025).

One key aspect is now made possible and accelerates rapidly around the understanding of big data and the growing scientific interest in cloud processes and life cycles. For example, cloud tracking is important, and observations will need to focus more on this aspect (Freeman et al. 2024). More generally, some methods for integrating models and observations are very systematic and comprehensive (e.g., data assimilation), and some are single purpose and customised (e.g., for model validation and evaluation). Future developments will enable advanced heterogeneous observation networks and models to improve predictions of the Earth system (atmosphere, land surface, oceans, cryosphere, and chemistry) from weather to climate. There is a clear need to take a coordinated approach to models, observations, and their uncertainties to maximize the benefit for Earth system prediction and impacts on society (Gettelman et al. 2022).

All the above is strictly linked to the development of high-performance computers (HPC) that permit high resolution Earth system modelling (e.g., Zhang et al. 2023). These models on HPCs allow for capturing major weather-climate extremes in the atmosphere and ocean, permitting clouds and ocean sub mesoscale eddies in modelling tropical cyclones and eddy-mean flow interactions. These studies pave the way to further model development to resolve finer scales at even higher resolution and characterised by more realistic physics, not to mention the huge development towards non-hydrostatic cloud and ocean sub mesoscale resolved ESMs.

## References

Abbott, B.W., Bishop, K., Zarnetske, J.P., Minaudo, C., Chapin III, F.S., Krause, S., Hannah, D.M., Conner, L., Ellison, D., Godsey, S.E., Plont, S., Marçais, J., Kolbe, T., Huebner, A., Frei, R.J., Hampton, T., Gu, S., Buhman, M., Sayedi, S.S., Ursache, O., Chapin, M., Henderson, K.D., Pinay, G.: Human domination of the global water cycle absent from depictions and perceptions. Nat. Geosci. **12**, 533–540 (2019). https://doi.org/10.1038/s41561-019-0374-y

Bonan, D.B., Kay, J.E., Feldl, N., Zelinka, M.D.: Mid-latitude clouds contribute to Arctic amplification via interactions with other climate feedbacks. Environ. Res.: Clim. **4**(1), 015001 (2025). https://doi.org/10.1088/2752-5295/ada84b

Ceppi, P., Nowack, P.: Observational evidence that cloud feedback amplifies global warming. PNAS. **118**(30), e2026290118 (2021). https://doi.org/10.1073/pnas.2026290118

Chao, L.-W., Zelinka, M.D., Dessler, A.E.: Evaluating cloud feedback components in observations and their representation in climate models. J. Geophys. Res. Atmos. **129**(2), e2023JD039427 (2024). https://doi.org/10.1029/2023JD039427

Douville, H., Raghavan, K., Renwick, J., Allan, R.P., Arias, P.A., Barlow, M., Cerezo-Mota, R., Cherchi, A., Gan, T.Y., Gergis, J., Jiang, D., Khan, A., Pokam Mba, W., Rosenfeld, D., Tierney, J., Zolina, O.: Water cycle changes. In: Masson-Delmotte, V., Zhai, P., Pirani, A., Connors, S.L., Péan, C., Berger, S., Caud, N., Chen, Y., Goldfarb, L., Gomis, M.I., Huang, M., Leitzell, K., Lonnoy, E., Matthews, J.B.R., Maycock, T.K., Waterfield, T., Yelekçi, O., Yu, R., Zhou, B. (eds.) Climate Change 2021: The Physical Science Basis. Contribution of Working Group I to the Sixth Assessment Report of the Intergovernmental Panel on Climate Change, pp. 1055–1210. Cambridge University Press (2021). https://doi.org/10.1017/9781009157896.010

Eyring, V., Cox, P.M., Flato, G.M., Gleckler, P.J., Abramowitz, G., Caldwell, P., Collins, W.D., Gier, B.K., Hall, A.D., Hoffman, F.M., Hurtt, G.C., Jahn, A., Jones, C.D., Klein, S.A., Krasting, J.P., Kwiatkowski, L., Lorenz, R., Maloney, E., Meehl, G.A., Pendergrass, A.G., Pincus, R., Ruane, A.C., Russell, J.L., Sanderson, B.M., Santer, B.D., Sherwood, S.C., Simpson, I.R., Stouffer, R.J., Williamson, M.S.: Taking climate model evaluation to the next level. Nat. Clim. Chang. **9**, 102–110 (2019). https://doi.org/10.1038/s41558-018-0355-y

Freeman, S.W., Brunner, K., Jones, W.K., Kukulies, J., Senf, F., Stier, P., van den Heever, S.C.: Advancing our understanding of cloud processes and their role in the Earth system through cloud object tracking. Bull. Am. Meteorol. Soc. **105**(1), E297–E299 (2024). https://doi.org/10.1175/BAMS-D-23-0204.1

Gettelman, A., Sherwood, S.C.: Processes responsible for cloud feedback. Curr. Clim. Chang. Rep. **2**, 179–189 (2016). https://doi.org/10.1007/s40641-016-0052-8

Gettelman, A., Geer, A.J., Forbes, R.M., Carmichael, G.R., Feingold, G., Posselt, D.J., Stephens, G.L., van den Heever, S.C., Varble, A.C., Zuidema, P.: The future of Earth system prediction: advances in model-data fusion. Sci. Adv. **8**(14) (2022). https://doi.org/10.1126/sciadv.abn3488

Gettelman, A., Fox-Kempe, B., Flato, G., Klocke, D., Stamer, D., Stevens, B., Vidale, P.L.: Kilometre-scale modelling of the Earth system: a new paradigm for climate prediction. WMO Bull. **72**(2) (2023) Available at https://wmo.int/media/magazine-article/kilometre-scale-modelling-of-earth-system-new-paradigm-climate-prediction. Last accessed 3 Apr 2025

Ghausi, S.A., Zehe, E., Ghosh, S., Tian, Y., Kleidon, A.: Thermodynamically inconsistent extreme precipitation sensitivities across continents driven by cloud-radiative effects. Nat. Commun. **15**, 10669 (2024). https://doi.org/10.1038/s41467-024-55143-8

Gleick, P.H.: Water resources. In: Schneider, S.H. (ed.) Encyclopedia of Climate and Weather, vol. 2, pp. 817–823. Oxford University Press (1996)

Hasselbring, W., Jung, R., Schnoor, H.: Software architecture evaluation of Earth system models. J. Softw. Eng. Appl. **18**(3), 113–138 (2025). https://doi.org/10.4236/jsea.2025.183008

Hofer, S., Hahn, L.C., Shaw, J.K., McGraw, Z.S., Bruno, O., Hellmuth, F., Pietschnig, M., Mostue, I.A., David, R.O., Carlsen, T., Storelvmo, T.: Realistic representation of mixed-phase clouds increases projected climate warming. Commun. Earth Environ. **5**, 390 (2024). https://doi.org/10.1038/s43247-024-01524-2

Lambert, F.H., Allan, R.P., Behrangi, A., Byrne, M.P., Ceppi, P., Chadwick, R., Durack, P.J., Fosser, G., Fowler, H.J., Greve, P., Lee, T., Mutton, H., O'Gorman, P.A., Osborne, J.M., Pendergrass, A.G., Reager, J.T., Stier, P., Swann, A.L.S., Todd, A., Vicente-Serrano, S.M., Stephens, G.L.: Changes in the regional water cycle and their impact on societies. WIREs: Clim. Change. **16**, e70005 (2025). https://doi.org/10.1002/wcc.70005

Launiainen, S., Futter, M.N., Ellison, D., Clarke, N., Finér, L., Högbom, L., Laurén, A., Ring, E.: Is the water footprint an appropriate tool for forestry and forest products: the fennoscandian case. Ambio. **43**, 244–256 (2014). https://doi.org/10.1007/s13280-013-0380-z

Lawson, R.P., Gettelman, A.: Impact of Antarctic mixed-phase clouds on climate. PNAS. **111**(51), 18,156–18,161 (2014). https://doi.org/10.1073/pnas.1418197111

Lee, J., Gleckler, P.J., Ahn, M.-S., Ordonez, A., Ullrich, P.A., Sperber, K.R., Taylor, K.E., Planton, Y.Y., Guilyardi, E., Durack, P., Bonfils, C., Zelinka, M.D., Chao, L.-W., Dong, B., Doutriaux, C., Zhang, C., Vo, T., Boutte, J., Wehner, M.F., Pendergrass, A.G., Kim, D., Xue,

Z., Wittenberg, A.T., Krasting, J.: Systematic and objective evaluation of Earth system models: PCMDI Metrics Package (PMP) version 3. Geosci. Model Dev. **17**, 3919–3948 (2024). https://doi.org/10.5194/gmd-17-3919-2024

Leroux, S., Bellon, G., Roehrig, R., Caian, M., Klingaman, N.P., Lafore, J.-P., Musat, I., Rio, C., Tyteca, S.: Inter-model comparison of subseasonal tropical variability in aquaplanet experiments: effect of a warm pool. J. Adv. Model. Earth Syst. **8**(4), 1526–1551 (2016). https://doi.org/10.1002/2016MS000683

McIlhattan, E.A., Kay, J.E., L'Ecuyer, T.S.: Arctic clouds and precipitation in the Community Earth System Model Version 2. J. Geophys. Res. Atmos. **125**(22), e2020JD032521 (2020). https://doi.org/10.1029/2020JD032521

Mülmenstädt, J., Salzmann, M., Kay, J.E., Zelinka, M.D., Ma, P.-L., Nam, C., Kretzschmar, J., Hörnig, S., Quaas, J.: An underestimated negative cloud feedback from cloud lifetime changes. Nat. Clim. Chang. **11**, 508–513 (2021). https://doi.org/10.1038/s41558-021-01038-1

Murray, B.J., Carslaw, K.S., Field, P.R.: Opinion: cloud-phase climate feedback and the importance of ice-nucleating particles. Atmos. Chem. Phys. **21**, 665–679 (2021). https://doi.org/10.5194/acp-21-665-2021

Radenz, M., Engelmann, R., Henning, S., Schmithüsen, H., Baars, H., Frey, M.M., Weller, R., Bühl, J., Jimenez, C., Roschke, J., Muser, L.O., Wullenweber, N., Zeppenfeld, S., Griesche, H., Wandinger, U., Seifert, P.: Ground-based remote sensing of aerosol, clouds, dynamics, and precipitation in Antarctica: first results from the 1-year COALA campaign at Neumayer Station III in 2023. Bull. Am. Meteorol. Soc. **105**(8), E1438–E1457 (2024). https://doi.org/10.1175/BAMS-D-22-0285.1

Reale, R., Giorgi, F., Solidoro, C., Di Biagio, V., Di Sante, F., Mariotti, L., Farneti, R., Sannino, G.: The regional Earth system model RegCM-ES: evaluation of the Mediterranean climate and marine biogeochemistry. J. Adv. Model. Earth Syst. **12**(9), e2019MS001812 (2020). https://doi.org/10.1029/2019MS001812

Rodell, M., Beaudoing, H.K., L'Ecuyer, T.S., Olson, W.S., Famiglietti, J.S., Houser, P.R., Adler, R., Bosilovich, M.G., Clayson, C.A., Chambers, D., Clark, E., Fetzer, E.J., Gao, X., Gu, G., Hilburn, K., Huffman, G.J., Lettenmaier, D.P., Liu, W.T., Robertson, F.R., Schlosser, C.A., Sheffield, J., Wood, E.F.: The observed state of the water cycle in the early twenty first century. J. Clim. **28**(21), 8289–8318 (2015). https://doi.org/10.1175/JCLI-D-14-00555.1

Rossow, W.B.: Evolution of the concept of cloud-climate feedbacks. J. Eur. Meteorol. Soc. **1**, 100004 (2024). https://doi.org/10.1016/j.jemets.2024.100004

Rossow, W.B., Knapp, K.R., Young, A.H.: International satellite cloud climatology project: extending the record. J. Clim. **35**(1), 141–158 (2022). https://doi.org/10.1175/JCLI-D-21-0157.1

Schiffer, R.A., Rossow, W.B.: The International Satellite Cloud Climatology Project (ISCCP): the first project of the World Climate Research Programme. Bull. Am. Meteorol. Soc. **64**, 779–784 (1983). https://doi.org/10.1175/1520-0477-64.7.779

Stanford, M.W., Fridlind, A.M., Silber, I., Ackerman, A.S., Cesana, G., Mülmenstädt, J., Protat, A., Alexander, S., McDonald, A.: Earth-system-model evaluation of cloud and precipitation occurrence for supercooled and warm clouds over the Southern Ocean's Macquarie Island. Atmos. Chem. Phys. **23**, 9037–9069 (2023). https://doi.org/10.5194/acp-23-9037-2023

Stephens, G.L.: Cloud feedbacks in the climate system: a critical review. J. Clim. **18**(2), 237–273 (2005). https://doi.org/10.1175/JCLI-3243.1

Stubenrauch, C.J., Bonazzola, M., Protopapadaki, S.E., Musat, I.: New cloud system metrics to assess bulk ice cloud schemes in a GCM. J. Adv. Model. Earth Syst. **11**(10), 3212–3234 (2019). https://doi.org/10.1029/2019MS001642

Sun, L., Fu, Y.: A new merged dataset for analyzing clouds, precipitation and atmospheric parameters based on ERA5 reanalysis data and the measurements of the Tropical Rainfall Measuring Mission (TRMM) precipitation radar and visible and infrared scanner. Earth Syst. Sci. Data. **13**, 2293–2306 (2021). https://doi.org/10.5194/essd-13-2293-2021

Trenberth, K.E., Fasullo, J.T.: Regional energy and water cycles: transports from ocean to land. J. Clim. **26**, 7837–7851 (2013). https://doi.org/10.1175/JCLI-D-13-00008.1

Trenberth, K.E., Smith, L., Qian, T., Dai, A., Fasullo, J.: Estimates of the global water budget and its annual cycle using observational and model data. J. Hydrometeorol. **8**(4), 758–769 (2007). https://doi.org/10.1175/JHM600.1

Vicente-Serrano, S.M., Tramblay, Y., Reig, F., González-Hidalgo, J.C., Beguería, S., Brunetti, M., Cindrić Kalin, K., Patalen, L., Kržič, A., Lionello, P., Lima, M.M., Trigo, R.M., El-Kenawy, A.M., Eddenjal, A., Türkes, M., Koutroulis, A., Manara, V., Maugeri, M., Badi, W., Mathbout, S., Bertalanič, R., Bocheva, L., Dabanli, I., Dumitrescu, A., Dubuisson, B., Sahabi-Abed, S., Abdulla, F., Fayad, A., Hodzic, S., Ivanov, M., Radevski, I., Peña-Angulo, D., Lorenzo-Lacruz, J., Domínguez-Castro, F., Gimeno-Sotelo, L., García-Herrera, R., Franquesa, M., Halifa-Marín, A., Adell-Michavila, M., Noguera, I., Barriopedro, D., Garrido-Perez, J.M., Azorin-Molina, C., Andres-Martin, M., Gimeno, L., Nieto, R., Llasat, M.C., Markonis, Y., Selmi, R., Ben Rached, S., Radovanović, S., Soubeyroux, J.-M., Ribes, A., Elmehdi Saidi, M., Bataineh, S., El Khalki, E.M., Robaa, S., Boucetta, A., Alsafadi, K., Mamassis, N., Mohammed, S., Fernández-Duque, B., Cheval, S., Moutia, S., Stevkov, A., Stevkova, S., Luna, M.Y., Potopová, V.: High temporal variability not trend dominates Mediterranean precipitation. Nature. **639**, 658–666 (2025). https://doi.org/10.1038/s41586-024-08576-6

Weigel, K., Bock, L., Gier, B.K., Lauer, A., Righi, M., Schlund, M., Adeniyi, K., Andela, B., Arnone, E., Berg, P., Caron, L.-P., Cionni, I., Corti, S., Drost, N., Hunter, A., Lledó, L., Mohr, C.W., Paçal, A., Pérez-Zanón, N., Predoi, V., Sandstad, M., Sillmann, J., Sterl, A., Vegas-Regidor, J., von Hardenberg, J., Eyring, V.: Earth System Model Evaluation Tool (ESMValTool) v2.0—diagnostics for extreme events, regional and impact evaluation, and analysis of Earth system models in CMIP. Geosci. Model Dev. **14**, 3159–3184 (2021). https://doi.org/10.5194/gmd-14-3159-2021

Zelinka, M.D., Zhou, C., Klein, S.A.: Insights from a refined decomposition of cloud feedbacks. Geophys. Res. Lett. **43**(17), 9259–9269 (2016). https://doi.org/10.1002/2016GL069917

Zhang, S., Xu, S., Fu, H., Wu, L., Liu, Z., Gao, Y., Zhao, C., Wan, W., Wan, L., Lu, H., Li, C., Liu, Y., Lv, X., Xie, J., Yu, Y., Gu, J., Wang, X., Zhang, Y., Ning, C., Fei, Y., Guo, X., Wang, Z., Wang, X., Wang, Z., Qu, B., Li, M., Zhao, H., Jiang, Y., Yang, G., Lu, L., Wang, H., An, H., Zhang, X., Zhang, Y., Ma, W., Yu, F., Xu, J., Lin, X., Shen, X.: Toward Earth system modeling with resolved clouds and ocean submesoscales on heterogeneous many-core HPCs. Natl. Sci. Rev. **10**(6), nwad069 (2023). https://doi.org/10.1093/nsr/nwad069

# Observing Clouds

# 2

What is a cloud and what is hidden behind different cloud types? Why certain clouds produce precipitation while others do not? Why diverse clouds produce different precipitation types? Seeking answers, we need to penetrate the intimate structure of clouds and their components. There are many ways to investigate clouds and their structure. In this chapter we will briefly classify clouds and describe some of the methods to study the interior of clouds and the hydrometeor types.

## 2.1 Cloud Classification

### 2.1.1 What Is a Cloud?

It may seem a trivial question at first sight, but defining a cloud is all but simple and universally accepted. In fact, we find different definitions from different institutions. The World Meteorological Organization (WMO 2018) defines a cloud as "an aggregate of very small water droplets, ice crystals, or a mixture of both, with its base above the Earth's surface, which is perceivable from the observation location". At the same time, a cloud for the American Meteorological Society (AMS 2020) is "a visible aggregate of minute water droplets and/or ice particles in the atmosphere above the Earth's surface". Thus, we realise that on one side the cloud is a *visible* atmospheric component while in the other definition it is *perceivable*. In fact, subvisual clouds, for example, are clouds on their own right like a towering thundercloud. The reason is very simple: we observe clouds not only visually with our own eyes, but also with a wealth of sensors, including lidars, radars and radiometers (Islam et al. 2017). Each one of these instruments sees the clouds differently and with its own sensitivity. It becomes then crucial to define observation uncertainties that have been elusive in the past (e.g., Stephens and Kummerow 2007).


V. Levizzani, C. Kidd, *Precipitation*, Geophysics and Environmental Physics, https://doi.org/10.1007/978-3-031-97096-2_2

Spänkuch et al. (2022) try to sort the matter out by examining pros and cons of the various definitions. Their conclusion is that no unique definition exists. This is no surprise since clouds are ephemeral and change constantly with the atmospheric conditions. The authors propose a new definition, which is based on physics and on observational conditions: "A meteorological cloud is an aggregate of minute particulate matter (solid, liquid, or mixed) in the atmosphere above the ground that becomes visible from the ground at a line-of-sight optical depth of at least about 0.03 at day and 0.05 at night".

No matter if a universally accepted definition (if any) will stem from the ongoing debate, we can safely state that a cloud is an aggregate of hydrometeors, i.e., droplets, drops, ice crystals, crystal aggregates, graupels, hailstones and obviously water vapour and aerosols. Although Spänkuch et al. (2022) admit that such a definition will "not cover all aspects of clouds", it will be appropriate for a meteorological framework where clouds are part of precipitating or non-precipitating systems.

From the cloud physics point of view, a more appropriate definition of a cloud is to consider it as a system of condensed water particles (drops and ice) suspended in the air. Such system is called colloid. The colloidal system is characterised by the size distribution of the particles, which is related to the stability of the system. If the cloud system is stable, it tends to keep its status, i.e. the cloud system. The particles will then remain dispersed and will not coagulate and eventually precipitate. If the system is unstable, then it will form rain and snow. The cloud will thus rapidly dissipate.

### 2.1.2 Cloud Types

It may seem that classifying clouds is a pointless exercise because clouds are ephemeral and constantly changing their shape, colour, altitude, and organisation. Clouds are not animals, plants, or minerals whose variety is well defined since the time of Carl Nilsson Linnaeus (1707–1778) although constantly changing with new discoveries in genetics, biochemistry, physics, chemistry, and geology. Clouds change in a matter of minutes or even seconds and thus it is difficult to describe them by means of fixed visual categories. This is why the first attempt to classify clouds dates back only to 1803 when the book *Essay on the modifications of clouds* of the British chemist and pharmacist Luke Howard (1772–1864) was first published (a modern edition is available, Howard 2011). Howard was a very methodical observer of the London skies together with his wife Mariabella and used to annotate his observations seeking for recurrent patterns. He concluded that clouds are indeed constantly changing but there are evident patterns that help defining categories for the recognition of cloud types. This way of thinking was borrowed from the classification method of Linnaeus and impressed greatly Johann Wolfgang von Goethe (1749–1832) who was investigating the order behind the external face of nature (Goethe 1820).

A long time has passed since Howard and Goethe first classified clouds. However, the classification of clouds, while evolving, remains basically the same. It is based mostly on their altitude and structure from a visual point of view. The official classification is contained in the WMO *International Cloud Atlas* (WMO 2017), which is a living document with the last introduction of a cloud type in March 2017. Global cloud observations are generally conducted from satellite, and we have already introduced the ISCCP (Rossow et al. 2022).

Observing clouds allows for obtaining information on temperature, moisture, and wind conditions at different heights in the atmosphere to be discussed. In one word, it prompts for observing the weather with special emphasis on precipitation.

The currently most used classification is based on their morphology as it appears from the perspective of an observer from the ground. It is a tree-like classification exactly as proposed by Linnaeus for plants and animals and it uses Latin names. The main group is called *genera* and is subdivided into *species* (shape and internal structure), which in turn contain *varieties* that describe specific features (transparency, shape of boundaries). *Supplementary features* are sometimes attached or partially merged with the cloud while *accessory clouds* are small clouds that may accompany the main cloud body being separate or partially merged with it. The classification also involves the category *mother clouds* that represent clouds from which other clouds generate instead of growing in clear air. Finally, *special clouds* form or grow because of local generating factor, either natural or resulting from human activities.

WMO defines the following concepts for describing a cloud:

- Height: Vertical distance from the point of observation on the Earth's surface to the point being measured.
- Altitude: Vertical distance from mean sea level to the point being measured.
- Height/Altitude of cloud base: For surface observations, height of the cloud base above ground level; for aircraft observations, altitude of the cloud base above mean sea level.
- Vertical extent: Vertical distance from a cloud's base to its top.

The cloud levels delimit the atmospheric layers where cloud genera most frequently develop. The classification avails itself of three levels described in Table 2.1. The levels have different height and depth depending on the region of the globe where the clouds form (Fig. 2.1); such regions may vary with season depending on the

**Table 2.1** Cloud levels, cloud genera belonging to each level, and approximate cloud base height

| Level | Genera | Polar region | Temperate region | Tropical region |
|---|---|---|---|---|
| High | Cirrus<br>Cirrocumulus<br>Cirrostratus | 3–8 km | 5–13 km | 6–18 km |
| Middle | Altocumulus<br>Altostratus<br>Nimbostratus | 2–4 km | 2–7 km | 2–8 km |
| Low | Stratus<br>Stratocumulus<br>Cumulus<br>Cumulonimbus | 0–2 km | 0–2 km | 0–2 km |

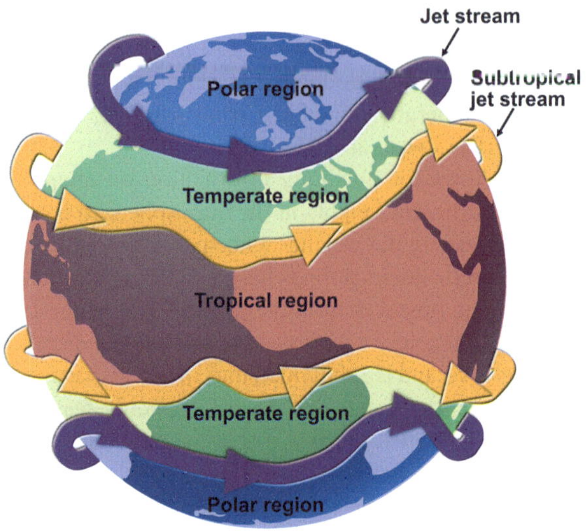


**Fig. 2.1** The three regions of the Earth's atmosphere separated by the polar and subtropical jet streams

position of the jet stream and of the subtropical jet stream. The vertical cloud levels also tend to overlap and thus clouds may be found at heights slightly different from those envisaged in the classification (Fig. 2.2). However, note that almost all clouds form in the troposphere, i.e., below 15 km height where almost all the water vapour is confined.

**High clouds** are also called cirriform clouds (cirrus in Latin is a curl of hairs). They form at temperatures below −40 °C and contain ice crystals. Cirriform clouds often characterise the upper level of a cold front and are thus often precursors of changing weather or upcoming storms. High-level cloud genera are three:

- **Cirrus (Ci)**. They are easily recognised as white filaments, white patches, or narrow bands. They display a fibrous or silky appearance. Precipitation from these clouds comes as thin bands of ice crystals not reaching the ground because of sublimation at high altitude.
- **Cirrocumulus (Cc)**. Thin, white patches, sheet, or layer of clouds without shading. They are composed of small elements organised as grains or ripples, merged or separate, regularly arranged.
- **Cirrostratus (Cs)**. Clouds with a transparent, whitish veil either fibrous or smooth. The Sun can often be seen through the cloud veil and sometimes a halo is produced. Precipitation mostly does not reach the ground.

**Medium clouds** consist of two genera:

- **Altocumulus (Ac)**. White or grey patch, sheet, or layer of cloud, generally with shading, composed of laminae, rounded masses, rolls, etc., which are sometimes partly fibrous or diffuse and which may or may not be merged; most of the regu-

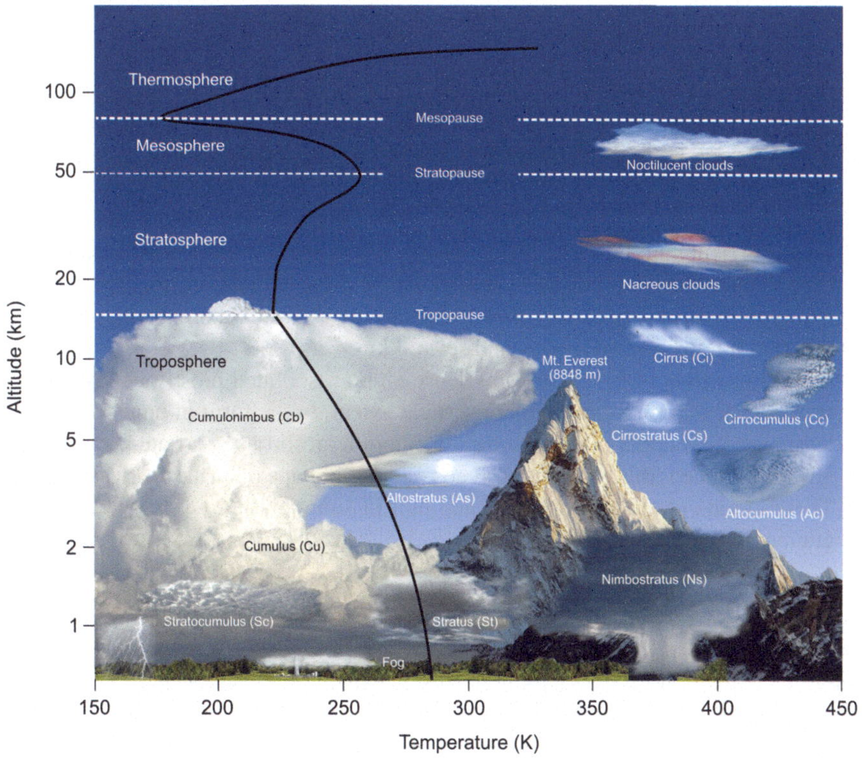


**Fig. 2.2** The ten genera of the WMO tropospheric cloud classification. Fog, nacreous clouds and noctilucent clouds are also shown. The curve of the vertical variation of temperature (K) of the atmosphere is in black. Note that the altitude scale is logarithmic

larly arranged small elements. Precipitation may reach the ground as raindrops or snow crystals.

- **Altostratus (As)**. Greyish or bluish cloud sheet or layer of striated, fibrous, or uniform appearance, totally or partly covering the sky. The Sun is revealed through their thinnest parts. Altostratus does not show halo phenomena unlike Cirrostratus.

**Low clouds** are essentially represented by two genera:

- **Stratocumulus (Sc)**. Grey or whitish patch, sheet, or layer of cloud with dark parts, composed of rounded masses, rolls, etc., which are non-fibrous, and which may or may not be merged; most of the cloud body is regularly arranged in small elements. Precipitation is weak or short-lived reaching the ground as rain or snow.
- **Stratus (St)**. Generally grey cloud layer with a uniform base, which may produce drizzle, snow, or snow grains. When the Sun is visible through the cloud, its outline is discernible, but no halo is produced except, possibly, at very low temperatures.

There are three remaining cloud genera that generally occupy thick layers spread among several levels (see Fig. 2.2):

- **Nimbostratus (Ns)**. Grey cloud layer, often dark, often characterised by continuously falling rain or snow, which, in most cases, reaches the ground. Nimbostratus is the precipitating cloud par excellence. They are thick enough throughout to obscure the Sun.
- **Cumulus (Cu)**. Detached clouds, dense and with sharp outlines, developing vertically as rising mounds, domes, or towers, of which the bulging upper part often resembles a cauliflower. The sunlit parts of these clouds are mostly brilliant white; their base is relatively dark and nearly horizontal. Cloud depth may vary between a few hundred metres and several km.
- **Cumulonimbus (Cb)**. Heavy and dense cloud associated mostly with thunderstorms. They show a considerable vertical extent, in the form of huge towers. At least part of its upper portion is usually smooth, or fibrous or striated, and nearly always flattened; this part often spreads out in the shape of an anvil or vast plume. These clouds produce heavy rain and hail in summer, and heavy snowfall and graupel in winter.

**Fog** is defined in the WMO International Cloud Atlas as a separate cloud type, i.e., a suspension of very small, usually microscopic water droplets in the air, reducing visibility at the Earth's surface. It is basically a very low altitude stratus cloud lying below 1 km and whose thickness may vary between a few metres and several hundred metres. Fog formation can be very diverse so that we observe radiation fog, advection fog, freezing fog, evaporation fog, upslope fog, hill fog, and frontal fog.

Only two types of high-altitude clouds are found in the stratosphere and in the mesosphere being visible during the night or at sunset. **Nacreous clouds** are also known as mother pearl clouds or polar stratospheric clouds. They typically occur at high latitudes and below −85 °C. They are wintertime ice clouds and contain solid hydrates or liquid solution droplets of nitric acid and water. **Noctilucent clouds** (or polar mesospheric clouds) are composed of very small water ice crystals formed on minute particles of dust, possibly of cosmic origin from micrometeors. They form only during the summer when temperatures in the mesosphere are lowest since temperatures lower than about −120 °C are required for their formation.

Certain clouds are generated by human activities such as **condensation trails**. These are cirrus clouds that form from the gases and particles of the exhaust of aircraft engines. Similar phenomena are the **ship tracks** that form from the warm air of the exhaust of ships that modifies the Sc cloud decks over the ocean leaving behind tracks in the cloud layer of enhanced albedo.

When a mountain chain is in the vicinity, we may be able to observe **orographic clouds** that develop on top of mountains due to uplift and wave trains. Spectacular cloud trains are generated, often organised in superimposed layers. Lenticular clouds are one of such scenic cloud formations.

There are several pictorial guides on clouds, and we will just mention a few of the most popular: Ludlam and Scorer (1957), Scorer (1975), Scorer and Verkaik (1989), Ludlum (2005), and Pretor-Pinney (2007, 2011, 2019). The *Cloud*

*Appreciation Society* (CAS, https://cloudappreciationsociety.org) is dedicated to cloud spotting and open to all interested individuals. The web site *What's This Cloud* (https://whatsthiscloud.com, last accessed 13 October 2023) aims at making cloud identification accessible to everyone.

### 2.1.3 Cloud Physical Classification

Classifying clouds only from their visual appearance (height, colour hues, vertical vs horizontal organisation) is a useful tool for cloud spotting, but it represents only one aspect of the classification. Certainly, it does not tell us all the details of cloud formation, their internal structure, and their capacity of producing precipitation. There are in fact physical characteristics that contribute to divide clouds in categories based on the physics of aerosol and water.

We will see in the next chapter that cloud droplets nucleate on aerosol particles and the reason behind this crucial fact. For the time being, it is enough to consider that water vapour condensates on aerosol particles whose size determines the initial size of the nucleated droplet. This initial size has an influence on the growth velocity of the droplet and thus on the precipitation potential of the cloud. This physical fact tells us that the aerosol content of the air mass where the cloud develops is crucial for determining the structure of the cloud, at least its initial genesis. This is known since the late 1950s when Squires (1958) first observed the size distribution of drops in clouds forming on maritime and continental air masses (Fig. 2.3) and later investigated by Twomey and Squires (1959). The size distribution of **maritime clouds** is peaked on large drops and their total number is relatively small compared with the very high number of drops in **continental clouds** whose size distribution peaks on small drops. So, the water in maritime clouds is divided up in a relatively small number of large drops while it is distributed in many smaller drops in continental clouds: this is due to the size distribution of aerosol in the two air mass types. Maritime clouds will then produce the first precipitation in a shorter time compared with continental clouds whose precipitation potential is clearly smaller. All these considerations are valid for **water clouds**.

When clouds reach and grow beyond the freezing level the structure of the cloud changes considerably. Liquid water continues to exist, but ice forms in various ways depending on temperature, saturation conditions, pressure/height, and aerosol type. We are then talking about **ice clouds** and **mixed-phase clouds**.

A more physical cloud classification consists of three major classes:

- **Liquid water clouds**. All hydrometeors are in the liquid state. Low and middle level clouds normally fall under this category, but some ice may be present when the environmental conditions permit. Sometimes the term **warm cloud** is used to identify this type of clouds. However, one must consider that such clouds may be at temperatures well below 0 °C and consist of supercooled water as we will see later. In this case the term can be misleading.

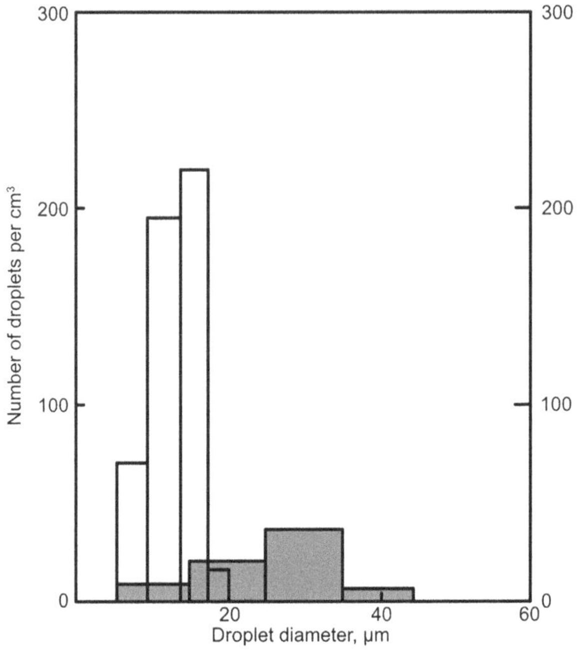


**Fig. 2.3** Number of droplets per cm$^3$ and their size distribution for maritime (filled boxes) and continental (hollow boxes) cumuli. The liquid water content is 0.5 g m$^{-3}$ in the maritime case and 0.35 g m$^{-3}$ in the continental case. (Adapted from Squires 1958. © The Author under CC-BY-4.0 license)

- **Ice clouds**. All hydrometeors are in the ice state. This is the category describing all high clouds, although highly concentrated aqueous solution drops may be present in some cases.
- **Mixed-phase clouds**. Both liquid and ice hydrometeors form these clouds. Summertime cumulonimbus is a prominent example of such a cloud where liquid and ice hydrometeors coexist depending on height/temperature.

Dividing clouds in classes based on any principle one can devise may seem a sterile exercise. However, it has some clear applications, apart from the obvious recognition of clouds in the meteorological observational praxis. An example are cloud-radiation interactions. Water and ice interact differently with radiation and thus defining the cloud composition is extremely valuable for weather forecasting and climate models. However, the process-level understanding of aerosol-cloud interactions remains insufficient to date (Kreidenweis et al. 2019).

## 2.2 Cloud Observations

Visual cloud observations are important and so is cloud sampling from mountain top observatories or, even better, from aircraft. There are several other techniques that allow for in depth cloud observations for the accurate description of cloud structure, genesis, and evolution. However, measuring physical quantities in the atmosphere is a very difficult enterprise and special instruments and techniques are required that offer the necessary precision and rapid response (Foken 2021).

### 2.2.1 Observatories

Several of the early cloud physics discoveries were made from observations in cloud observatories due to their unique observational perspectives.

Cloud observations from the ground are mostly conducted by observers and instruments located at high altitude to allow direct in-cloud sampling of hydrometeors. Mountain tops are preferred sites for such laboratories being often surrounded by cloud decks. Other observing sites are conceived to sample clouds from observing windows at various heights on a mountain slope. This latter strategy allows hydrometeor observations during orographic cloud formation.

Several observatories exist worldwide, and we will just mention a few of them to give an idea of the different observation strategies that depend on the location and associated cloud systems.

The Observatoire de Physique du Globe is located on top of the volcanic mountain Puy de Dôme and is run by the Laboratoire de Météorologie Physique of the Université de Clermont Auvergne in Clermont-Ferrand, France. The laboratory is located on the mountain top at 1464 m asl (Fig. 2.4) and the first meteorological

**Fig. 2.4** The observatory on top of the Puy de Dôme in the Massif Central, Auvergne, France. (Public domain)

station was established in 1875. Several window openings on the slope of the mountain allow for collection of samplings at several cloud levels. The history of the site goes considerably back in time. In 1648 Florin Périer was charged by Blaise Pascal to experiment the theory of Evangelista Torricelli that the barometric measurements were influenced by the air weight: to carry on the research Périer did his measurements at three different heights.

The Jungfraujoch Sphinx Observatory is the highest mountain observatory in Europe at 3571 m asl in the Swiss Alps. It hosts a meteorological station and an astronomical observatory. It was completed in 1937 and since then research has been carried out continuously on glaciology, physics, astronomy, and medicine, and specifically on cosmic rays and neutrino physics.

The observatory of Holme Moss in the Southern Pennines is situated close to the top of the Black Hill at 525 m asl in the Peak District in the UK. It hosts studies on cloud-aerosol interactions and cloud processing of aerosol. Facilities are installed on masts with a small cabin for external instruments. Measurements include cloud and rainwater chemical composition and an infrared visibility meter for resolving precipitation type and rate.

One of the oldest observatories in Europe is the meteorological observatory Milešovka at 837 m asl on top of a volcanic cone, which is higher than all the surrounding terrain in the České středohoří Mountains. It is the oldest observatory in the Czech Republic founded in 1905 in an area known for its windy and thunderstorm rich atmosphere. The mountain top is often covered by low stratiform clouds. Aerosol and cloud physics instruments are hosted with a Ka band cloud profiler.

In the United States several observatories have a long history of observations. One of the longest observational records of alpine climate belongs to the Mt. Washington Observatory in New Hampshire. The Blue Hill Meteorological Observatory was established in 1875 at the summit of the Great Blue Hill on the Blue Hills Reservation at the junction of Routes 93 and 138 about 16 km south of Boston in Milton, Massachusetts. Its mission is to produce, maintain, and analyse a meticulous, consistent record of weather and climate observations. The Elk Mountain facility is located at the northern end of the Medicine Bow Mountain range in south-eastern Wyoming. The facility consists of an observatory, located at the head of Halleck Creek in a small semi-circular basin (3300 m asl), and the Schaefer Hut located atop the western peak of Elk Mountain (3350 m asl). Meteorological and cloud physics sensors include ambient temperature, dew point temperature, wind direction and speed, and cloud liquid water content. There is also a nephelometer and several atmospheric chemistry monitors at the observatory.

Arctic cloud observatories are deployed at Barrow and Atqasuk, Alaska; Eureka, Nunavut, Canada; Ny-Ålesund, Svalbard Islands, Norway; Summit, Greenland. An all year round observatory is deployed in the permanent ice pack in the western Arctic Ocean. Each observatory employs some combination of cloud radar, lidar, ceilometer, and/or infrared interferometer for a unique perspective on Arctic clouds.

In Antarctica several countries maintain permanent or seasonal observatories both on the coast or on the Antarctic Plateau. The Dome C (Dome Concordia) at 3233 m asl is operated by France and Italy (Fig. 2.5). The site is used for astronomical, physical, atmospheric, and medical science observations and experiments. The station measured the lowest Earth temperature on record of −84.7 °C on 13 August 2010.

Observatories in the Tropics are not numerous, but they are very important to ensure observations of monsoon and ocean cloudiness. The most famous observatory is perhaps the Mauna Loa Observatory (Fig. 2.6) located on the north flank of

**Fig. 2.5** The French-Italian base of Dome Concordia on the Antarctic Plateau. (Public domain)

**Fig. 2.6** NOAA's Mauna Loa observatory in Hawaii, a benchmark site for monitoring global climate change. In the background a marine stratocumulus cloud deck. (Public domain)

the Mauna Loa Volcano on the Big Island of Hawaii at 3444 m asl and run by the National Oceanic and Atmospheric Administration (NOAA). It is a premier atmospheric research facility that has been continuously monitoring and collecting data related to atmospheric change since the 1950s. Continuous $CO_2$ measurements started in March 1958 directed by Charles D. Keeling producing the well-known Keeling curve, a foundation of studies on the behaviour of greenhouse gases in the atmosphere. The Barbados Cloud Observatory was established on the Island of Barbados to enhance the understanding of clouds, circulation, and climate sensitivity through sustained observation at the edge of the ITCZ. The Indian Government is establishing several cloud physics observatories including one on the Western Ghats in Munnar, Kerala. This is a high-altitude observatory to observe cloud and rain processes over that region with state-of-the-art observations. It enables a better understanding of rainfall distribution and allows for a characterization of rainfall processes in the numerical models for the prediction of monsoon rainfall.

### 2.2.2 Aircraft

The most direct way of observing clouds is via penetration using instrumented aircraft. Given the high speed of the aircraft and the rapid evolution of the cloud system, the sensors must be specifically designed for fast acquisition and sampling, much faster than that of any ground-based instrument. Moreover, in general the aircraft are equipped for high-speed flight because the pilot sometimes needs to change route to cope with dangerous evolutions of the cloud system.

Instruments mounted on aircraft to sample cloud systems are numerous and designed to collect measurements of the cloud environment (temperature, humidity, wind, and pressure) and of the hydrometeors (cloud hydrometeor size, concentration, habits, and total water content). Fast-response hygrometers, liquid water probes, cloud condensation and ice nuclei counters, aerosol probes, and imaging spectrometers for hydrometeor sampling are among the most important instruments that equip research aircraft for cloud studies. The most important cloud hydrometeor sensors evolved from the Knollenberg probe designed for count and measure the sizes of particles within all types of clouds. This type of sensors nowadays includes Cloud Droplet Probe, Cloud Imaging Probe (CIP), Precipitation Imaging Probe, and the Particle Habit and Imaging and Polar Scattering Probe to measure number and size of cloud particles. The Nevzorov Probe and the Johnson-Williams Hot Wire Probe measure cloud liquid and total water content. The aircraft itself is equipped with Pitot tubes, gyroscopes and other ancillary sensors normally used for navigation and providing the necessary measurements of the environment around the aircraft during flight. Sampling instruments are normally mounted on the far side of the wings where turbulence is lower, and the environment is less perturbed by the aircraft mass (Fig. 2.7). Note that it is never true that the instruments measure in unperturbed conditions because the aircraft always modifies the cloud environment to some extent.

**Fig. 2.7** The Cessna Citation II Research Aircraft of Weather Modification Inc. Note the cloud imaging probes mounted under the wing tip. (Public domain)

CIPs sample dimensions and shapes of liquid and ice hydrometeors during cloud penetration. They record cloud particle images by capturing the shadow of the particles as they pass through a laser beam. They also provide information on the particle size distribution and cloud liquid water content. The CIP has a typical resolution of 25 μm and can measure particles in the size range 12.5 μm–1.55 mm. These sensors are also instrumental for investigating the ice crystal habits and the type of hydrometeors in response to the cloud environment. Examples of the vertical distribution of ice crystal shapes in a cirrus cloud sampled by the Cloud Particle Imager (CPI) are shown in Fig. 2.8. See also Lawson et al. (2019).

A few aircraft have been designed to fly at high altitude for the observations of high clouds, hurricanes, and severe storms. They are equipped with passive and active remote sensing instruments and often with dropsonde launchers. One of these aircrafts is the Gulfstream G 550 High Altitude and Long Range Research Aircraft (HALO, Fig. 2.9) of the German Aerospace Center (Deutsches Zentrum für Luft- und Raumfahrt, DLR). The downward-looking radiometers of the aircraft have bands in several spectral ranges: $H_2O$ (water vapour, liquid water path), $O_2$ (temperature, liquid water path), window (ice/liquid water condensate path), bandpass $O_2$ line (temperature profiling), and bandpass $H_2O$ line (water vapour profiling and ice). The passive instruments are complemented by a nadir pointing cloud and precipitation radar at Ka band (35.5 GHz).

A dropsonde launcher is normally included. The dropsonde (Fig. 2.10), developed by the National Center for Atmospheric Research (NCAR), is designed to be dropped from an aircraft at altitude to measure and track storm conditions as the device falls to the surface. The sonde contains a global positioning system (GPS) receiver, along with pressure, temperature, and humidity sensors to capture atmospheric profiles and thermodynamic data. Dropsondes are the only current method to measure the winds and barometric pressure through the atmosphere and down to the sea surface within the core of tropical cyclones far from land-based weather radars.

Several other aircraft belong to research centres worldwide. The list is long, but we will mention a few that have significantly contributed to the advancement of our

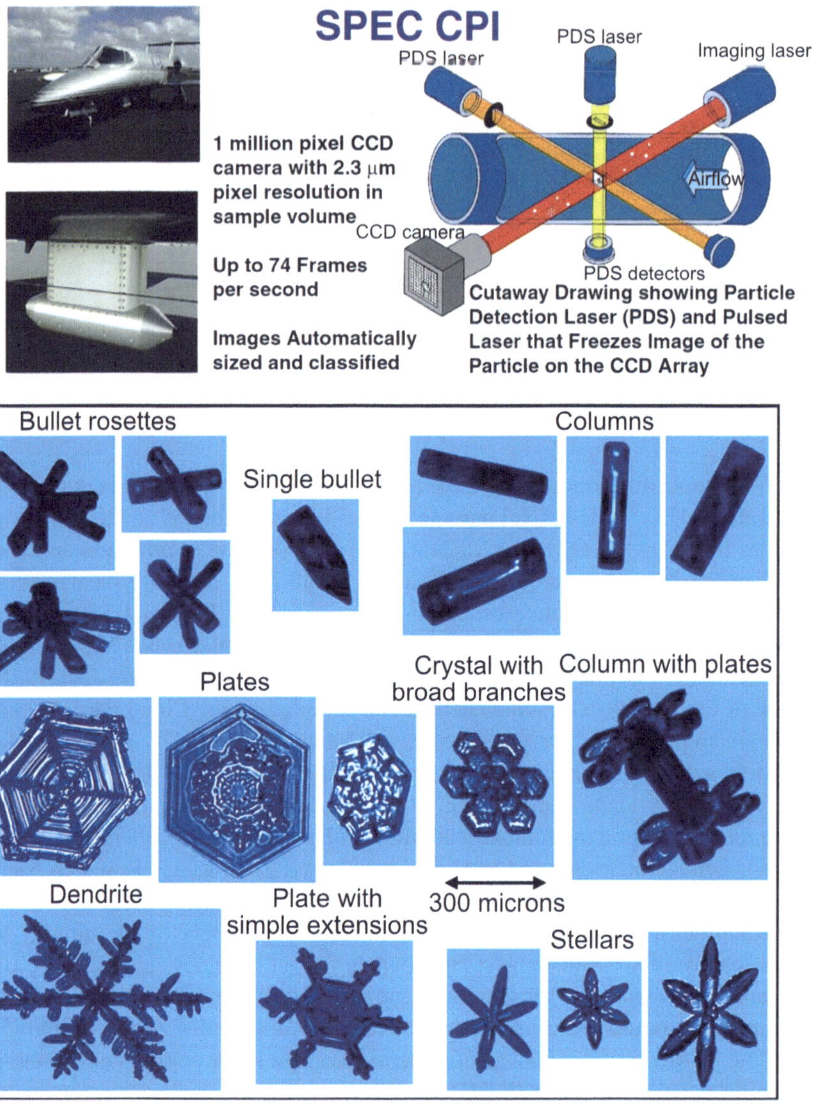


**Fig. 2.8** Top: CPI instrument recording at high-resolution (2.3 μm pixel size) digital images of particles that pass through the sample volume at speeds up to 200 m $s^{-1}$ (recently replaced by 3 V-CPI). Bottom: sample of ice crystal images from in cloud penetrations (Source: NASA)

knowledge of cloud physics. NCAR Earth Observing Laboratory (EOL) operates two research aircraft (Fig. 2.11). The most recent one is the High-performance Instrumented Airborne Platform for Environmental Research (HIAPER) funded by the National Science Foundation (NSF) and based on a Gulfstream V platform. It hosts a wing mounted cloud radar which allows high resolution views into snow

**Fig. 2.9** The Gulfstream G 550 High Altitude and Long Range Research Aircraft (HALO) of the German Aerospace Center. (Public domain)

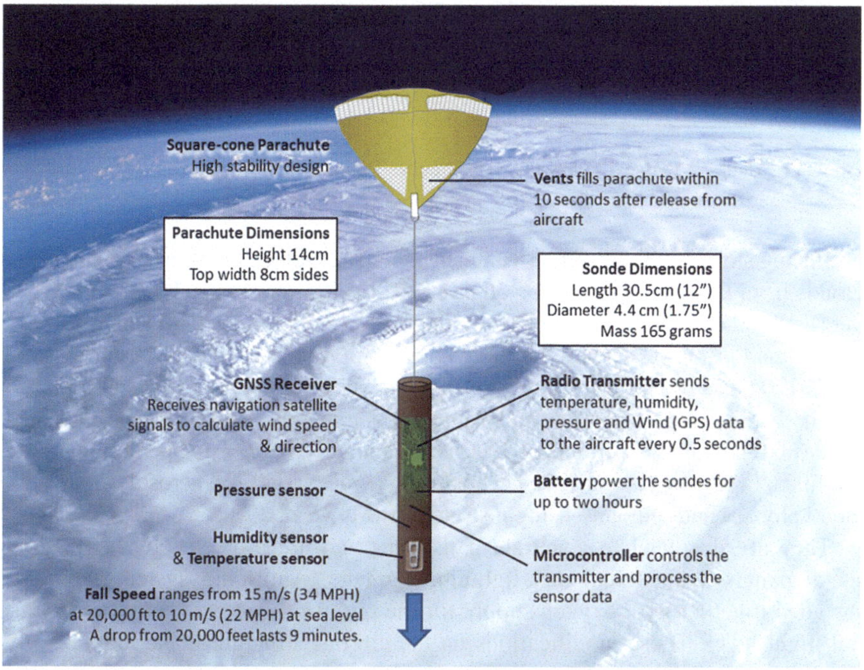


**Fig. 2.10** The NCAR GPS dropsonde to be deployed by manned and unmanned research aircraft. (Public domain)

**Fig. 2.11** The two NSF/NCAR research aircraft: the HIAPER (left) for stratospheric flights and the Hercules C-130 (right). (Public domain)

**Fig. 2.12** The FAAM Bae-146 UK research aircraft. (Public domain)

producing storms. The aircraft is designed and instrumented to observe and measure clouds from the stratosphere. The other NSF/NCAR aircraft is a Hercules C-130, a versatile and capable research platform that carries a wide variety of scientific payloads: in addition to standard thermodynamic, microphysics and radiation sensors, the C-130 has a roomy fuselage area and many versatile inlets and optical ports.

The Facility for Airborne Atmospheric Measurements (FAAM) was established by the UK Natural Environment Research Council (NERC). It operates a highly modified BAe-146 four-engine jet hosting a wide variety of aerosol, chemistry, cloud physics and remote sensing instruments (Fig. 2.12).

They are designed to penetrate hailstorms and are protected by heavy-duty metal panels against hail and lightning damage. Anti-icing systems are also installed together with oxygen supply for the pilot to fly above the aircraft nominal flight level. These modifications on the aircraft structure make it safe to fly inside clouds with about 8 cm diameter hailstones. A special aircraft is the armpured T-28 ai that belongs to the South Dakota School of Mines and Technology (SDSMT) and has been modified in time since its initial development in 1970 (Fig. 2.13).

**Fig. 2.13** The T-28 armoured research aircraft of the South Dakota School of Mines and Technology. (Public domain)

**Fig. 2.14** The NASA Global Hawk Grumman Northrop unmanned aircraft for high-altitude, long-duration missions. (Source: NASA)

Unmanned aerial vehicles (UAV) have been originally conceived for long range warfare operations. However, as frequently happens, applications of these military technologies have revealed important for science: UAVs have been shown relevant for the exploration of the atmosphere. The National Aeronautics and Space Administration (NASA) operates two Northrop Grumman Global Hawk unmanned aircraft for high-altitude, long-duration missions (Fig. 2.14). The Global Hawk aircraft operates autonomously and executes a flight plan loaded to the aircraft prior to flight. Nominally these aircraft are flown on missions of up to 24 h, during which they travel approximately 16,000 km. Although autonomous, the aircraft's flight is managed, and systems are monitored through satellite and line-of-site communication links using a ground control station.

### 2.2.3 Radar

The inception of radar technology between the two World Wars and the advent of radar meteorology during and shortly after the Second World War have been fundamental chapters of a success story in cloud physics. Radar reflectivity has immediately revealed as one of the most direct methods to investigate the structure of clouds and precipitation (Battan 1973; Atlas 1990; Wakimoto and Srivastava 2003). However, using the simple reflectivity factor to infer type and intensity of precipitation soon demonstrated difficult and prone to several known errors: e.g., the scattering from hydrometeors is not sufficient to fully characterize the formation of precipitation and more information are needed.

The first one comes from the measurement of the Doppler effect to identify flow fields in the cloud from the displacement of hydrometeors (Doviak and Zrnić 2006). The Doppler effect is the displacement between the emitted and the received frequency of the radar pulse. From the spectra of pulse-Doppler radars one can derive the motion of precipitation. Since the end of the 1950s these radars underwent a decade of experimentation in the United States and the first operational Doppler weather radar was turned on in Norman, Oklahoma in 1971 (Fig. 2.15) at NOAA's

**Fig. 2.15** The first NOAA/NSSL Doppler weather radar in Norman, Oklahoma. (Public domain)

**Fig. 2.16** A Doppler on Wheels radar deployed to observe severe storms and tornado. (Public domain)

National Severe Storms Laboratory (NSSL), leading to the Next-Generation Radar (NEXRAD) network being deployed in the 1980s.

Mention must be given to the mobile radar systems that allow to conduct field campaigns in remote sites not covered by existing radar networks. One of the most well-known is the Doppler on Wheels (DOW), a fleet of X- and C-band (7–12.5 and 4–8 GHz, respectively) managed by the University Illinois Urbana-Champaign. These radars are used mostly for severe weather research in the US Great Plains (Fig. 2.16).

However, radars can do much more for cloud physics when their emitted radiation is polarised (Bringi and Chandrasekar 2001; Fabry 2015; Chandrasekar et al. 2023). The antenna of the radar system is designed to transmit and receive electromagnetic waves of a specific polarization. Components in two orthogonal or basis polarisations are needed to create a wave with a given polarisation. The two most common polarisations are horizontal linear or H and vertical linear or V. Circular polarisations are also used, and, in this case, they are called Right Hand Circular (R) and Left Hand Circular (L). The scatterer can change the polarisation of the incident wave and thus the antenna is designed to receive the different polarisation components simultaneously.

Dual-polarimetric radars transmit and receive pulses in both a horizontal and vertical orientation. As a result, the returning frequencies provide measurements of the horizontal and vertical dimensions of targets, supplying forecasters with better estimates of the size, shape, and variety of targets (Fig. 2.17). Water droplets are usually oblate as they grow and thus the horizontally polarised wave will have a

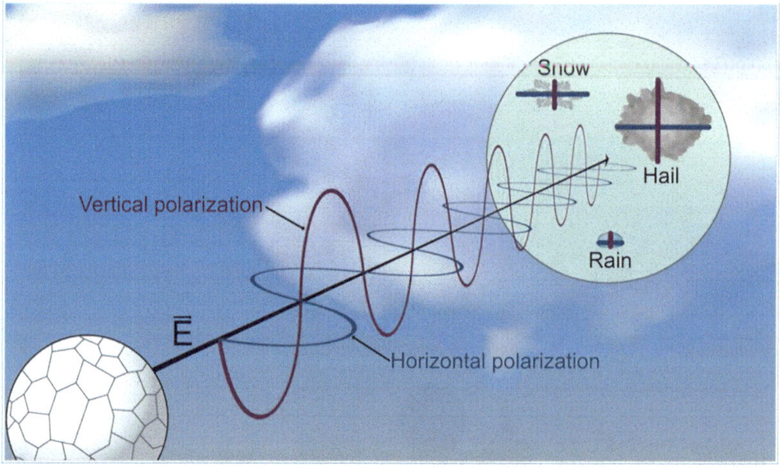


**Fig. 2.17** Dual-polarisation radar waves provide information on precipitation type, geometry and falling habit. (Modified from original of the NWS, public domain)

stronger return than the vertical one. Hail on the other hand has no preferred axis. Comparing the V and H returns will thus give information on the shape of the weather echoes.

The two signals are compared in different ways, such as for example: differential reflectivity $Z_{DR}$ (proportional to the ratio of the reflected horizontal ZH and vertical ZV power returns), correlation coefficient $\rho_{HV}$ (statistical correlation between the reflected horizontal and vertical power returns), linear depolarization ration $L_{DR}$ (ratio of a vertical power return from a horizontal pulse or a horizontal power return from a vertical pulse), and differential phase $\Phi_{DP}$ (comparison of the returned phase difference between the horizontal and vertical pulses). Dual-polarization quantities, added to reflectivity, velocity, and the spectrum of Doppler signals, have allowed development of new algorithms to differentiate precipitation types and non-meteorological targets, and to produce better accumulation estimates. An example of vertical profiles of various quantities is shown in Fig. 2.18.

A further step ahead was made with the design of millimetre-wave cloud radars, also known as cloud radars, operating at frequencies between 24 and 110 GHz. The wavelengths (1 mm to 1.11 cm) are ten times shorter than those used in conventional precipitation radars. The main use of these radars is in the investigation of cloud properties and evolution made possible by the detection of small-size hydrometeors directly in cloud. These radars operate typically at Ka-band (35 GHz, 8 mm) and W-band (94 GHz, 3 mm). Note that longer wavelengths are less attenuated by drizzle and rain, while shorter wavelengths are more sensitive to smaller particles. Accordingly, radar signals are less attenuated in Ka-band than in W-band, while W-band radars detect smaller particles.

Airborne weather radars at all wavelengths are hosted on board of research aircraft for field campaigns. However, radars are normally mounted under the nose of

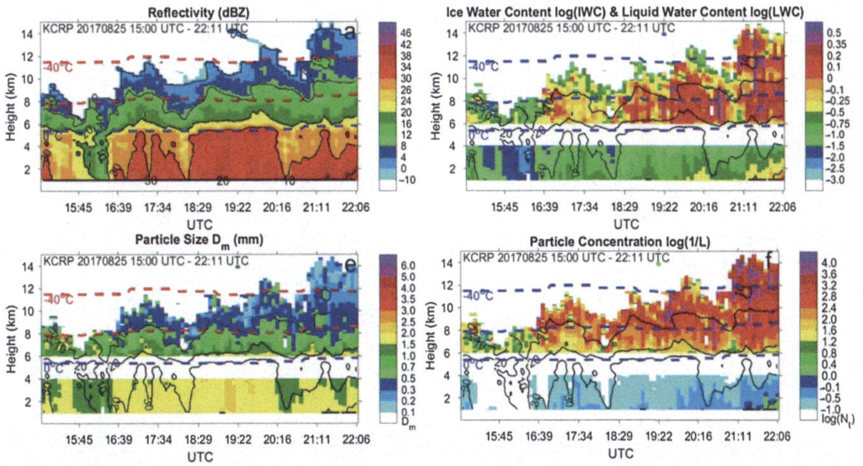


**Fig. 2.18** Columnar vertical profiles of ZH, Ice Water Content (IWC)/Liquid Water Content (LWC), particle size $D_m$, and particle concentration $N_t$ in the eyewall of hurricane Harvey from the perspective of the KCRP WSR-88D radar in Corpus Christi, Texas. Overlaid are contours of ZH every 10 dBZ. (Ryzhkov et al. 2020. © The Author(s) under CC-BY-4.0 license)

**Fig. 2.19** A flat plate airborne radar for detecting microbursts housed in the nose of the NASA 737. (Source: NASA)

commercial aircrafts for weather surveillance, collision avoidance, target tracking and more. An example is shown in Fig. 2.19.

### 2.2.4 Lidar

The atmospheric lidar is an instrument for determining range by targeting an object with a laser and measuring the time for the reflected light to return to the receiver (see one of the first article on the subject, Collis 1966). Lidar is mostly operated

**Fig. 2.20** A vertically pointing lidar of the Andes Lidar observatory. (© NOIRLab-NSF under CC-BY-4.0 license)

along a fixed direction (normally vertically pointing, Fig. 2.20), but it can also scan multiple directions (scanning lidar). It is normally operated during night-time.

Lidars have been used for the classification of cloud types (i.e., cumuli vs cirrus). Cloud boundaries can be retrieved from a ground-based lidar operating at a visible and/or near-infrared band. Cloud-base height can be identified, and the laser beam is attenuated when it penetrates through the clouds. However, when a powerful laser (e.g., Nd:YAG laser with high energy per pulse) is used, cloud tops can be retrieved too. Another physical parameter that can be retrieved is the cloud phase.

Particularly useful for profiling water vapour, clouds and aerosol are the Raman lidars. Raman lidar systems detect, in addition to backscattering at the laser wavelength (elastic backscattering), signals at different wavelengths. Such signals emerge from scattering by molecules, which absorb a part of the photon's energy or add an amount of energy to the photon's energy (inelastic scattering), whereby these molecules change their vibrational and/or rotational state (Raman process) causing a frequency shift due to its change of energy.

Lidars, as well as radars, can be hosted on board of mobile vans, research aircraft and satellites. An example of such lidar systems is the Cloud Physics Lidar (CPL), a NASA backscatter lidar designed to operate at three wavelengths: 1064, 532 and 355 nm. Its purpose is to provide multi-wavelength measurements of cirrus, subvisual cirrus and aerosols at high spatial and temporal resolution. It is hosted on board NASA's research aircrafts and of the Global Hawk. The High Spectral Resolution

Lidar (HSRL) is a more recent lidar system developed at NASA Langley Research Center utilizing HSRL techniques at 532 nm and standard backscatter techniques at 1064 nm to obtain profiles of aerosol backscatter, extinction, and depolarisation. The vertical resolution of the profile is 30 m with a horizontal resolution of 1 km. The CPL has been the predecessor of the Cloud-Aerosol Transport System (CATS), launched in January 2015, a lidar to investigate atmospheric aerosol and clouds from the International Space Station (ISS) focusing on clouds, volcanic ash, pollutants, dust, and other aerosols (Fig. 2.21).

### 2.2.5 Satellite

Clouds and precipitation are global phenomena and their observation from the ground or from airborne platforms can only be partial especially in remote areas and over the oceans. Moreover, long-term observations for climate studies are needed at the global scale. The top-down observations from space offer a unique perspective, which is complementary to that from other sensors and networks (Kidder and Vonder Haar 1995).

Satellites fly in a variety of different orbits that ensure a global coverage while maintaining the necessary space-time resolution for observing cloud and precipitation processes. Observation strategies depend on the application, but also on the specific phenomenon and/or type of information to be derived. For example, geostationary orbits (36,000 km height) are suited for observations of dynamic aspects of

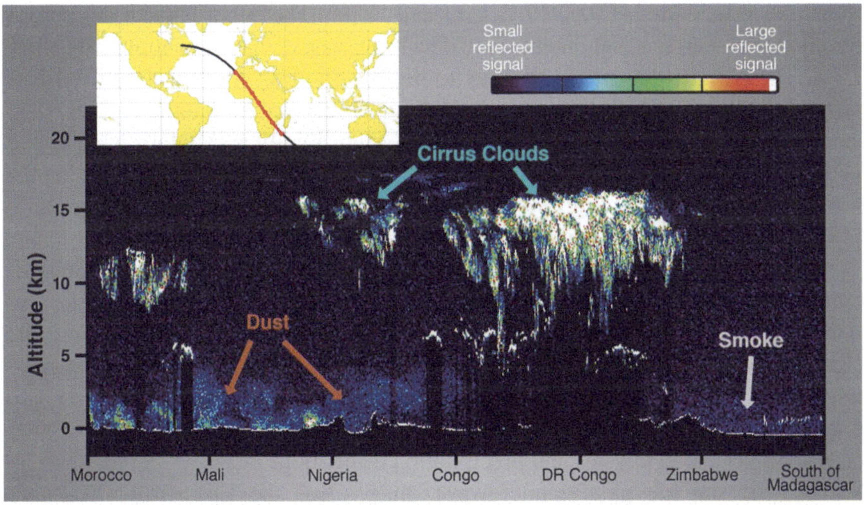


**Fig. 2.21** Cross-section of the atmosphere over Africa showing clouds, dust, and smoke from fires, as well as topography returned by the Cloud-Aerosol Transport System instrument aboard the International Space Station on 11 February 2022. (Source: NASA)

precipitating systems while lower orbits serve smaller scale observing purposes. The orbit at which a particular sensor can be launched depends also on the remote sensing principle of its construction that limits its sensitivity, space-time resolution and observing viewing angle. Finally, processes can be observed directly by measuring their emitted or scattered radiation field, but there are also indirect observations that can provide a wealth of information on ongoing cloud processes.

Clouds and precipitation have been the most important target since the early stages of the satellite era. The first meteorological satellite to provide images was the Television Infrared Observation Satellite (TIROS), which was launched on 1 April 1960. The very first satellite was the Vanguard-2 launched on 17 February 1959, but it failed to collect useful data due to several orbital reasons. Since then, many satellite missions were designed and successfully launched by various countries contributing to a global observation strategy, which is still in place and provides data whose value constantly increases. We tend to distinguish between meteorological and environmental satellites, but the distinction does not serve any practical purpose, and data are nowadays used in synergy using the whole set of sensors, even of those not specifically designed for clouds and precipitation observations. We are now in the era of the so-called "big data" and data processing methods are subject to a very rapid evolution involving artificial intelligence and machine learning.

The first two decades of cloud observations from satellite were characterised by the launch of visible and infrared sensors. Perhaps the most outstanding example of a satellite sensor and a true champion of longevity is the Advanced Very High Resolution Radiometer (AVHRR) operated by NOAA, which is active in several versions since the 1970s. The AVHRR measures reflectance and emission in five spectral bands centred at 0.6, 0.9, 3.5, 11 and 12 μm. The importance of this sensor for climate research is significant since data have been collected continuously since 1981. The AVHRR has paved the way to the design of another successful visible-infrared sensor, the Moderate-resolution Imaging Spectroradiometer (MODIS) launched aboard NASA's Earth Observing System satellites. MODIS is equipped with 36 spectral bands and observes the Earth every 1–2 days. At the time of launch it represented a breakthrough for clouds and precipitation observations (Fig. 2.22) as well for numerous other fields.

The second fundamental chapter of satellite meteorology for clouds and precipitation started in the 1970s with the advent of sensors in the microwave spectral range. In fact, visible and infrared radiometers are suited only for observing the cloud top but fail penetrating the interior of clouds or provide only indirect information. Microwave emission and scattering from hydrometeors, on the contrary, are directly linked to the cloud level where the hydrometeors develop and thus are powerful tools for cloud physics and precipitation estimation.

The first satellite passive-microwave imager was the Electrically Scanning Microwave Radiometer (ESMR) launched on NASA's Nimbus 5 satellite in December 1972, preceded by a non-scanning radiometer launched on the Russian

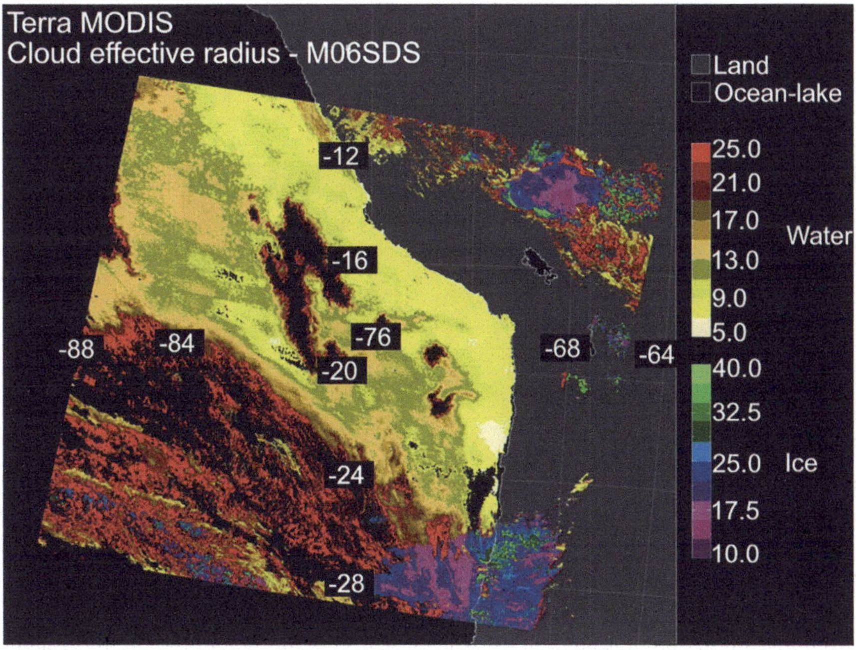


**Fig. 2.22** Sample of MODIS Level 2 data of effective radii for water and ice clouds. (Source: NASA)

Cosmos satellite series. The launch of the Scanning Multichannel Microwave Radiometer (SMMR) in 1978 represented a milestone because it was the first conically scanning radiometer used in space. The launch of this mission gave the opportunity to image the Earth at a constant incidence angle that is important as surface emissivity is angle dependent. In the 1980s, new multi-frequency, dual-polarization radiometric instruments were developed. Technology helped considerably in the following years to the present with microwave sensors being successful in cloud exploration from space. Perhaps the most important of these sensors was the Special Sensor Microwave/Imager (SSM/I) aboard the satellite of the Defense Meteorological Satellite Program (DMSP), which contributed to developing new algorithms and new insights on cloud development. It was later substituted by the Special Sensor Microwave Imager/Sounder (SSMIS).

The most important breakthrough came in the 1990s with the launch of the Tropical Rainfall Measuring Mission (TRMM; Kummerow et al. 1998; Fig. 2.23) launched on 27 November 1997 as a joint program of NASA and the Japan Aerospace Exploration Agency (JAXA). This revolutionary satellite hosted the TRMM Microwave Imager (TMI), the first Precipitation Radar (PR) in space, the Lightning Imaging Sensor (LIS), the Visible and Infrared Scanner (VIRS), and the

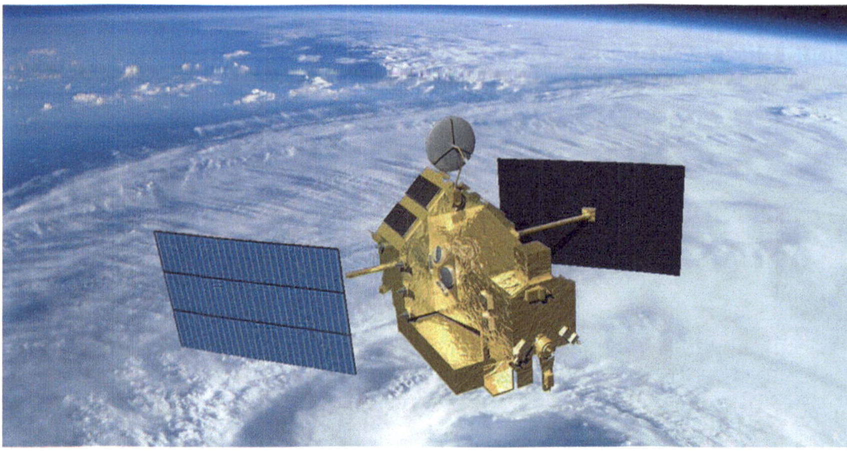

**Fig. 2.23** The Tropical Rainfall Measuring Mission (TRMM). (Source: NASA)

Clouds and the Earth's Radiant Energy Sensor (CERES). The satellite operated until 2015 and has considerably improved our understanding of the distribution and variability of precipitation within the tropics as part of the water cycle in the climate system. TRMM provided important precipitation information that increased our understanding of the interactions between water vapour, clouds, and precipitation.

TRMM was followed by the Global Precipitation Measurement (GPM) mission (Hou et al. 2014; Skofronick-Jackson et al. 2017) consisting of a core satellite and a constellation of international microwave sensors. The Core Observatory (Fig. 2.24), launched on 27 February 2014, hosts the Dual-frequency Precipitation Radar (DPR) and the multi-channel GPM Microwave Imager (GMI). The dual-frequency radar represents another breakthrough in clouds and precipitation observations with two frequency bands at Ka (35.5 GHz) and Ku (13.6 GHz). The greater latitudinal coverage and the cloud radar frequencies allow better observations of low-intensity rainfall and some insights into snowfall processes, even though limitations remain due to the attenuation and power of the radar.

The mission has generated a considerable improvement in our capability of measuring precipitation from space. The Integrated Multi-satellitE Retrievals for GPM product combines precipitation observations using infrared and microwave sensors from the GPM constellation satellites, united by the GPM Core Observatory, to provide near real-time half-hourly precipitation estimates at 10 km resolution for the entire globe (Fig. 2.25).

Two other missions have represented significant steps ahead in observing clouds and precipitation from space; they both advance the concept of using active sensors, radar and lidar. NASA's CloudSat satellite (Stephens et al. 2002)

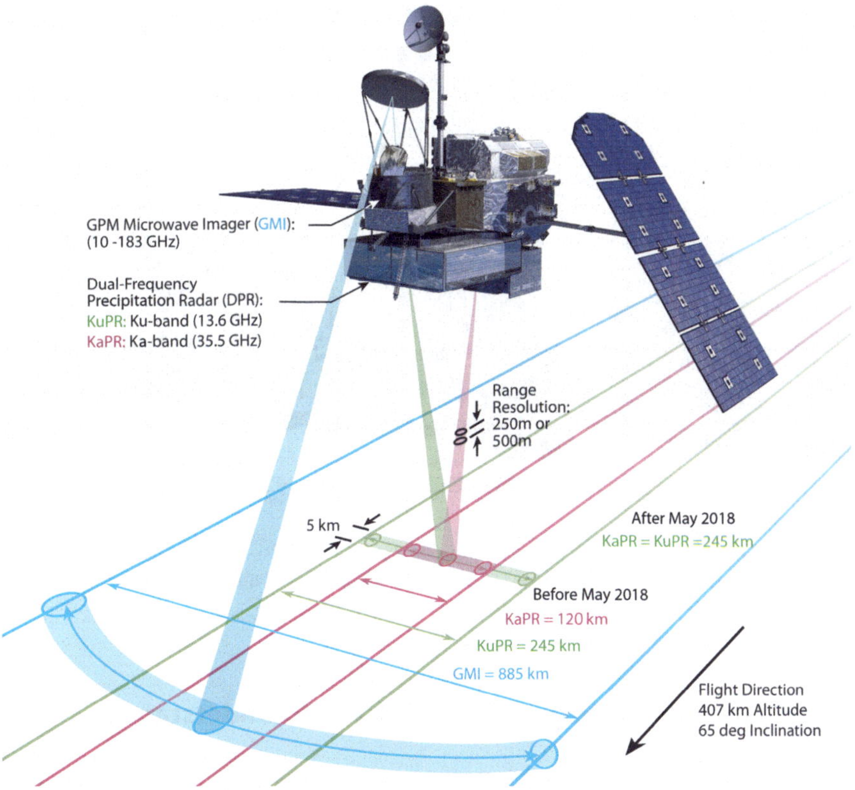


**Fig. 2.24** Swath dimensions and scan geometry of the Global Precipitation Measurement (GPM) Core Observatory's Dual-frequency Precipitation Radar (DPR) and GPM Microwave Imager (GMI) instruments. As of May 2018, the KaPR swath width is also 245 km. (Source: NASA)

is a satellite designed to measure the vertical structure of clouds from space using a W-band (94 GHz) nadir-pointing cloud radar. CloudSat observations have opened new avenues in cloud structure observations especially as regards snow (Fig. 2.26) and warm rain. The other mission that brought a novel observing perspective is the Cloud-Aerosol Lidar and Infrared Pathfinder Satellite Observation (CALIPSO, Winker et al. 2010) mission launched on 28 April 2006 together with CloudSat as a joint mission between NASA and the French Centre National d'Études Spatiales (CNES).

They are part of a novel type of constellation, the A-Train (Fig. 2.27), that observes clouds from satellites flying a few minutes apart thus providing an idea of cloud evolution from different platforms and wavelengths (Stephens et al. 2018).

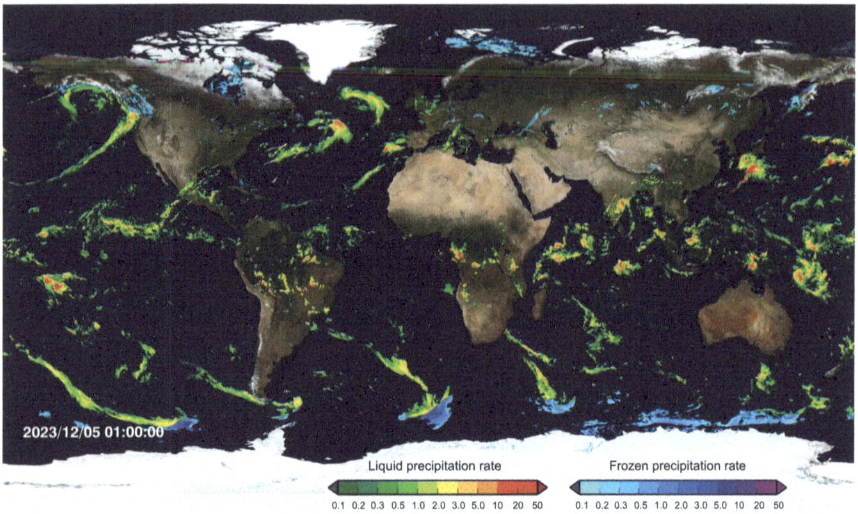


**Fig. 2.25** Global precipitation estimation from the Integrated Multi-satellitE Retrievals for GPM (IMERG) algorithm on 5 December 2023 0100 UTC. Note the snowfall classification including a small snowstorm over northern Italy at the time. (Source: NASA)

## 2.3 What We Know and What We Need to Understand

The knowledge of cloud physics and cloud structure has made giant steps ahead in the second half of the 1900s. We now understand how clouds form and the details of hydrometeor formation thanks to in situ and laboratory observations. The use of polarimetry (Fig. 2.28) and of new frequencies in radarmeteorology has allowed for in depth-investigations whose results are starting to enter the operational practice.

New technologies become available making it possible to observe clouds in situ to fill some gaps in our knowledge of the physics, especially that of glaciated clouds. A recent development in airborne technology for the atmosphere is the one proposed by UAVOS Inc., which has designed the stratospheric aircraft HiDRON. The HiDRON is a balloon-launched unmanned glider capable of autonomous soaring flight modes and designed for collecting high-altitude atmospheric data. The key objective of the HiDRON is to provide a new dynamic method for measuring and evaluating atmospheric phenomena at various altitudes in a variety of geographical locations. During the initial test flights, the HiDRON reached altitudes up to 25 km. The vehicle is launched to the stratosphere via a helium-filled balloon. Once launched the autopilot and integrated atmospheric measurement system transmit data in real-time to a ground station. After reaching the desired altitude (or as directed by an operator on the ground) the HiDRON is released from the balloon and glides back to a pre-specified landing site.

Remote sensing techniques in turn are instrumental to advance our knowledge of cloud dynamics and cloud formation in general.

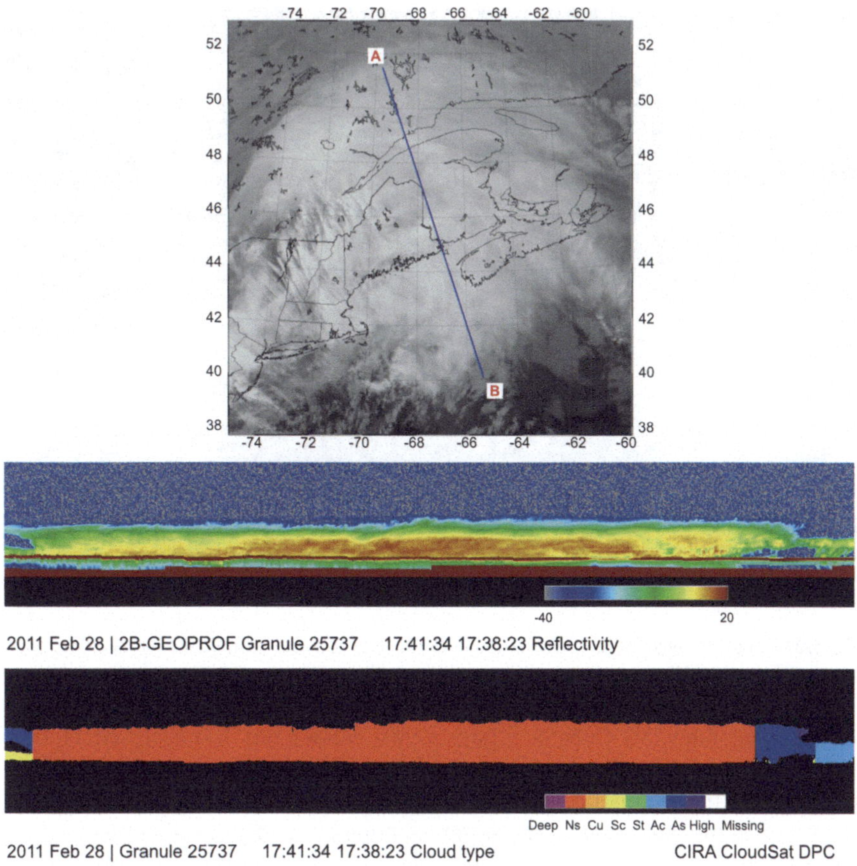


**Fig. 2.26** CloudSat overpass over a snowstorm in New Brunswick on 28 February 2011 1740 UTC (top panel). Cloud tops are shallow as most of the precipitation is snow over the land areas but over the ocean some of precipitation is falling as rain; the freezing level (or melting layer) is distinguishable along the southern section as the snow changes to rain (reflectivity map in middle panel). The CloudSat cloud classification (2B-CLD-CLASS) product identifies most of the clouds as nimbostratus (lower panel). The storm brought around 20 cm of snowfall to the area. (Source: NASA)

The discussion we have conducted so far indicates that constellations of satellites are gaining increasing importance in observing clouds and precipitation. While the existing constellations concentrate more on improving the coverage of satellite precipitation products, the future ones will dwell on the most recent advances of miniaturisation of spacecraft together with those of radar technology in space. The advent of micro- and nanosatellites has in fact opened opportunities for launching constellations of small, light-weighted, and relatively inexpensive satellites, that have the potential to follow the evolution of meteorological systems at high time frequencies. An advanced proof of concept in this direction is the Time-Resolved Observations of Precipitation structure and storm Intensity with a Constellation of Smallsats (TROPICS; Blackwell et al. 2018). TROPICS is a NASA constellation of

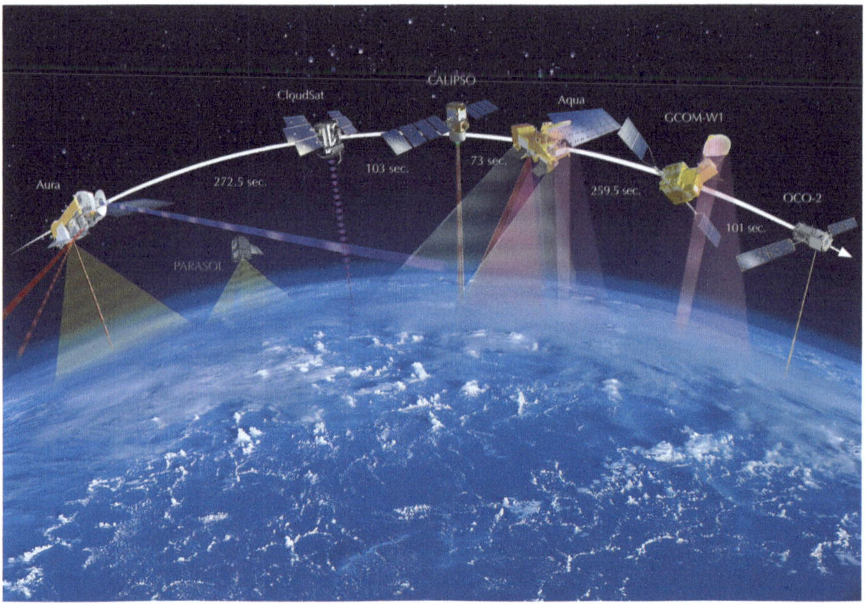


**Fig. 2.27** The A-train satellite constellation in 2013. Note that the constellation has been modified in time. (Source: NASA)

six small satellites, 3U CubeSats (https://www.cubesat.org, last accessed 11 Dec. 2023), that measure temperature and moisture profiles and precipitation in tropical systems with high temporal frequency. The data document the dynamic processes that occur in the inner core of the storm resulting in rapid genesis and intensification. The millimetre-wave radiometers sound mid-tropospheric water vapour to better understand its role in the growth and organization of convection in various large-scale environments adding to existing data sources.

Constellations of small radars aboard CubeSats have recently been explored. The antennas of these small satellites are deployable and make the payload small and cheap. The Ka band nadir-pointing radar of the Radar in a CubeSat (RainCube) mission is a technology demonstration mission to enable Ka band precipitation radar technologies on a low-cost, quick-turnaround platform (Fig. 2.29). The RainCube project has developed a 35.75 GHz radar payload operated on a 6U CubeSat form factor. Studies were conducted on cross validating TEMPEST radiometer observations with nearly simultaneous observations from RainCube (Radhakrishnan et al. 2022).

Advances in W band (94 GHz) Doppler radar technology blended with polarisation diversity and a radiometric mode is paving the way to a future mission of the European Space Agency (ESA), the Wind Velocity Radar Nephoscope (WIVERN; Illingworth et al. 2018). WIVERN aims to provide global measurements of in-cloud winds using the Doppler-shifted radar returns from hydrometeors using a conical scanning strategy. The novel concept is to derive the horizontal line of sight

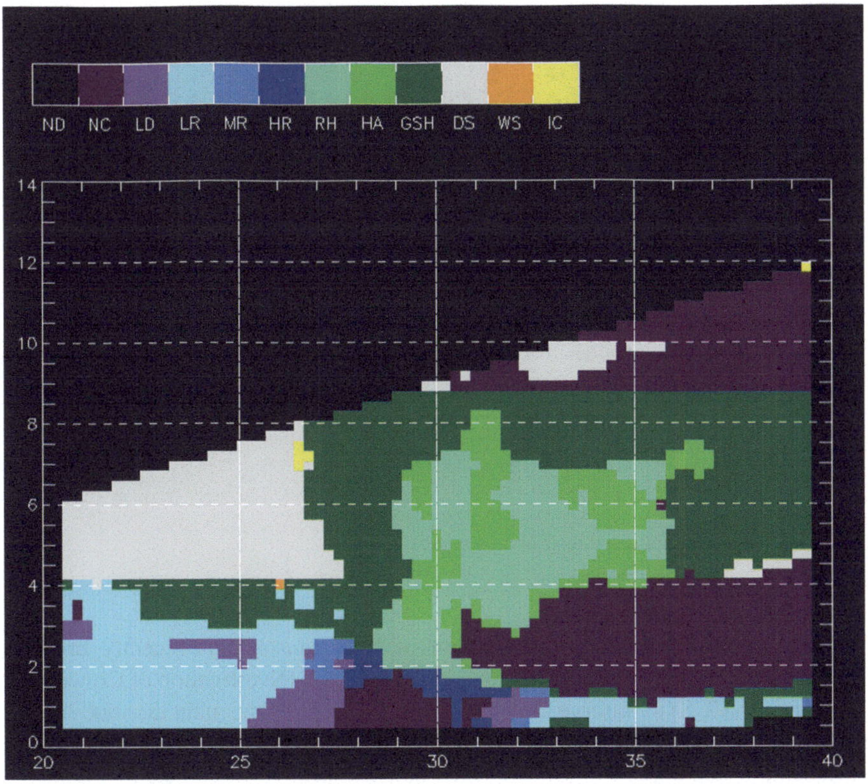


**Fig. 2.28** Polarimetric classification of hydrometeors in a hailstorm on 3 September 2002 1234 UTC. The classes refer to no data (ND), non-classified (NC), large drops (LD), light rain (LR), moderate rain (MR), heavy rain (HR), rain/hail mixture (RH), hail (HA), Graupel-small hail (GSH), dry snow (DS), wet snow (WS) and ice crystals (IC). (Printed under permission, P. P. Alberoni and V. Levizzani, ARPAE-SIMC and CNR-ISAC, Bologna)

component of the wind, which is representative of large-scale flow over distances of 20 km or more. Moreover, transient convective motions detected by the Doppler shift will be available at the 1 km scale thus making it possible to identify convective regions. From reflectivity and brightness temperature observations the liquid water path (LWP), the IWC and rain and snow rates will be also observed. The overall aim is to make available products for model data assimilation in the perspective of the development of new models by several meteorological services.

One more mission that is worth mentioning is the joint ESA-JAXA Earth Cloud Aerosol and Radiation Explorer (EarthCARE), which aims at advancing our understanding of the role played by clouds and aerosols in the climate system. It hosts an atmospheric lidar, a cloud profiling radar, a multispectral imager, and a broadband radiometer. The mission collects information about thicker clouds and observes the vertical velocities of cloud particles. This single-satellite mission enhances the coverage on clouds and aerosols of the previous CloudSat and CALIPSO missions.

**Fig. 2.29** The RainCube satellite. (Public domain)

Finally, the enhanced miniaturisation of MW sensors has recently brought Haddad and Sy (2024) to propose the mission concept INvestigation of Convective UpdraftS (INCUS) to observe systematically and globally the main thermodynamical process in convective storms: the vertical transport of air and moisture from the surface up to the upper troposphere. This new perspective could help significantly in improving the quality of precipitation measurements from space in convective regimes by making available observations on the vertical transport of air and moisture from the surface to the upper troposphere.

## References

AMS: Cloud. Glossary of Meteorology. http://glossary.ametsoc.org/wiki/cloud. Last accessed 5 Oct 2023 (2020)

Atlas, D.: Radar in Meteorology. American Meteorological Society (1990)

Battan, L.J.: Radar Observation of the Atmosphere. The University of Chicago Press (1973)

Blackwell, W.J., Braun, S., Bennartz, R., Velden, C., DeMaria, M., Atlas, R., Dunion, J., Marks, F., Rogers, R., Annane, B., Leslie, R.V.: An overview of the TROPICS NASA Earth Venture Mission. Q. J. R. Meteorol. Soc. **144**(S1), 16–26 (2018). https://doi.org/10.1002/qj.3290

Bringi, V.N., Chandrasekar, V.: Polarimetric Doppler Weather Radar: Principles and Applications. Cambridge University Press (2001)

Chandrasekar, V., Beauchamp, R.M., Bechini, R.: Introduction to Dual Polarization Weather Radar: Fundamentals, Applications, and Networks. Cambridge University Press (2023)

Collis, R.T.H.: Lidar: a new atmospheric probe. Q. J. R. Meteorol. Soc. **92**, 220–230 (1966). https://doi.org/10.1002/qj.49709239205

Doviak, R.J., Zrnić, D.S.: Doppler Radar and Weather Observations, 2nd edn. Dover Publ (2006)

Fabry, F.: Radar Meteorology. Cambridge University Press (2015)

Foken, T. (ed.): Springer Handbook of Atmospheric Measurements. Springer Nature (2021)

Haddad, Z.S., Sy, O.O.: From RainCube to INCUS: using miniaturized microwave instruments to analyze the dynamics of tropical convection. IEEE J. Microw. **4**(4), 858–870 (2024). https://doi.org/10.1109/JMW.2024.3438126

Hou, A.Y., Kakar, R.K., Neeck, S.A., Azarbarzin, A., Kummerow, C.D., Kojima, M., Oki, R., Nakamura, K., Iguchi, T.: The global precipitation measurement mission. Bull. Am. Meteorol. Soc. **95**, 701–722 (2014). https://doi.org/10.1175/BAMS-D-13-00164.1

Howard, L.: Essay on the Modifications of Clouds. Cambridge University Press (2011)

Illingworth, A.J., Battaglia, A., Bradford, J., Forsythe, M., Joe, P., Kollias, P., Lean, K., Lori, M., Mahfouf, J.-F., Melo, S., Midthassel, R., Munro, Y., Nicol, J., Potthast, R., Rennie, M., Stein, T.H.M., Tanelli, S., Tridon, F., Walden, C.J., Wolde, M.: WIVERN: a new satellite concept to provide global in-cloud winds, precipitation, and cloud properties. Bull. Am. Meteorol. Soc. **99**(8), 1669–1687 (2018). https://doi.org/10.1175/BAMS-D-16-0047.1

Islam, T., Hu, Y., Kokhanovsky, A., Wang, J. (eds.): Remote Sensing of Aerosols, Clouds and Precipitation. Elsevier (2017)

Kidder, S.Q., Vonder Haar, T.H.: Satellite Meteorology: An Introduction. Academic Press (1995)

Kreidenweis, S.M., Petters, M., Lohmann, U.: 100 years of progress in cloud physics, aerosols, and aerosol chemistry research. Meteorol. Monogr. **59**(1), 11.1-11.72 (2019). https://doi.org/10.1175/AMSMONOGRAPHS-D-18-0024.1

Kummerow, C., Barnes, W., Kozu, T., Shiue, J., Simpson, J.: The Tropical Rainfall Measuring Mission (TRMM) sensor package. J. Atmos. Ocean. Technol. **15**, 809–817 (1998). https://doi.org/10.1175/1520-0426(1998)015<0809:TTRMMT>2.0.CO;2

Lawson, R.P., Woods, S., Jensen, E., Erfani, E., Gurganus, C., Gallagher, M., Connolly, P., Whiteway, J., Baran, A.J., May, P., Heymsfield, A., Schmitt, C.G., McFarquhar, G., Um, J., Protat, A., Bailey, M., Lance, S., Muehlbauer, A., Stith, J., Korolev, A., Toon, O.B., Krämer, M.: A review of ice particle shapes in cirrus formed in situ and in anvils. J. Geophys. Res. Atmos. **124**(17–18), 10049–10090 (2019). https://doi.org/10.1029/2018JD030122

Ludlam, F.H., Scorer, R.S.: Cloud Study: A Pictorial Guide. John Murray (1957)

Ludlum, D.: National Audubon Society Field Guide to North American Weather. Knopf (2005)

Pretor-Pinney, G.: The Cloudspotter's Guide: The Science, History, and Culture of Clouds. Tarcher Perigee (2007)

Pretor-Pinney, G.: The Cloud Collector's Handbook. Sceptre (2011)

Pretor-Pinney, G.: A Cloud a Day. Batsford (2019)

Radhakrishnan, C., Chandrasekar, V., Berg, W., Reising, S.C.: Cross validation of TEMPEST-D and RainCube observations over precipitation systems. IEEE J. Sel. Top. Appl. Earth Obs. Remote Sens. **15**, 7826–7838 (2022). https://doi.org/10.1109/JSTARS.2022.3199402

Rossow, W.B., Knapp, K.R., Young, A.H.: International satellite cloud climatology project: extending the record. J. Clim. **35**(1), 141–158 (2022). https://doi.org/10.1175/JCLI-D-21-0157.1

Ryzhkov, A.V., Snyder, J., Carlin, J.T., Khain, A., Pinsky, M.: What polarimetric weather radars offer to cloud modelers: forward radar operators and microphysical/thermodynamic retrievals. Atmosphere. **11**(4), 362 (2020). https://doi.org/10.3390/atmos11040362

Scorer, R.S.: Clouds of the World: A Complete Color Encyclopedia. Stackpole Books (1975)

Scorer, R.S., Verkaik, A.: Spacious skies. David & Charles UK (1989)

Skofronick-Jackson, G., Petersen, W.A., Berg, W., Kidd, C., Stocker, E.F., Kirschbaum, D.B., Kakar, R., Braun, S.A., Huffman, G.J., Iguchi, T., Kirstetter, P.E., Kummerow, C., Meneghini, R., Oki, R., Olson, W.S., Takayabu, Y.N., Furukawa, K., Wilheit, T.: The Global Precipitation Measurement (GPM) mission for science and society. Bull. Am. Meteorol. Soc. **98**, 1679–1695 (2017). https://doi.org/10.1175/BAMS-D-15-00306.1

Spänkuch, D., Hellmuth, O., Görsdorf, U.: What is a cloud? Toward a more precise definition. Bull. Am. Meteorol. Soc. **103**(8), E1894–E1929 (2022). https://doi.org/10.1175/BAMS-D-21-0032.1

Squires, P.: The microstructure and colloidal stability of warm clouds. Part I: the relation between structure and stability. Tellus. **10**(2), 256–261 (1958). https://doi.org/10.3402/tellusa.v10i2.9228

Stephens, G.L., Kummerow, C.D.: The remote sensing of clouds and precipitation from space: a review. J. Atmos. Sci. **64**(11), 3742–3765 (2007). https://doi.org/10.1175/2006JAS2375.1

Stephens, G.L., Vane, D.G., Boain, R.J., Mace, G.G., Sassen, K., Wang, Z., Illingworth, A.J., O'Connor, E.J., Rossow, W.B., Durden, S.L., Miller, S.D., Austin, R.T., Benedetti, A., Mitrescu, C., the CloudSat Science Team: The CloudSat mission and the A-Train - A new dimension of space-based observations of clouds and precipitation. Bull. Am. Meteorol. Soc. **83**(12), 1771–1790 (2002). https://doi.org/10.1175/BAMS-83-12-1771

Stephens, G., Winker, D., Pelon, J., Trepte, C., Vane, D., Yuhas, C., L'Ecuyer, T., Lebsock, M.: CloudSat and CALIPSO within the A-Train: ten years of actively observing the Earth system. Bull. Am. Meteorol. Soc. **99**(3), 569–581 (2018). https://doi.org/10.1175/BAMS-D-16-0324.1

Twomey, S., Squires, P.: The influence of cloud nucleus population on the microstructure and stability of convective clouds. Tellus. **11**(4), 408–411 (1959). https://doi.org/10.3402/tellusa.v11i4.9331

Goethe, J.W. von: Die Form der Wolken nach Howard. In: Morphologie. Bildung und Umbildung d. organischen Naturen (1820)

Wakimoto, R.M., Srivastava, R. (eds.): Radar and Atmospheric Science. A Collection of Essays in Honor of David Atlas. American Meteorological Society (2003)

Winker, D.M., Pelon, J., Coakley Jr., J.A., Ackerman, S.A., Charlson, R.J., Colarco, P.R., Flamant, P., Fu, Q., Hoff, R., Kittaka, C., Kubar, T.L., LeTreut, H., McCormick, M.P., Megie, G., Poole, L., Powell, K., Trepte, C., Vaughan, M.A., Wielicki, B.A.: The CALIPSO Mission: a global 3D view of aerosols and clouds. Bull. Am. Meteorol. Soc. **91**(9), 1211–1230 (2010). https://doi.org/10.1175/2010BAMS3009.1

WMO: International cloud atlas. Manual on the observation of clouds and other meteors. WMO-No. 407. https://cloudatlas.wmo.int/en/home.html (2017)

WMO: Measurement of meteorological variables, vol. 1, Guide to Meteorological Instruments and Methods of Observation, WMO-8, World Meteorological Organization, 548 pp. https://library.wmo.int/doc_num.php?explnum_id=10616 (2018)

# Cloud Physics 3

Clouds and precipitation are part of a complex Earth system and represent essential components of meteorology, hydrology, and climate. This fact may seem obvious, but it is not. If we want to unambiguously place clouds within the mechanisms of the water cycle, we need to be specific on what is the physics behind their formation, evolution, and dissipation. Once we will have grasped the details of the cloud structure, we will address precipitation in all its countless forms. Excellent books populate the scientific literature on atmospheric sciences including chapters on clouds and precipitation (Houghton 1985; Wallace and Hobbs 2006; North et al. 2014). More specifically, numerous textbooks provide insights on cloud physics (Fletcher 1962; Byers 1965; Mason 1971; Hobbs 1974; Levin and Litvinov 1977; Rogers and Yau 1979; Kokhanovsky 2010; Pruppacher and Klett 2010; Hobbs and Deepak 2012; Wang 2013; Lohmann et al. 2016; Andronache 2018), dynamics (Cotton and Anthes 1992; Wang 2002; Cotton et al. 2011; Houze 2014; Baumgardner et al. 2018; Wang 2021), thermodynamics (Wegener 1911; Dufour and Defay 1963; Curry and Webster 1999; Khvorostyanov and Curry 2014), electricity (Rakov and Uman 2007; Uman 2012), and chemistry (Junge 1963; Lamb and Verlinde 2011; Seinfeld and Pandis 2016; Brasseur and Jacob 2017). Several books concentrate on aerosol particles as fundamental components of clouds in their formation (Friedlander 1977; Hobbs 1993; Kondratyev et al. 2006; Lenoble et al. 2013; Boucher 2015; Tomasi et al. 2016; Carslaw 2022). Other books examine the contribution of clouds and precipitation in shaping the Earth's climate (Crutzen and Ramanathan 1996; Randall 2012; Oke et al. 2017; Pincus et al. 2018; Ropelewski and Arkin 2019; Siebesma et al. 2020; Sullivan and Hoose 2023). They all contribute to define what we know of clouds and precipitation while keeping in mind that we are far from knowing all the details of cloud structure.



V. Levizzani, C. Kidd, *Precipitation*, Geophysics and Environmental Physics, https://doi.org/10.1007/978-3-031-97096-2_3

## 3.1 Cloud Hydrometeors

### 3.1.1 Atmospheric Meteors

The Greek term *metéoros* stands for "what is in the sky" and was used since ancient times to indicate missiles coming from the sky. The word meteorology was devised by Aristotle to describe the study of all the physical phenomena that happen in the terrestrial troposphere and are responsible for the weather. In meteorology meteors are all liquid and ice objects falling from clouds except those from space that are called meteoroids. In the Glossary of Meteorology of the AMS (AMS 2020) and in the WMO International Cloud Atlas (WMO 2017) the following meteors are identified:

Hydrometeors—wet: clouds, rain, snow, fog, dew, frost, hail, graupel, etc.
Lithometeors—dry: dust, sand, smoke, haze.
Igneous meteors—lightning, corona.
Electrometeors—lightning, thunder.
Luminous meteors—rainbows, halos, etc.

Only the term hydrometeor is used in meteorology while the other terms are rarely used and rapidly becoming outdated.

### 3.1.2 Cloud and Precipitation Particles

Condensed water particles in clouds are collectively identified as hydrometeors. They are the product of condensation or deposition of the atmospheric water vapour (AMS 2020) and can be classified in many ways. A physically based categorisation can be the following:

- Liquid or solid water particles formed and remaining suspended in the air, for example, damp haze, cloud, fog, ice fog, and mist. Cloud droplets belong to this category and have sizes between a few microns and ~400 μm.
- Liquid precipitation, for example, drizzle and rain. Raindrops have sizes between a few hundred microns up to ~3 mm. Drizzle droplets have radii smaller than 250 μm.
- Freezing precipitation, for example, freezing drizzle and freezing rain. Dimensions are similar to those of their liquid versions.
- Ice crystals. They are ice particles predominantly of hexagonal shape. Their dimensions range between a few tens and a few hundred microns. Sometimes they are called cloud ice.
- Snowflakes. Relatively large crystalline ice particles falling as single crystals (snow crystals) or aggregates (snowflakes or snow aggregates). The size range is between a few hundred microns for snow crystals to a few centimetres for large aggregates. Hundreds of individual crystals may aggregate to form a large snowflake.

- Graupel. Particles formed when supercooled droplets collide with an ice or snow crystal at low temperatures. Rime forms and a rimed crystal is produced. If the rime thickens, the particle loses its rimed crystal characteristics and becomes a graupel (also called soft hail or snow pellet). Due to turbulent motions around their centre of gravity, graupels often have a conical shape. Graupels normally has a diameter less than 5 mm; above 5 mm they are classified as hail. However, we will see later that the difference among the two precipitation types goes beyond size.
- Hail. Graupel eventually evolve in hail formation if the updraft is sufficiently strong showing a stratification of different ice density due to wet and dry growth. Hailstones have normally a diameter larger than 5 mm but can attain diameters of 15 cm or more.
- Ice pellets. A form of precipitation consisting of small, hard, translucent balls of ice not to be mistaken for graupels. Sleet is a regionally variant term for ice pellets, for rain and snow mixed or glaze (smooth coating of ice formed on objects by freezing rain).
- Falling hydrometeors that evaporate before reaching the ground, for example, virga when the atmosphere below the cloud base is very dry.
- Liquid or solid water particles lifted by the wind from the Earth's surface, for example, drifting snow, blowing snow, and blowing spray.

Precipitation is a term from chemistry indicating sedimentation of a solid material (a precipitate) from a liquid solution. In principle this would apply to all precipitating hydrometeors as "precipitate" of water vapour. However, in meteorology one way of distinguishing between cloud hydrometeors and precipitation particles is to consider the strength of the updraft in a cloud. When the updraft is strong enough to maintain the hydrometeors suspended in air, they are named cloud particles. When the updraft is not strong enough to prevent the hydrometeors from falling, they are named precipitation particles. No clear-cut dimensional boundary is identified between cloud and precipitation particles apart from their weight against the strength of the updraft. Empirical boundaries are fixed; for example, the one between cloud droplets and precipitation drops (Fig. 3.1) is normally set at 100 μm (0.1 mm).

The boundary between cloud ice particles and precipitation particles is less dependent on size since the density of the hydrometeor is crucial. Moreover, the supersaturation determines form and density of pristine crystals (Fig. 3.2) thus influencing their aggregation and precipitation habits.

Ice crystal shapes are numerous and have been studied extensively since the 1950s (Nakaya 1954) and 1960s (Magono and Lee 1966). However, the topic is far from being completely explored and recent studies have brought new laboratory (Kikuchi et al. 2013) and in situ (Um et al. 2015) insights into shapes, habits, and sizes of natural ice crystals.

Cloud droplets and ice crystals undergo growth processes in cloud that transform them into precipitation hydrometeors. The time they spend to reach precipitation size and mass varies between liquid drops and ice crystals (Fig. 3.3). For cloud droplets the radius is sufficient to document their growth while for ice crystals their

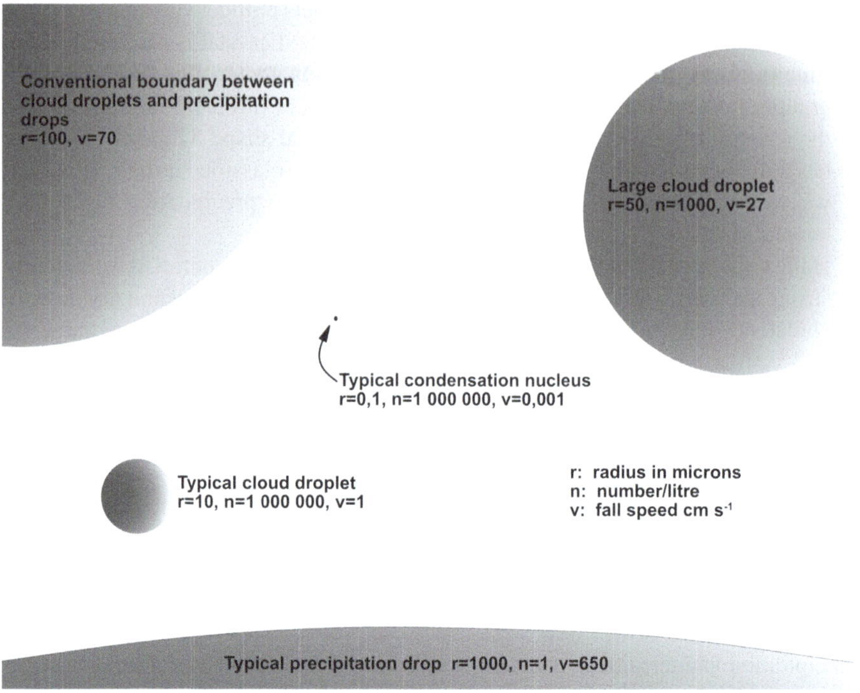


**Fig. 3.1** Size (microns), concentration (number of particles per litre of air) and fall speed (cm $s^{-1}$) of cloud condensation nuclei, cloud droplets, and rain drops

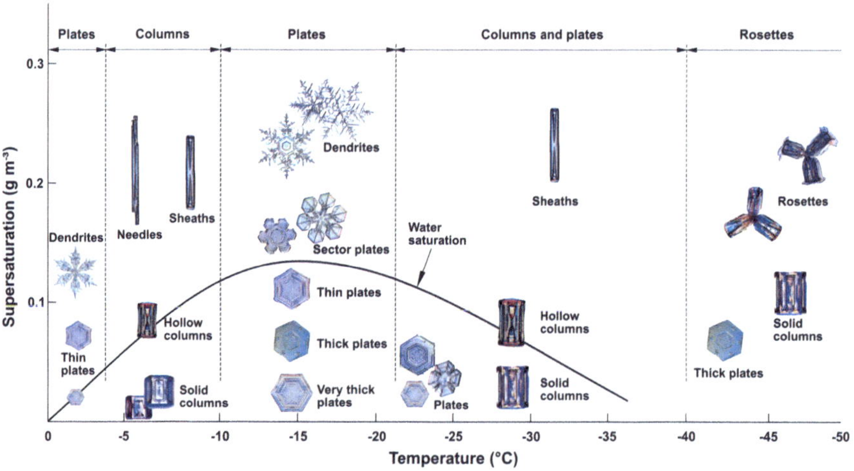


**Fig. 3.2** Ice crystal shapes. Needles are long, narrow, solid columns and sheaths are hollow needles. Dendrites are star-shaped crystals with thick or thin branches depending on the temperature

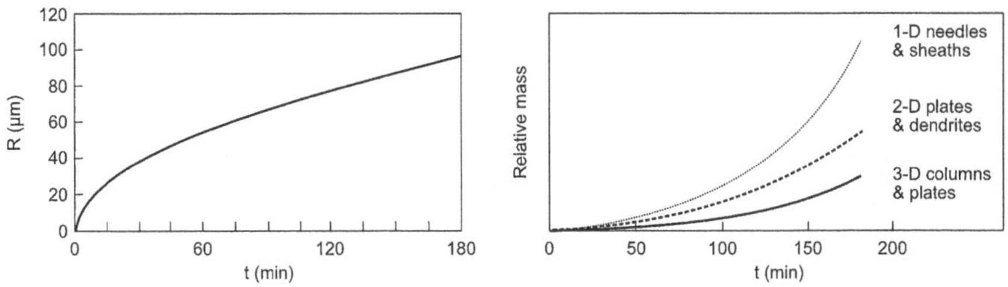


**Fig. 3.3** Growth time of water droplets (left) and ice crystals (right) in cloud assuming only diffusional growth

shape is necessary because of the great variety of crystals in a cloud. The growth time differs among crystals depending on their morphology: 1-D crystals are those that will precipitate faster and the reason is that their growth is only in one direction with respect to 2-D and 3-D crystals, thus offering less resistance to air friction.

Note that a cloud droplet or an ice crystal would require around 3 h to be large enough to become precipitation particles, and this is not compatible with the time of the first appearance of precipitation since the initial cloud formation, which is around 30 min. Other mechanisms are involved, and we will examine them in detail in the following pages.

## 3.2 Hydrometeor Size Distributions

### 3.2.1 Cloud Droplet Size Distributions

Cloud droplets are typically characterised by sizes between 5 and 50 μm in diameter and they remain suspended in a cloud before becoming precipitation drops. Since the population of droplets is composed of countless tiny members, it is important to describe it in terms of size distributions that are relevant to cloud processes. For example, the larger the variety of sizes the faster will be the formation of rain-size drops as we will see when introducing the formation of precipitation particles. At the same time, a greater variety of sizes of cloud droplets will translate in a slower evaporation of the cloud along its external boundaries even if the mean droplet size is the same. Thus, it is important to understand droplet distributions in clouds before they start to fall as precipitation.

We have already seen while commenting on Fig. 2.3 that larger mean drop size distributions tend to correspond to lower number of drop concentrations (e.g., maritime clouds). This is obviously because the liquid water is distributed among a relatively small number of large droplets. The opposite is true for narrower size distributions (mostly in continental clouds) where the water is distributed among many small droplets. Narrower spectra are in general highly stable and thus are less associated with precipitation production and the opposite is true for broader spectra. As we have previously noted, the tendency to produce narrow or broad spectra is

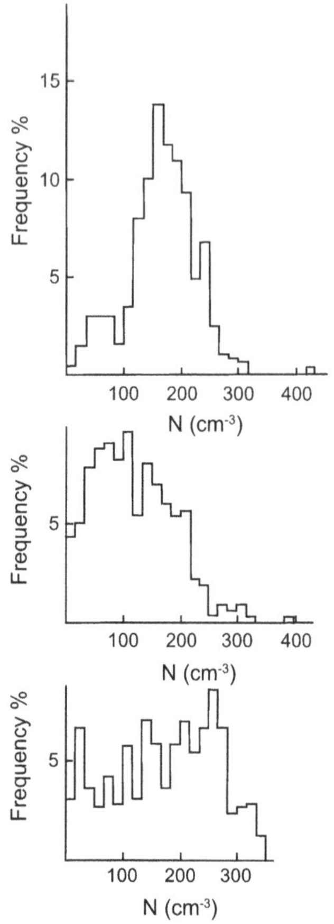


**Fig. 3.4** Frequency of the total number concentrations of droplets in a fair weather cumulus during three flight penetrations of the cumulus tower on 24 February 1975. (Adapted from Ichimura et al. 1980. © The Authors under CC-BY-4.0 license)

due to microphysical processes and to the dynamic environment of the cloud. A high number of very small aerosol nuclei will induce a narrow spectrum with high number concentrations while smaller amounts of large nuclei will do the opposite.

Droplet distributions vary considerably in time and space in a growing cloud. Flight penetrations have clearly demonstrated that cloud locations a few hundred metres apart are characterised by considerably different frequencies of the droplet number per unit volume as exemplified in Fig. 3.4.

Different clouds and different stages of cloud development are associated with different size distributions that reflect the microphysical processes in the cloud. For example, droplet sampling by Ichimura et al. (1980) in stratocumulus and fair-weather cumulus clouds (Fig. 3.5) have documented broader spectra of $N_D$ (dN/dlogD) in stratocumulus than in cumulus although a marginal peak is found around the second peak of the stratocumulus distributions. Moreover, the various stages of cloud development (growing, mature and decaying) correspond to different size distributions reflecting the action of the microphysical processes that we will describe later.

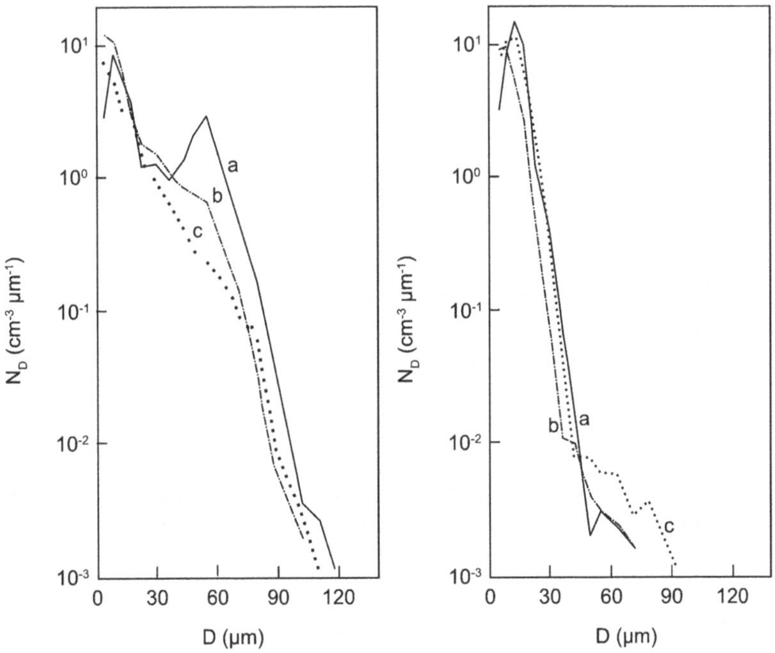


**Fig. 3.5** Size distributions obtained from sampling during flight penetrations in stratocumulus (left) and fair-weather cumulus (right) on 25 February 1975. Three penetrations are shown: growing stage (a), mature stage (b), and decaying stage (c). (Adapted from Ichimura et al. 1980. )

For the same reason the size distributions of cloud droplets also vary with height from cloud base under the action of collision and coalescence processes. This is evident in warm rain clouds in the monsoon area as noted by Prabha et al. (2011): the authors note the broadening of the spectrum with increasing height from cloud base. The effect is measurable also in continental non-raining cumulus clouds. Miles et al. (2000) have also demonstrated that the general assumption of constant cloud droplet concentration with height is appropriate for marine, but not continental, stratus.

The drop spectra in non-raining cumulus clouds were measured by Paul (2000) in clouds of higher thickness (1067–2134 m) and of lower thickness (305–1067 m). The maximum size 80 μm for thicker clouds was greater than that of 66 μm for thinner clouds. The mean distributions were bimodal (primary and secondary peaks at 7 and 75 μm diameter), and unimodal (peak at 7 μm), respectively.

### 3.2.2 Cloud Droplet Size Distribution Functions

The observed cloud droplet size distributions we have introduced in the previous section are described by mathematical functions to be used in models and calculations. As we have seen, it is not easy to introduce a function that describes all modes of the observed size distributions. Among the empirical functions widely used in cloud physics are the following (see Yangang et al. 1995 for an overview): (1) the modified gamma (Deirmendjian 1963; Petty and Huang 2011), (2) the Khrgian and Mazin (1952) and (3) the log-normal.

The Khrgian-Mazin function has the following expression:

$$N(r) = Ar^2 \exp(-Br) \tag{3.1}$$

where $N(r)dr$ represents the number concentration of droplets of radius between $r$ and $r + dr$ and A and B are two adjustable constants to fit the data. Note that the Khrgian-Mazin function is a special case of the modified gamma distribution:

$$N(r) = N_0 r^{\mu} \exp\left(-\lambda r^q\right) \tag{3.2}$$

where $N_0$, $\mu$, $\lambda$ and $q$ are adjustable parameters to fit the data ($\mu$, $\lambda$ and $q$ control the shape and $N_0$ the scaling of the distribution). One more function derives from the modified gamma, the Weibull distribution $N(r) = N_0 r^{q-1} \exp(-\lambda r^q)$ that is normally transformed into $N(r) = N_0 r^2 \exp(-\lambda r^3)$ called the maximum likelihood distribution. This latter form of the Weibull is used under the assumption of dynamically conservative total cloud droplet number and mass.

From (3.1) the main statistical properties of the cloud drop size spectrum can be derived:

- mean drop size

$$\overline{r} = \frac{\int_0^{\infty} rn(r)dr}{\int_0^{\infty} n(r)dr} = \frac{3}{B} \tag{3.3}$$

- liquid water content (LWC)

$$W_L = \frac{4}{3}\pi\rho_w \int_0^{\infty} r^3 n(r)dr \tag{3.4}$$

- total droplet concentration

$$N = \int n(r) = \frac{2A}{B^3} \tag{3.5}$$

where $\rho_w$ is the density of water.

The log-normal size distribution has the following form:

$$N(r) = \frac{N}{\sqrt{2\pi} r\sigma} \exp\left[-\frac{1}{2}\left(\frac{\ln r - \ln r_m}{\sigma}\right)^2\right] \tag{3.6}$$

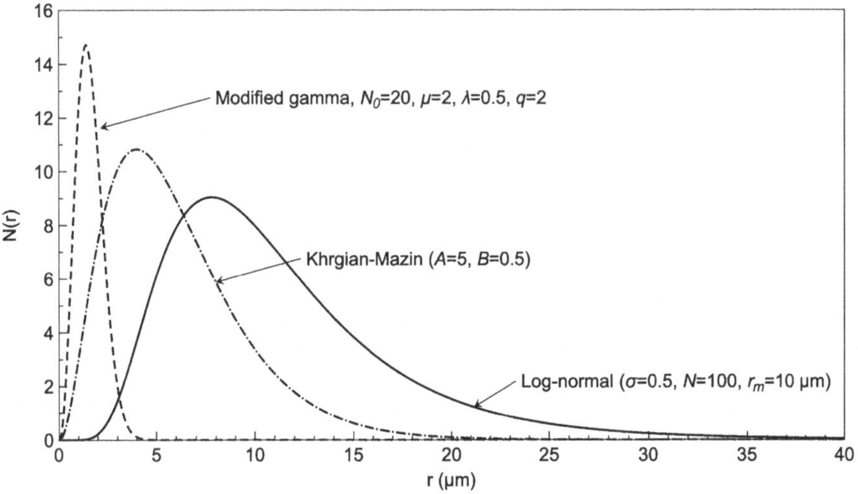


**Fig. 3.6** The Khrgian-Mazin, log-normal and modified gamma functions for three sets of the functional parameters (see text for details on the parameters)

where $r_m$ is the modal radius and $\sigma$ the logarithmic standard deviation (spectrum width).

Both Khrgian-Mazin and log-normal distribution functions are skewed towards the larger drop sizes (positively skewed), but they are unimodal (Fig. 3.6) and cannot adequately describe the observed size distributions that are distinctively bimodal. Thus, using these forms of the size distribution functions introduces errors in the calculations. However, the log-normal curve is perhaps best suited to describe observed distributions since it does not extend to size zero, which is correct for the measured size spectra in clouds.

### 3.2.3 Raindrop Size Distributions

Size distributions are an important way of describing raindrop size spectra as well as they are for cloud droplets. Raindrop sizes range from 500 μm to more than 3 mm for large drops. Because of the air pressure during fall large raindrops lose their spherical shape (Fig. 3.7) and become flat at their base around 2 mm diameter assuming the typical "hamburger bun" shape. With increasing diameter, the air pressure transforms the drops creating bags of water that eventually explode into several smaller droplets (Villermaux and Bossa 2009) thus determining the evolution of the size distribution.

The description of the size distribution of raindrops that considers their nonspherical shape normally adopts the concept of "equivalent size", i.e. the size of a spherical drop that has the same volume of the deformed drop. Raindrop size spectra have been observed extensively since the 1940s and found to be exponentially distributed with size (Laws and Parsons 1943; Marshall and Palmer 1948). Marshall and Palmer (1948) proposed the well-known formula based on observations of

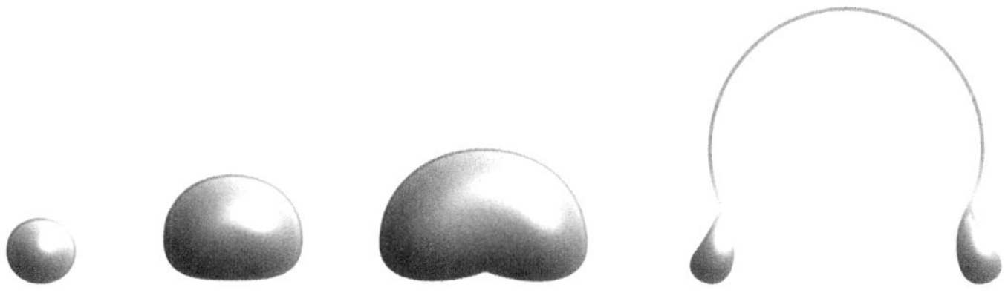


**Fig. 3.7** Typical evolution of raindrop shapes from the initial spherical shape to the final bag shape that results in the drop disruption

raindrops captured on a dyed filter in Ottawa. They correlated the distribution with the radar echo return of rain and introduced the following exponential distribution that is still being used:

$$N(D) = N_0 \exp(-\Lambda D) \tag{3.7}$$

where $D$ is the equivalent diameter of the drop and $N(D)$ (in $m^{-3}$ $mm^{-1}$) is the drop concentration of equivalent diameter D. The factor $\Lambda$ (in $mm^{-1}$) is a function of the rainfall rate R (in mm $h^{-1}$):

$$\Lambda = 4.1 R^{-0.21} \tag{3.8}$$

This is an empirical formula whose values do not vary. $N_0$ is the intercept of the curve with the y axis and has the value of $8 \times 10^3$ $m^{-3}$ $mm^{-1}$ (see the original graph of Marshall and Palmer 1948 in their Fig. 2). In the semi-logarithmic plot the distributions are straight lines that converge to the value of $N(D) = 8000$ for $D = 0$. For each fixed R, the decrease of the diameter is monotonic, and the slope decreases with increasing rainfall intensity.

The Marshall and Palmer size distribution has been widely used but obviously presents problems when compared to observed distributions. In fact, it describes the distribution of larger drops well while tending to overestimate that of smaller drops. Moreover, the function is the result of averaging over many observed spectra, which implies that any real instantaneous size distribution may deviate significantly from the proposed average. Size distributions vary considerably during a rainfall event as shown by Paul (2000). The rainfall intensity may vary very little, but the size distribution may change dramatically.

One more consideration needs to be made on the Marshall and Palmer expression. The observations of Marshall and Palmer lacked data for $D < 1$ mm because of instrumental limitations. Successive observations by several authors have shown that normally a maximum is detected at the very small size end and thus the Marshall and Palmer distribution cannot be used for such sizes: a modified gamma or a log-normal function stand better chances to describe small size spectra.

The variation of the shape of the size distributions of cloud and raindrops during a single rain episode obviously translates in a wealth of size distributions that

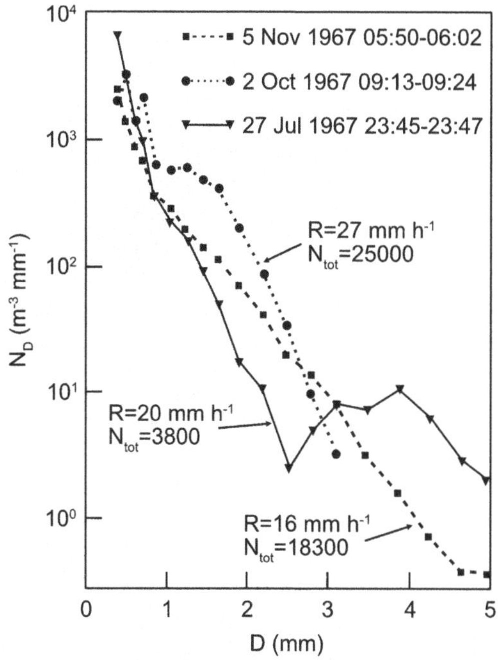


**Fig. 3.8** Examples of measured size distributions of raindrops for different rain episodes. Indicated for each rain episode are the duration, the average rainfall intensity and the total number of drops counted ($N_{tot}$). The distributions of October and November were measured during nearly constant stratiform rain while the distribution of July was recorded during a thunderstorm. (Adapted from Joss et al. 1968. © The Authors and IAMAS)

characterise the specific rain showers of the various seasons. An example is shown in Fig. 3.8 for two different stratiform rain episodes during autumn and a summer thunderstorm. Note the monomodal shape that becomes bimodal with the increase of the rainfall rate and the inception of convection.

Such differences in the raindrop size distributions depend on various factors still under investigation, but they relate to the microphysics of rain formation as demonstrated by the conceptual model of Dolan et al. (2018) (Fig. 3.9). More studies are needed to better understand how raindrops are distributed globally also in view of their parameterisations in numerical weather and climate models.

### 3.2.4 Composite Size Distribution Functions

From the above considerations on the effectiveness of the various mathematical forms in fitting observational raindrop spectra data it becomes evident that simple forms such exponential or gamma do not fully describe the complete spectrum of drops. In particular, the bimodal or multimodal shapes are far from being adequately described. More evidence comes from Fig. 3.10 (You et al. 2022) where the inadequacy of the Marshall and Palmer exponential function fit is evident. The graph shows that the inadequacy increases with increasing rainfall rate. This is obviously linked to the microphysical processes behind the rainfall production in a cloud, especially complex in thunderstorm clouds.

One way of fitting bimodal size distributions is the double gamma, which makes use of two gamma distributions added together. Another method involves adding together an exponential and a gamma function:

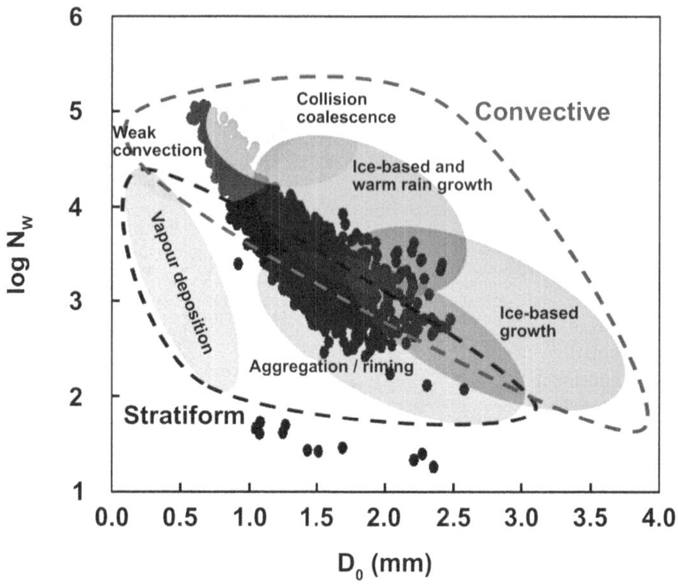


**Fig. 3.9** Estimation of dominant cloud microphysical processes using the conceptual model by Dolan et al. (2018) using observations at Daegwallyeong (Korea) from 2010 to 2011 in log$N_w$–$D_0$ space ($N_w$ is the intercept of the size distribution and $D_0$ the median diameter). (Adapted from Cha et al. 2023. © The Authors under CC-BY-4.0 license)

$$N(D) = N_1 \exp(-\Lambda_1 D) + N_2 D^{\mu_2} \exp(-\Lambda_2 D) \tag{3.9}$$

An example is displayed in Fig. 3.11 where the composite function of Eq. 3.9 is instrumental in shaping a distribution that fits well the experimental data of Okazaki et al. (2023).

### 3.2.5 Snowflake Size Distributions

If it is difficult to describe raindrop size distributions because of drop deformations, it becomes even more difficult to write empirical functions for the size distributions of snowflakes. Snowflakes have a complex variety of geometrical shapes far from being spherical and thus the diameter of the hydrometeor depends upon its orientation in space. The way around the problem is to describe the distribution in terms of the diameter of the drop of equivalent volume resulting from the melting of the snowflake. This is the technique used in one of the first works on observing size distributions of snowflakes conducted at the McGill University by Gunn and Marshall (1958). The authors sampled snowflakes on an angora wool surface and then brought the wool surface inside to allow melting of the crystals. The crystals melted without wetting the wool and formed drops that wetted a filter paper dyed with a violet dye. In this way the diameter of the equivalent volume drops could be determined and measured. The resulting distributions are shown in Gunn and Marshall (1958, their Figure 5).

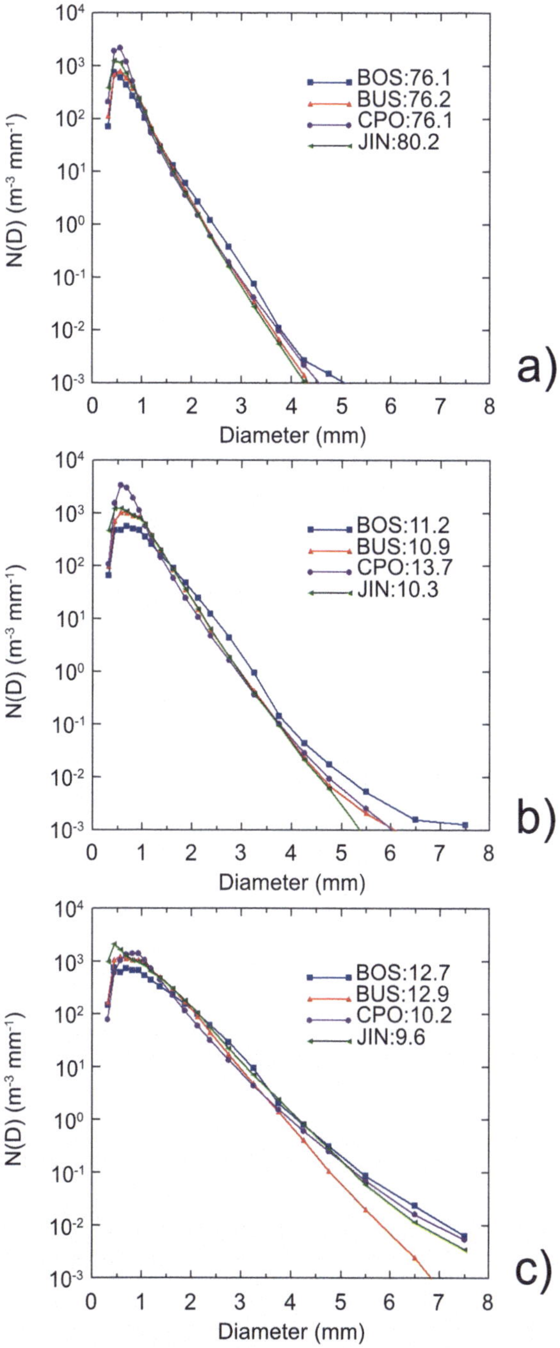


**Fig. 3.10** Average raindrop size distributions with respect to rain rate (**a**) 0.1 < R < 5.0 mm h$^{-1}$, (**b**) 5.0 < R < 10.0 mm h$^{-1}$, and (**c**) R > 10.0 mm h$^{-1}$. Data were collected at four different sites: Daegwallyeong (CPO), Jincheon (JIN), Boseong (BOS) and Busan (BUS). The numbers in the legend represent the percentage of each rain rate. (You et al. 2022. © The Author(s) under CC-BY-4.0 license)

The average size distribution with an equal snowfall intensity is an exponential distribution as shown in Fig. 3.12 maintaining the exponential form with increasing snowfall intensity. Harimaya et al. (2000) found that these size distributions agree

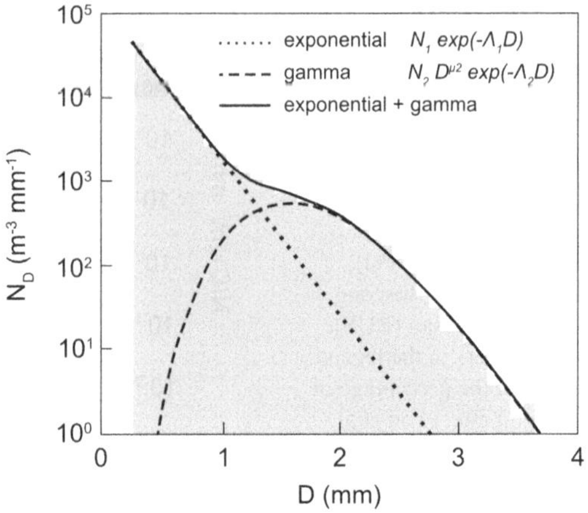


**Fig. 3.11** Example of fit a composite exponential + gamma distribution (solid line) to a raindrop spectrum measured by a 2D video disdrometer (grey histogram). The final composite function fit has the form described in Eq. 3.9. $N_1$ is 1.34 × 105, $\Lambda_1$ 4.27, $N_2$ 5.72 × 105, $\mu_2$ 12.5 and $\Lambda_2$ 9.65. (Adapted from Okazaki et al. 2023. © The Authors under CC-BY-4.0 license)

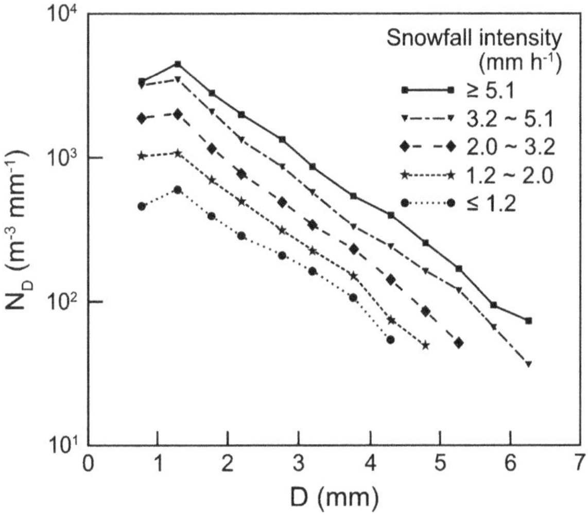


**Fig. 3.12** Size distributions of snowflakes grouped per snowfall intensity. (Adapted from Harimaya et al. 2000. © The Authors under CC-BY-4.0 license)

with those by Gunn and Marshall (1958) regarding the exponential distribution, but they differ from them in the fact that $N_0$ and $\Lambda$ do not decrease with increasing snowfall intensity. The Japanese group found that the density and riming proportion of snowflakes determine the slope. In fact, the slope of the size distribution becomes less pronounced when snowflakes have low density and are not composed of rimed snow crystals.

The same Japanese research group noted that the intercept parameter $N_0$ in different regions of Japan (Harimaya et al. 2004), becomes larger according to the distance from the coast, even under the condition of equal snowfall intensity. They also found that the slope value of the size distributions in the case of snowfalls in

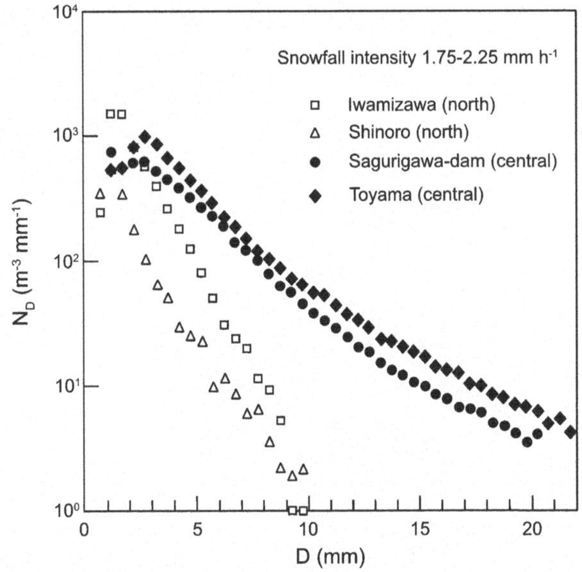


**Fig. 3.13** Snowflake size distributions collected at stations in central and northern Japan under the condition of snowfall intensity of 1.75–2.25 mm $h^{-1}$. (Adapted from Harimaya et al. 2000. )

central Japan (Toyama and Sagurigawa-dam) is smaller than the slope values in northern Japan (Shinoro and Iwamizawa) (Fig. 3.13). The authors argue that this is due to the sticking strength of snow crystals which is stronger under warmer environmental conditions.

Recent studies have tackled the problem of the primary modes of snowfall PSDs as a crucial aspect of remote sensing snowfall retrievals (King et al. 2024a, b).

## 3.3 Atmospheric Aerosol

Normally atmospheric aerosol particles are perceived as part of atmospheric pollution to be possibly removed to purify air. This is not correct when we consider clouds and precipitation formation.

Aerosol particles are essential for the formation of droplets and ice crystals in a cloud. This fact was discovered already in the last quarter of the nineteenth century. The notion of water vapour that condenses when it reaches saturation is in fact wrong in the atmosphere where cloud condensation nuclei (CCN) are necessary to form droplets. At the same time, ice nucleating particles (INP) are the substrate to form ice crystals unless, as we will see later, the temperature reaches −40 °C or even below this limit. It is thus fundamental to discuss the properties of the atmospheric nuclei for the formation of precipitation. More details can be found in Friedlander (1977), Pruppacher and Klett (2010) and Lenoble et al. (2013). A simple scheme of aerosol sources and processing in the atmosphere is shown in Fig. 3.14.

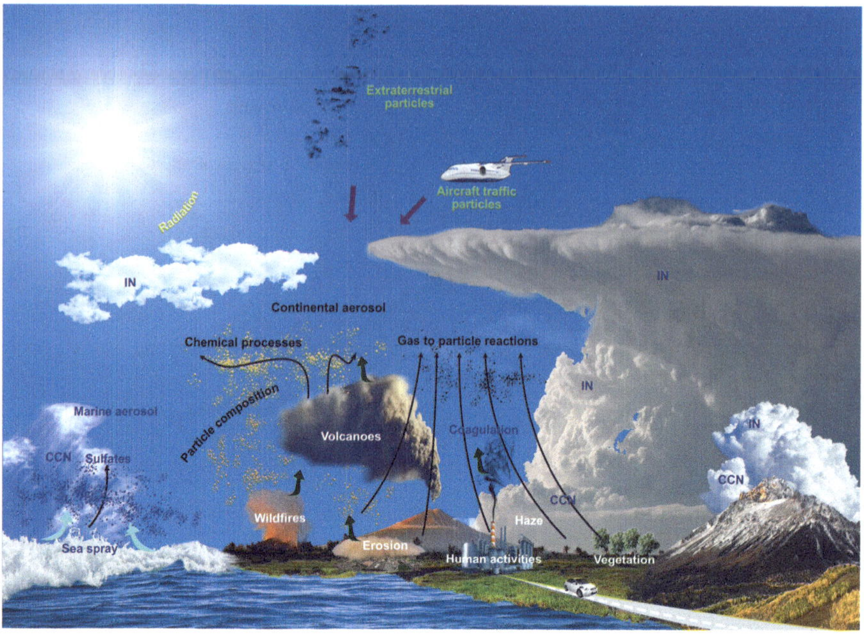


**Fig. 3.14** Simple scheme of atmospheric aerosol sources and their processing as condensation and ice nuclei for cloud formation

### 3.3.1 Sizes

Aerosol particle sizes span a wide range of dimensions from very large dust particles in the 0.1–1 mm range that remain close to the surface and have a relatively short residence time in the atmosphere to much smaller particles that reach higher levels. The aerosol size classification dates to Junge (1963) and consists of three main radius categories:

1. *Aitken nuclei* $r < 0.1\ \mu m$
2. *Large particles* $0.1 \leq r \leq 1\ \mu m$
3. *Giant particles* $r > 1\ \mu m$

Aitken particles are named after John Aitken (1839–1919) a Scottish meteorologist, physicist and marine engineer who was among the first to measure the number of dust and fog particles in the atmosphere. A general overview of aerosol sizes and their main sources and sinks is provided in Fig. 3.15. Note that large particles are divided into at least two categories depending on their generation mechanisms. Large particles are important scatterers of solar radiation since their size is in the same range of the visible wavelengths (0.4–0.7 μm). Moreover, they have an impact on human health since they tend to be trapped in our lungs with notable implications on respiratory diseases.

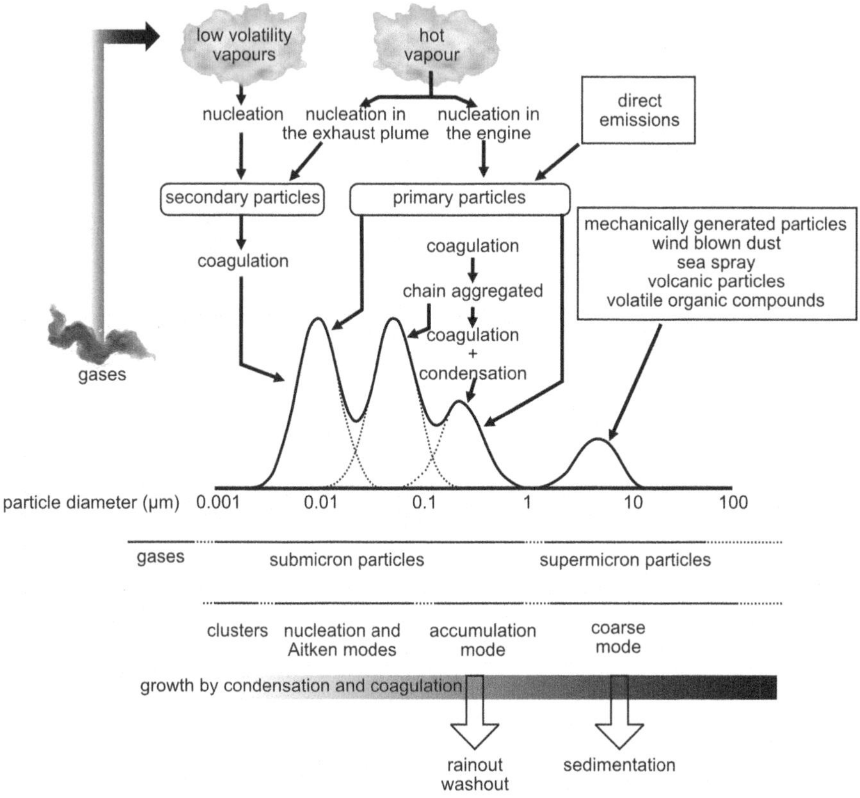


**Fig. 3.15** Aerosol sizes together with main sources and sinks

Aerosol particles that act as nuclei to nucleate cloud hydrometeors belong to all sizes, but not all aerosol particles serve as nuclei for cloud formation. One more useful aerosol classification is due to Whitby (1978) and is shown in Fig. 3.15. The Whitby classification considers the aerosol generation mechanisms, and its modes may overlap with those of other generation mechanisms (Fig. 3.16).

The modal size of the distribution defines the classification naming scheme:

1. *Nuclei mode*—Particles with diameter $D < 0.1$ μm. The mode is further divided into nucleation mode ($D < 0.03$ μm) and Aitken mode ($0.03 \leq D \leq 0.1$ μm).
2. *Accumulation mode*—Particles with $0.1 \leq D \leq 2.5$ μm.
3. *Coarse mode*—Particles with $D \geq 2.5$ μm.

The diameter is intended to be the aerodynamic diameter, i.e. the diameter of a sphere of density 1 g cm$^{-3}$ that has the same terminal settling velocity in the gas as the particle under consideration. In other words, it is an equivalent spherical diameter.

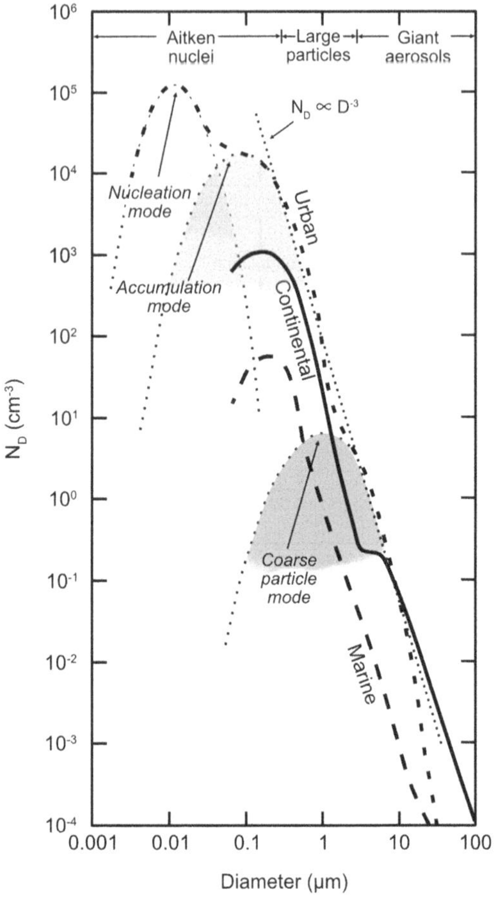


**Fig. 3.16** Aerosol classification following Junge (1963) (top) and Whitby (1978) (modal naming, shaded curves). Superimposed are typical distribution spectra of continental, marine and urban particles. An exponential distribution fit ($D^{-3}$, the Junge size distribution, dotted straight line) is also plotted for reference. (Adapted from public domain)

Particles with D < 1 μm are often called *fine* particles (FP) and those with D < 0.1 μm *ultrafine* particles (UFP). The naming depends on the application.

Different applications may also call for a different way of dividing aerosol dimensions. In fact, the diameter may not be a unique characterisation parameter. In Fig. 3.17 we see that smaller particles dominate the number concentration while larger particles contribute mainly to mass (volume). This is a very important consideration when approaching aerosol as nucleating agent since the nucleation power depends on the external surface of the particle and its mass content. Note that the characterisation of the size distribution provides information on the type of atmospheric processes resulting in the distributions (Morawska et al. 1999).

It is clear from the spectra that we have examined so far that aerosol size distributions tend to be bimodal and sometimes even trimodal. Thus, their mathematical description needs to avail itself of sums of log-normal, gamma and exponential functions rather than trying to describe the spectra with single functions.

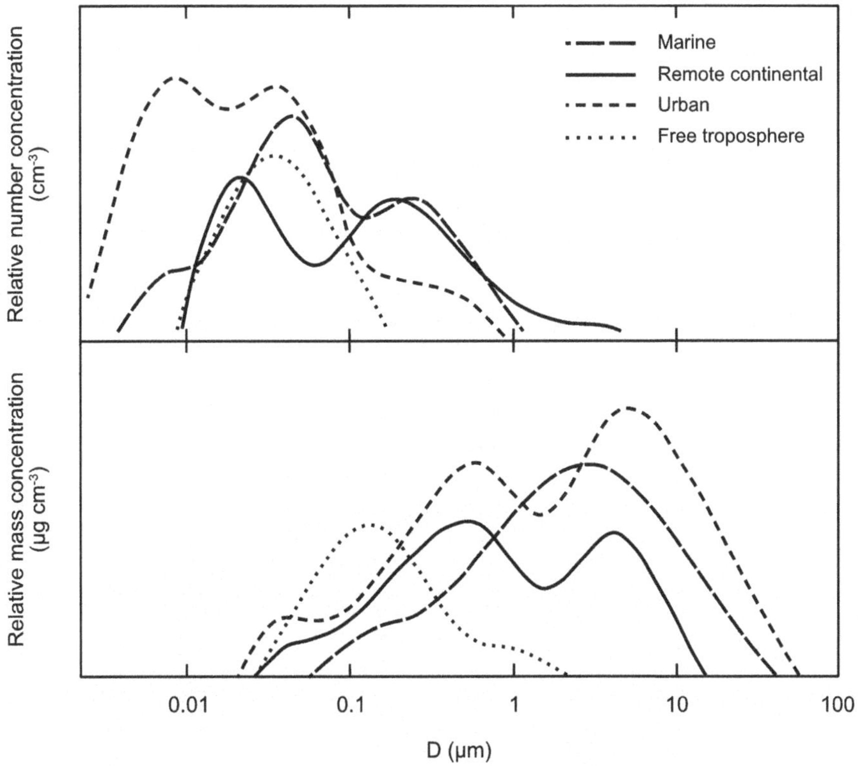


**Fig. 3.17** Typical size (top) and mass (bottom) distributions of atmospheric aerosol particles from different sources. (Adapted from public domain)

## 3.3.2 Global Concentration and Variation with Height and Time

Aerosol number concentration varies considerably with location depending on the proximity of sources. Variations span from a few hundred particles per $cm^{-3}$ over the oceans to over $10^6$ $cm^{-3}$ in highly polluted areas near cities or industrial plants. Over the mountains the concentration is typically from $10^2$ to $10^3$ $cm^{-3}$, usually decreasing with height. Mass concentration varies from 1 µg $m^{-3}$ over ocean to about 10 µg $m^{-3}$ in urban areas. In dust storm conditions mass concentration can be of several hundred µg $m^{-3}$. A global overview of the aerosol concentration can be derived from Fig. 3.18 in terms of the aerosol optical depth (AOD), a measure of the amount of light that aerosol scatter and absorb in the atmosphere. An AOD of 0.05 indicates clear skies whereas a value of 1 indicates hazy conditions.

Aerosol concentrations vary also with height. As we have seen, the atmosphere is a primary source of aerosol particles, but their major source is the Earth's surface. It is not surprising then that aerosol concentration is at its maximum close to the surface and decreases with height. This decrease is not homogeneous and monotonic since sources and sinks are present within the atmosphere.

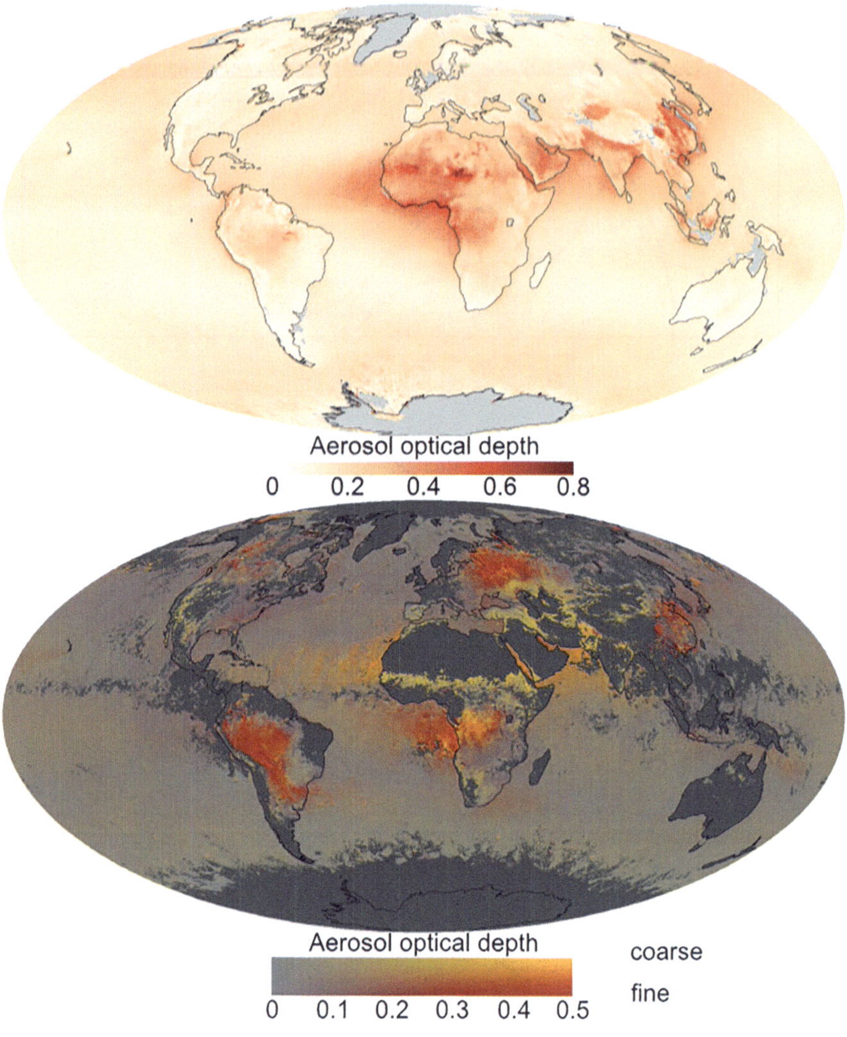


**Fig. 3.18** Top: average aerosol optical depth from June 2000 through May 2010 measured by the Multi-angle Imaging Spectroradiometer (MISR). Red indicates high aerosol concentration while beige low concentration. Bottom: average pattern of atmospheric aerosol from MODIS data. Yellow areas indicate predominantly coarse particles (dust) and red areas fine aerosols. (Source: NASA)

Junge (1961) was the first to show that fine aerosol particles have maximum number concentration close to the surface, which is then the source. The concentration decreases nearly exponentially up to 6 km altitude. From to 6 to 10 km the concentration remains nearly constant and above 10 km the Aitken nuclei decrease again with height. The variation with height of large particles is substantially different. The decrease with height of the number concentration continues through the troposphere until we reach the tropopause. Above the tropopause the concentration increases again until the middle troposphere depending on location and season. The stratospheric maximum is well known as the *Junge layer* or *stratospheric aerosol layer* that has been recently confirmed by new observations that locate it between 20 and 40 km altitude. Finally, giant particles are mostly confined in the lower troposphere due to their sources and dimensions.

Most aerosol particles are very small and are thus influenced by Brownian motion. Particles smaller than 0.5 μm are subject to such motions and collide and coagulate to form larger particles; this is the Brownian coagulation, which ceases to influence particles larger than 0.5 μm because they are too massive. This process implies that new particles are formed from particles of diverse chemical composition. Junge (1955) considered a cloud of freshly formed aerosol particles with an initial distribution in the UFP size range. In time the particles smaller than 0.5 μm are subject to Brownian coagulation and become progressively larger. At the same time the coagulation of particles larger than 0.5 μm is slower. After a certain time, the distribution becomes quasi-steady state and the process is called *ageing of aerosols*. The aerosol aging causes the size distribution to become quasi-straight in log-log coordinates and Junge suggested thus a power-law formulation of the aerosol size distribution (the well-known *Junge size distribution*, see also Fig. 3.16).

$$\frac{dN}{d\log r} = Ar^{\beta} \tag{3.10}$$

where $\beta \sim -3$. Once again, this distribution only represents a section of the overall spectrum and not the entire distribution.

### 3.3.3 Sources

Three most relevant pathways for the formation of aerosol particles are (see also Fig. 3.19):

1. *Pulverisation of bulk matter*—For example, breaking of rocks or dirt, breaking of bubbles at the surface of ocean waves, ejection of droplets from combustion engine exhaust.
2. *Chemical reactions*—Reaction in one phase (homogeneous reactions) or in different phases (heterogeneous reactions). When gaseous species are involved to form aerosol particles the reaction is called gas-to-particle conversion (GPC). Also drop-to-particle conversion (DPC) is possible and occurs frequently in clouds. Gases and particulate are mostly soluble in water and droplets become aqueous solutions. When the droplet evaporates the dissolved material frequently crystallizes becoming a solid particle.
3. *Condensation*—Either the homogeneous condensation of gas to form liquid droplets or solid particles or the freezing of liquid into solid, or heterogeneous condensation where gases condense into liquids or solids, or liquids

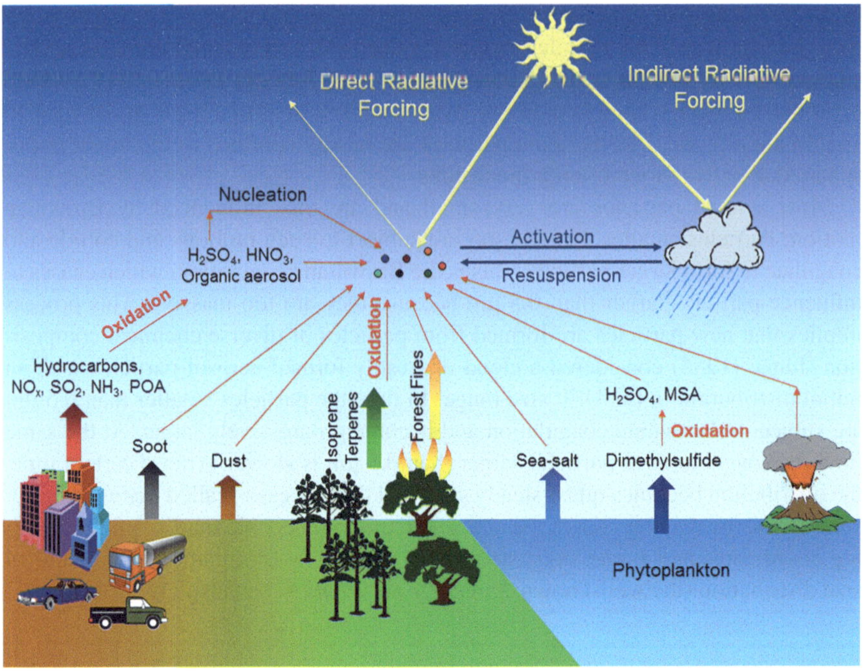


**Fig. 3.19** Atmospheric aerosol sources and major chemical mechanisms. (Graphic courtesy of Pacific Northwest National Laboratory)

freeze on solids. An example of homogeneous condensation occurs during biomass burning.

Aerosol particles directly injected into the atmosphere without undergoing GPC or DPC processes are called *primary aerosol particles*. Particles that are processed in the atmosphere via GPC or DPC are called *secondary aerosol particles* (SAP).

The major sources of aerosol particles in the Earth's atmosphere are manifold. Let's mention some of them.

1. *Fragmentation of the land surface*—Saltation of dust from the surface of deserts during dust storms (clay particles, $SiO_2$, $Al_2O_3$, $Fe_2O_3$ and MgO): mostly giant particles with a relatively low atmospheric residence time. Creation of particles through biomass burning mostly belonging to the accumulation mode with a median diameter of 0.15 μm with smaller amounts of coarse and nuclei modes. Particles from volcanic eruptions: sulphate particles and gases ($CO_2$ and $SO_2$) that can be precursors for the formation of SAPs. Aerosols from industrial activities: burning of fuels (wood, coal, etc.), sulphates, black carbon, and photochemical smog mainly due to vehicular traffic.
2. *Ocean surface processes*—Particles produced at the surface of the ocean are mainly salt particles, NaCl, but there are smaller amounts of $K^+$, $Mg^{2+}$, $CO_3^{2-}$,

$SO_4^{2-}$ and biological material. The production of this aerosol happens via the *bubble burst mechanism* (also called *sea spray mechanism*) under the action of the wind stress on the ocean waves. Dimensions of marine aerosol vary from less than 0.1 to more than 10 μm.

3. *Biogenic aerosol*—Plants are the major sources of all biological aerosol. Pollens are a typical example with particles larger than 10 μm. Then we must consider plant fragments, spores, viruses, and bacteria from a few microns to less than 1 μm. Several of these particles act as IN for freezing hydrometeors. An example is represented by cellulose particles recently investigated.
4. *Extraterrestrial particles*—The Earth, like any other planet, is constantly hit by meteoroids, which are subject to ablation when they reach the atmosphere. The explosion of meteoroids due to the extreme heating caused by the air friction generates particles of sand dust to pebble size that evaporate in the atmosphere. However, some larger particles may break up into smaller fragments and become aerosol particles. These processes happen in the mesosphere at heights between 80 and 95 km. Meteoric aerosols contain Fe, Si, Mg, S, Ca, Ni, Al, Cr, Mu, Cl, K, Ti, and Co. For some time after their discovery, the extraterrestrial particles were thought to have an influence on the formation of precipitation, but studies have not confirmed these speculations. More recent measurements have shown that the downward transport from the mesosphere of nanometer-size particles may make available condensation nuclei to form sulphuric acid and water in the stratosphere, perhaps contributing to the formation of polar stratospheric clouds (PSC).

## 3.4 Nucleation of Cloud Droplets

Condensation of water vapour is normally thought to happen when the saturation is exceeded. However, experiments tell us that in clean air the vapour pressure can significantly exceed the saturation threshold without condensation. The phenomenon is called *supersaturation*. The process of phase change from vapour to liquid water drops is called nucleation and two are the categories of nucleation:

1. *Homogeneous nucleation*. The aqueous phase starts from pure vapour without the intervention of any foreign substance.
2. *Heterogeneous nucleation*. The aqueous phase is produced from pure vapour with the involvement of particles.

Homogeneous nucleation does not happen in the terrestrial atmosphere except in some very limited cases. When the humidity is relatively low, water vapour molecules are far from each other while the opposite is true when we are near saturation. In saturation conditions water vapour molecules interact strongly and form molecular clusters of varying number of molecules. When the molecule is isolated, it is called a *monomer*, a *dimer* when two molecules stick together, a *trimer* with three molecules, and generally an *i-mer* with *i* molecules. In saturation conditions these clusters of molecules may collide and coagulate to form quasi-droplets. They may also disintegrate again because of thermal agitation. These clusters, not yet activated to become droplets, are called *embryos*. When the activation takes place, the embryo transforms into a *germ*. These quasi-droplets have physical properties like those of water droplets, including curvature. When the relative humidity reaches several hundred percent, the quasi-droplets stabilise and form real droplets. By convention the threshold defining a significant rate of homogeneous nucleation is 1 $cm^{-3}$ $s^{-1}$ and this is called the critical saturation ratio $S_c$. Miller et al. (1983) have shown that $S_c$ decreases with increasing temperature. Approximate values are 4.3 at 273 K, 6.3 at 250 K and 3.5 at 290 K.

However, while this result can be attained in the lab, it is generally not observed in the atmosphere during cloud formation since supersaturation rarely exceeds 1 or 2%. We will not expand on the homogeneous nucleation process, which is well described by Wang (2013).

### 3.4.1 Cloud Condensation Nuclei (CCN)

Nucleation is in fact normally heterogeneous and happens through vapour deposition on suitable aerosol particles that we have already introduced, cloud condensation nuclei (CCN). As we will see shortly hereafter, an aqueous solution drop will possibly survive even in an unsaturated environment due to the solute effect while a pure water droplet will evaporate even when the relative humidity is 100%. Thus, an aerosol particle made of material that can dissolve in water to form an aqueous solution is a suitable CCN candidate. A well know example is the sea salt (NaCl).

Since not all aerosol particles dispersed in the atmosphere can act as CCN, the most important question is: how many particles can serve as CCN? The answer is difficult due to sparse coverage of in situ measurements and difficulties associated with retrievals from satellites. A global appraisal of magnitude, source and temporal and spatial distribution cannot be easily obtained. The answer clearly depends on the environmental humidity conditions.

Direct measurements of CCNs are performed through instruments called condensation particle counters (CPC) (Fig. 3.20). They are based on a chamber where supersaturation is reached in controlled conditions: a wet surface is maintained as a source of water vapour and the temperature is allowed to vary along the path of the incoming air from the environment. The air experiences a relative humidity change due to temperature changes. When the air is subsaturated it is generally transparent. When supersaturation is reached some aerosol particles act as CCN and form droplets that are illuminated by a laser beam and detected by a photomultiplier. The droplet count then corresponds to the number of CCN in the air sample. Obviously, the number of CCN will increase with increasing supersaturation.

However, counting particles with a CPC is a point measurement and does not help much in drawing global maps of CCN. Recent studies have produced global datasets based on input from emission inventories using aerosol observations, with a strong tie to satellite retrieved AOD, which is assimilated as a constraining factor in reanalyses. Block et al. (2024) have produced a unique dataset of 3-D CCN

**Fig. 3.20** Schematic of a condensation particle counter, operated by diffusional thermal cooling

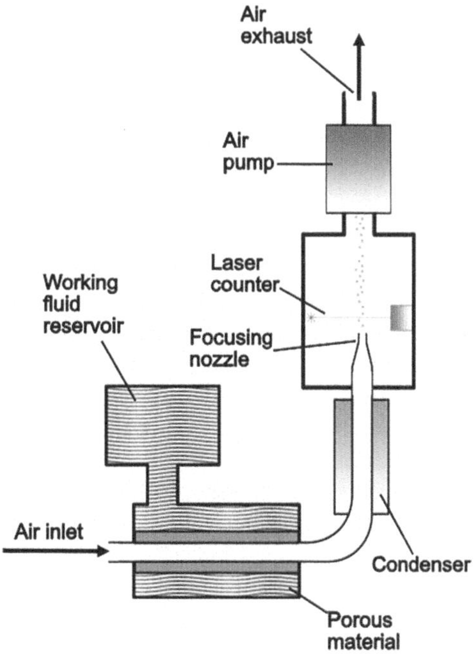


number concentrations with global coverage for various supersaturations and aerosol species covering the years 2003–2021 with daily frequency. Time is coming when global models can be evaluated, and aerosol-cloud interactions better studied.

What are the concentrations of CCNs? Typical concentrations are approximately $10–10^2$ cm$^{-3}$ over ocean and $10^2–10^3$ cm$^{-3}$ over land. It is thus evident that the concentration over land is approximately one order of magnitude greater than over ocean. This is consistent with the concentration of droplets over the two environments that we have discussed in Sect. 2.1.3. Note that, however, in polluted and industrial areas the concentration of nuclei can be 2–3 orders of magnitude greater than that in general overland conditions.

The composition of the CCN population is quite diverse. Mostly they are hygroscopic particles generated by many sources. The most common are NaCl (sea salt) and $(NH_4)_2SO_4$ (ammonium sulphate). There are also insoluble particles that can act as CCN, for example those coming from the degradation of soils. It is worth mentioning that biological particles have a nucleating power such as bacteria (e.g., Lazaridis 2019). Their distribution around the globe varies with their composition as shown by Block et al. (2024).

Is chemical composition the only parameter one must consider while measuring the effectiveness of an aerosol particle as a CCN? No, size was recently found as another, perhaps even more important characteristic (Dusek et al. 2006). Large particles, for example, are found to be more effective than Aitken nuclei. Thus, we need to know the chemical composition, but also the size distribution to assess the nucleation efficiency of an aerosol population.

The difference between the CCN population over ocean and over land is described in Fig. 3.21. Note the order of magnitude difference in the CCN concentration between oceanic and land air masses from the observations of Jayachandran et al. (2020) (see also earlier observations of Twomey and Wojciechowski 1969) demonstrating that over ocean the number of CCN is substantially lower, although their dimensions are larger. The effective hygroscopicity parameter *k*, which describes the influence of diameter and chemical composition on the CCN activity of aerosol particles (Petters and Kreidenweis 2007, 2008), simulated globally by Pringle et al. (2010) and observed by Pöhlker et al. (2023) shows how active nuclei are on a global mean in both environments. Marine nuclei are more hygroscopic than continental nuclei, but their hygroscopic power is more concentrated in the lower atmosphere due to the presence of large nuclei whose concentration decreases with height. Small continental nuclei are associated with a constant hygroscopicity throughout the atmosphere since they consist of small particles that are not subject to sedimentation.

A rapid way of expressing CCN activity is by its *activity spectrum*, i.e. the number of particles per unit volume that are activated to become cloud droplets, as a function of supersaturation. The spectra are measured in a cloud chamber through a CPC instrument. Counts are taken at steps of increasing supersaturation in the range from 0.3% to 1%. The nucleus count is approximated by a power law relation:

$$N_c = Cs^k \tag{3.11}$$

where $s = (S - 1) \times 100\%$ is the percent supersaturation ($S$ is the saturation ratio), $N_c$ the number of nuclei per unit volume activated at supersaturation less than $s$, and $C$ and $k$ are parameters that depend on the air mass type. Typical values are $C = 30$

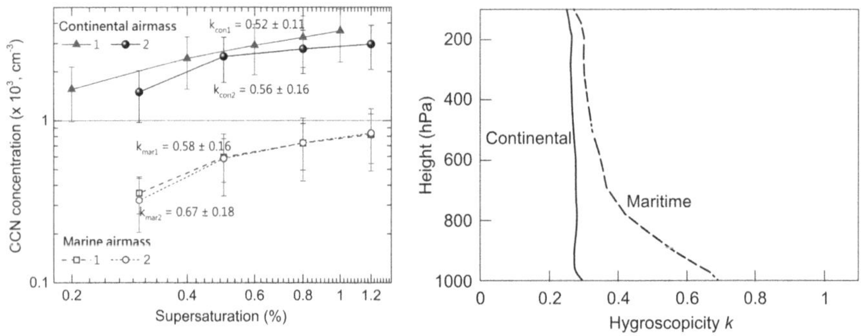


**Fig. 3.21** Mean CCN concentration for different supersaturations (the power law fit value of the hygroscopicity parameter *k* of each spectrum is also given) for two continental and maritime air masses by Jayachandran et al. (2020) (left panel). The global mean of the vertical distribution of *k* in continental and maritime air masses from model runs by Pringle et al. (2010) (right panel) is also shown. (Left: from Jayachandran et al. 2020. Right: adapted from Pringle et al. 2010. © The Authors under CC-BY-4.0 license)

to 3000 $cm^{-3}$ and $k = 0.3$ to 1. For maritime air, and $C = 300$ to 3000 $cm^{-3}$, and $k = 0.2$ to 2 for continental air.

The activation power law is derived in several works from the Junge-type power law. New parameterisations have been recently proposed reconciling the use of log-normal distributions with the original power law (e.g., Khvorostyanov and Curry 2006).

CCN of some sorts are always present in the atmosphere in considerable amounts and clouds form when there is vertical air motion and sufficient moisture. Precipitation is more likely to occur when the CCN population is formed by a few large particles instead of many small particles. The comparison between the observed droplet spectra with the activity spectra discussed above have demonstrated a relation between the CCN population and the population of the cloud droplets deriving from them. The development of a cloud after its initial formation stage and of its precipitation power depend on several large-scale phenomena such as the updraft speed and the moisture supply. However, at least to some extent, the initial microstructure of a cloud determines how it is susceptible to producing precipitation, and how much time is required before the first precipitation develops.

### 3.4.2 Heterogeneous Nucleation

How can we determine the chance that collisions and aggregations of water molecules lead to the formation of an embryo that will be stable and continue to grow as a cloud droplet during the initial stage of cloud development?

#### 3.4.2.1 The Kelvin Equation

The embryo will be stable as it exceeds a given critical size under which it will soon evaporate. The critical size is determined by the balance between forces that cause growth and decay. In the case of homogeneous nucleation, the growth rate depends on the partial pressure of water around the droplet because it determines the rate at which water molecules hit the growing surface. The decay depends strongly on temperature and on the surface tension of the droplet.

The molecules at the surface must have sufficient energy to overcome the binding force that trap them into the liquid. If an equilibrium is reached, then evaporation and condensation rates will be balanced, and the vapour pressure is equal to the equilibrium or saturation vapour pressure.

The equilibrium vapour pressure over a droplet depends upon its curvature and follows the equation derived by Thomson (1871) (Lord Kelvin, thus the name of Kelvin equation):

$$e_s(r) = e_s(\infty)\exp\left(\frac{2\sigma}{rR_v\rho_L T}\right) \tag{3.12}$$

where $e_s(r)$ is the saturation vapour pressure over a spherical droplet of radius r with surface tension $\sigma$, $\rho_L$ at the temperature $T$. $R_v$ is the gas constant for water vapour and $e_s(\infty)$ the saturation vapour pressure over bulk water. This equation was derived to explain the form of the meniscus of the water raising inside capillary tubes.

The surface tension $\sigma$ is the free energy per unit surface area of the liquid, i.e. the work per unit area required to extend the surface of the liquid at constant temperature. At meteorological temperatures, the surface tension of water is approximately $7.5 \times 10^{-2}$ N m$^{-1}$.

The growth rate of a droplet is proportional to the difference $e—e_s(r)$, where $e$ is the actual ambient vapour pressure. When $e—e_s(r) < 0$ the droplets will tend to decay and when $e—e_s(r) > 0$ will tend to grow. The critical size $r_c$ is defined as the radius for which $e—e_s(r_c) = 0$, or $e_s(r) = e_s(\infty)\exp\left(\frac{2\sigma}{r_c R_v \rho_L T}\right)$. Then, the critical size is

$$r_c = \frac{2\sigma}{R_v \rho_L T lnS} \tag{3.13}$$

where $S = e/e_s(\infty)$ is the saturation ratio. For a droplet forming by random collisions, it will be stable when growing larger than the critical radius.

High supersaturations are required for the stability of very small droplets (see for example Rogers and Yau 1979, their table 6.1). When supersaturation is 1% ($S = 1.01$) droplets with radii smaller than 0.121 μm will tend to evaporate. Thus, homogeneous nucleation is impossible at a relative humidity of 100% and a droplet cannot even start to form in these conditions. An illustration of the relative humidity and radius required for droplet to grow in a cloud is provided by Stull (2017) and plotted in Fig. 3.22.

In homogeneous nucleation, critical size droplets are formed by random collisions of water molecules. If one more molecule impinges the critical droplet, the

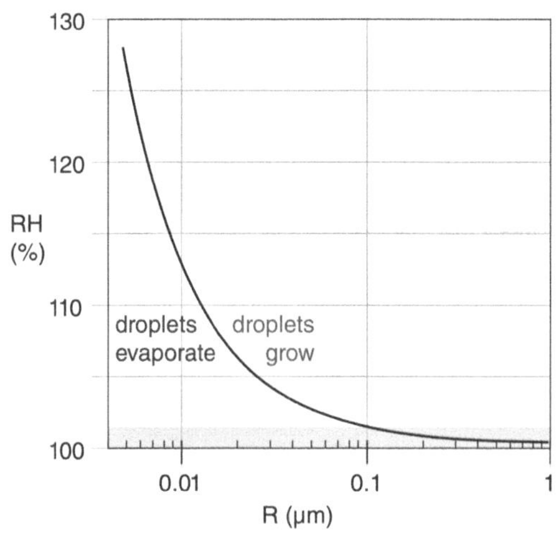


**Fig. 3.22** Curvature effect, showing the relative humidity (RH) needed in air for a droplet of radius R to grow. The curve is given by Kelvin's equation, for pure water at 0 °C. The shaded region highlights the range of RH typically found in clouds. (From Stull 2017. © The Author under CC-BY-4.0 license)

droplet will become supercritical, i.e. with increasing size $e_s(r)$ decreases and the growth rate (proportional to $e$-$e_s(r)$) increases. Supercritical droplets will then grow spontaneously. The rate of nucleation is thus the rate at which supercritical droplets are formed.

It is shown that, for a given temperature, the nucleation rate increases from very small values to extremely large values over a very narrow range of $S$. Once again, these conditions are attainable in the lab but not found in the atmosphere.

#### 3.4.2.2 The Raoult's Law

The nucleation of droplets on particles requires, as we have seen, hygroscopic particles. The rate of droplet formation depends on the number of CCNs present and not on collision statistics. Aerosols are classified in general as hygroscopic, neutral, and hydrophobic. Nucleation on a neutral particle requires a higher supersaturation than homogeneous nucleation. Supersaturation would be even higher for the nucleation on a hydrophobic particle, i.e. a particle that resists wetting.

A non-volatile dissolved substance contributes to lower the equilibrium vapour pressure of a liquid (Fig. 3.23). Very roughly, solute molecules are evenly distributed in the water (solvent) and, therefore, some of them occupy surface sites that would otherwise be occupied by water molecules. Thus, the solute prevents water molecules from evaporating from those sites. Adding more solute means that more surface sites would be occupied by solute molecules and water vapour would have even less opportunity to break hydrogen bonds and escape the liquid. The real view is more complicated by the electrostatic interactions between water and solute molecules that cause an attraction between them, but the basic result stands against experimental verification.

The evaporation rate is lowered, and this means that there will be net condensation until the water vapour flux to the surface matches the water vapour flux leaving the surface. When the equilibrium is established between the lower evaporation and condensation, the condensation will be less, which means that the saturation vapour

**Fig. 3.23** Scheme of the vapour pressure decrease from pure water (left) to a solution of water and a non-volatile substance (right). The molecules of the dissolved material are indicated by purple spheres

pressure will be lower. The equilibrium vapour pressure is less than $e_s$, the saturation vapour pressure over a flat surface of pure water.

As the amount of solute is increased, the equilibrium vapour pressure of the solu tion will decrease further. The effect drastically lowers the equilibrium water vapour pressure over a droplet and the solution droplet can be in equilibrium with the environment at a much lower supersaturation than a pure water droplet of the same size.

For a plane water surface the reduction of vapour pressure in presence of a non-volatile solute can be expressed as

$$\frac{e'}{e_s(\infty)} = \frac{n_0}{n + n_0} \tag{3.14}$$

where $e'$ is the equilibrium water vapour pressure over a solution consisting of $n_0$ molecules of water and $n$ molecules of solute. This is the well-known Raoult's Law. For dilute solutions $n \ll n_0$ and we have

$$\frac{e'}{e_s(\infty)} = 1 - \frac{n}{n_0} \tag{3.15}$$

For solutions in which the dissolved molecules are dissociated (ionic solutions) $n$ must be multiplied by the factor $i$, the degree of ionic dissociation or van't Hoff factor, which is determined experimentally. Tabulated values exist for several electrolytes, such as sodium chloride and ammonium sulphate, which we know are important CCN substances. $i \approx 2$ is a good approximation for such materials.

The number of effective ions in a solute of mass $M$ is given by

$$n = iN_0 M / m_s \tag{3.16}$$

where $N_0$ is the Avogadro's number (number of molecules per mole) and $m_s$ the molecular weight of the solute. The number of water molecules in mass m can be written likewise as

$$n_0 = N_0 m / m_v \tag{3.17}$$

where $m$ and $m_v$ are the mass and the molecular weight of water, respectively.

The mass of water is expressed by $m = \frac{4}{3}\pi r^3 \rho_L$. Thus, Eq. 3.15 can be expressed as

$$\frac{e'}{e_s(\infty)} = 1 - b / r^3 \tag{3.18}$$

where $b = 3im_v M / 4\pi\rho_L m_s$ and we note that $N_s = Mm_v/m_s$ is the number of moles of the solute.

#### 3.4.2.3 The Köhler's Equation

The equation of Kelvin and that expressing the solution effects are now combined to produce $e_s'(r)$ for the equilibrium water vapour pressure of a solution droplet

$$\frac{e_s'(r)}{e_s(\infty)} = \left[1 - \frac{b}{r^3}\right] e^{a/r} \tag{3.19}$$

where $a = 2\sigma/\rho_L R_v T$. For r not too small a good approximation of this equation is

$$\frac{e_s'(r)}{e_s(\infty)} = 1 + \frac{a}{r} - \frac{b}{r^3} \tag{3.20}$$

This is the equation of Köhler from the name of its original developer, the Swedish chemist Hilding Köhler (Köhler 1936). In this form, $a/r$ is the *curvature term*, which expresses the increase of saturation ratio over a droplet as compared with a plane surface. The term $b/r^3$ is called the *solution term* and shows the reduction of the vapour pressure due to the presence of a dissolved substance. For given $T$, $M$ and $m_s$, Eq. 3.20 describes the dependence of saturation ratio on the size of a solution droplet. The resulting curve is the Köhler's curve as shown in Fig. 3.24.

The curve interpretation is as follows. The solution (chemical) effect dominates when the radius of the droplet is small: a very small solution droplet is in equilibrium with vapour at relative humidity less than 100%. If a small amount of humidity is added, the droplet will grow until the equilibrium is reached again. This process

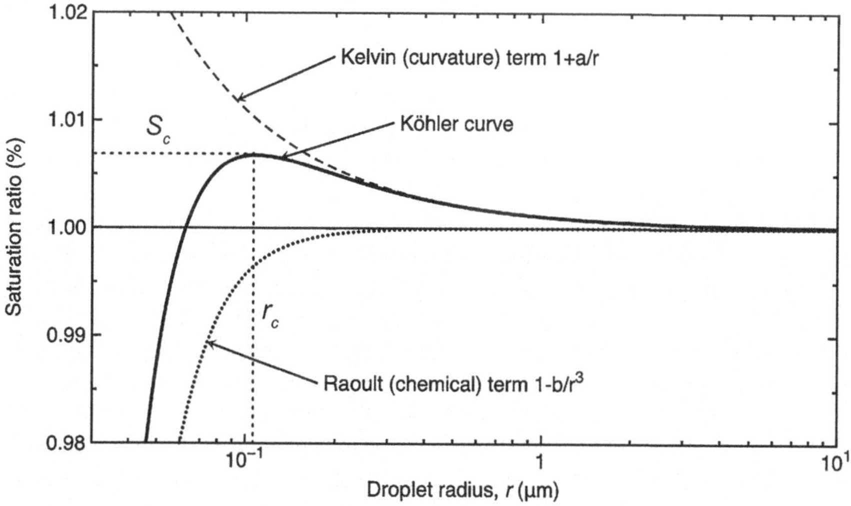


**Fig. 3.24** Equilibrium saturation ratio of a solution droplet with $a$ = 1.1× $10^{-1}$ μm and $b = 4.3 \times 10^{-6}$ μm$^3$. The number of moles is assumed $N_S = 10^{-18}$. $S_c$ and $r_c$ are the critical saturation and critical radius, respectively

of increasing the ambient humidity and allowing the droplet to grow to an equilibrium size can continue up to a 100% relative humidity or slightly above. Then, the critical saturation $S_c$ is reached corresponding to the peak of the Köhler curve (in the example in Fig. 3.24 $S_c$ is about 0.6% corresponding to a critical radius $r_c$ of approximately 0.12 μm). Until this critical point, the relative humidity had to be increased to allow the droplet to grow. When the relative humidity slightly exceeds the peak $S_c$ and the droplet radius is greater than $r_c$, the equilibrium saturation ratio falls below $S_c$. the vapour will thus continue to diffuse to the droplet that will continue to grow without the need of further increasing the ambient saturation ratio. Up to the size $r_c$ (for most droplets between 0.1 and 1 μm) the droplet is in stable equilibrium with the environment: changes of the saturation ratio will either cause the droplet to grow or evaporate until reaching equilibrium. When the droplet is smaller than $r_c$ and in stable equilibrium, it is called a *haze* particle. The CCN is said to be activated (see Sect. 3.4.1) when the droplet forming on its surface reaches the size $r_c$. Once the droplet exceeds the critical radius it will continue to grow to cloud droplet size if the ambient saturation remains above the equilibrium curve. In real clouds the growth does not continue indefinitely because many droplets are present that compete for the available vapour present in the updraft and tend to lower the saturation ratio once the condensation becomes more rapid than the production of supersaturation.

The critical values of the supersaturation and of the radius can be derived from Eq. 3.20:

$$r_c = \sqrt{3b / a} \tag{3.21}$$

$$S_c = 1 + \sqrt{4a^3 / 27b} \tag{3.22}$$

Critical conditions are plotted in Fig. 3.25 for various masses of different chemicals. $S_c - 1$ is inversely related to $r_c$, so as solute mass increases, smaller supersaturations are required to reach the critical point, and at that point the droplets will be larger. Also, different chemicals will grow to different sizes, which is one factor causing a range of drop sizes in the cloud.

The Köhler curve is not a single curve for each solute. It is rather a family of curves, which depend on the molar content of solute as shown in Fig. 3.26. The hygroscopic growth is thus sensitive to the chemical term in the Köhler equation, but the same is true for the curvature term (Wex et al. 2008). The Raoult term is dominant for smaller radii whereas the Kelvin term dominates at larger radii, with both terms acting in opposite directions as shown by Wex et al. (2008). This is due to the $r^{-3}$ versus the $r^{-1}$ dependency of the Raoult and Kelvin term, respectively.

One can imagine all kinds of scenarios that can happen when there is a distribution of CCN of different sizes and different amounts of solute. Drops with more solute have lower values for the critical supersaturation and therefore are likely to nucleate first at lower supersaturation values. We can imagine the larger CCN taking up the water first and taking up so much water that the ambient supersaturation drops below the critical supersaturation for the smaller CCN. As a result, the larger

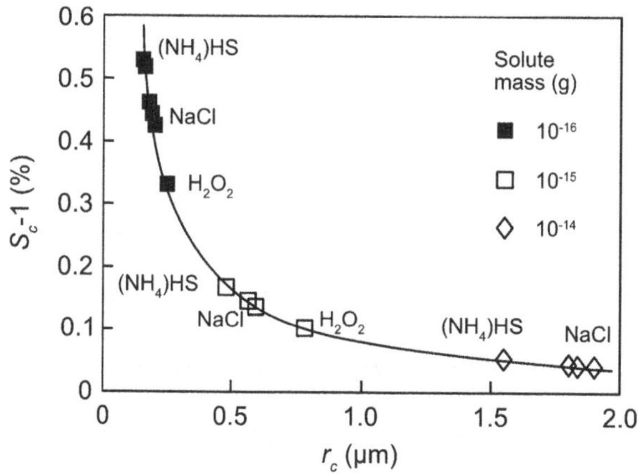


**Fig. 3.25** Critical values of supersaturation vs. critical drop radius, for various solute masses and chemicals: ammonium sulphide, sodium chloride and hydrogen peroxide. Nitric and sulfuric acid are near the sodium chloride data points. (Adapted from Stull 2017. © The Author under CC-BY-4.0 license)

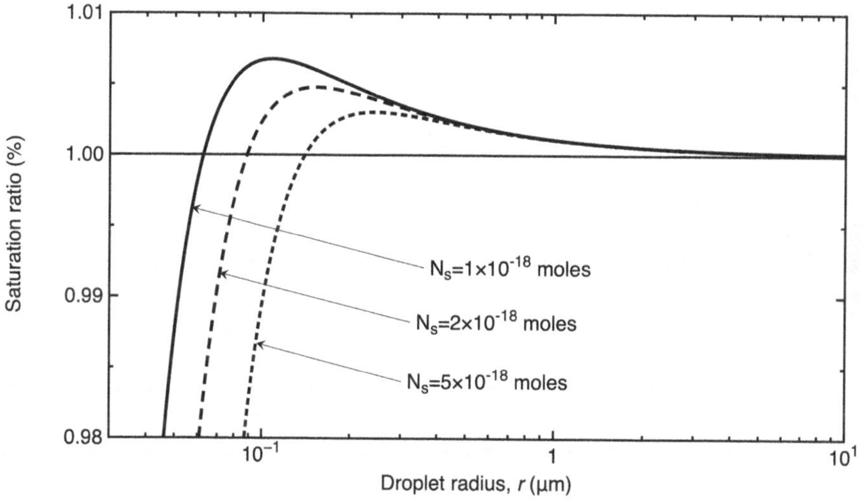


**Fig. 3.26** Köhler curves for three droplets with different molar content. The increasing $N_s$ contributes to decrease the supersaturation required to nucleate a larger droplet

CCNs nucleate cloud drops while the smaller CCNs turn into haze. These considerations support the notion that the first instants in the formation of a cloud are crucial for its further development.

However, things are not as simple as we have described so far. Recent studies have shown that soluble gases (Laaksonen et al. 1998), slightly soluble solutes (Shulman et al. 1996), and surface tension depression by organic substances

(Charlson et al. 2001) influence the formation of cloud droplets. Ruehl et al. (2016) measured droplet diameters at the point of CCN activation for particles composed of secondary organic aerosol and ammonium sulphate. The diameters were 40% to 60% larger than predicted if the organic was assumed to be dissolved within the bulk droplet. The idea is that the organic molecules move to the surface of the growing droplet. These molecules reduce the surface tension, altering the relationship between water vapour supersaturation and droplet size, leading to the larger diameters observed at activation.

One more important aspect needs to be considered. Precipitating clouds are major sinks for particulate matter (due to scavenging processes), but recent studies suggest that precipitation may also act as a particle source. The magnitude of the sources versus sinks, particularly for CCN numbers, remain unquantified. Khadir et al. (2023) have analysed multi-year in situ observations from tropical and boreal forests, as well as Arctic marine environment, finding links between precipitation and enhanced particle concentrations, including CCN particles. In some cases, the magnitude of precipitation-related source equals or surpasses corresponding removal effect. These findings highlight the importance of cloud-processed material in determining near-surface particle concentrations. New particle formation happens also near the surface and is an important source of CCN particles whose vertical distribution is under investigation with new recent findings (see for example O'Donnell et al. 2023). Long term observations from the ground, from satellite, and multiscale modelling need to be used seamlessly as recommended by Kulmala et al. (2023).

## 3.5 Warm Clouds

The first physical type of clouds that stems from the formation of cloud droplets described above is warm clouds or, if we prefer, non-freezing clouds. Warm clouds are clouds that develop in the lower troposphere normally at temperatures above the freezing level. This doesn't necessarily mean that all the cloud body develops at temperatures above the freezing level. A warm cloud is a cloud whose microphysical processes take place "mainly" at temperatures above the freezing level. For example, tropical storm clouds reach altitudes well below the freezing level, but the mechanism of rain formation is that of a warm cloud with little or no presence of ice processes.

Precipitation formation in warm clouds follows a series of processes (see Fig. 3.27): CCN activation, droplet condensation, collision/coalescence, evaporation, raindrop formation and breakup. Note that not all cloud droplets end up forming raindrops after collision/coalescence. In-cloud aerosol activation (through the cloud walls and outside the updraft) and evaporation contribute to the formation of small droplets that may continue to feed the collision/coalescence process as secondary droplets. The process is normally very rapid, and this is why tropical showers develop relatively quickly after the initial cloud formation.

We will now examine in detail each step of the warm rain formation process.

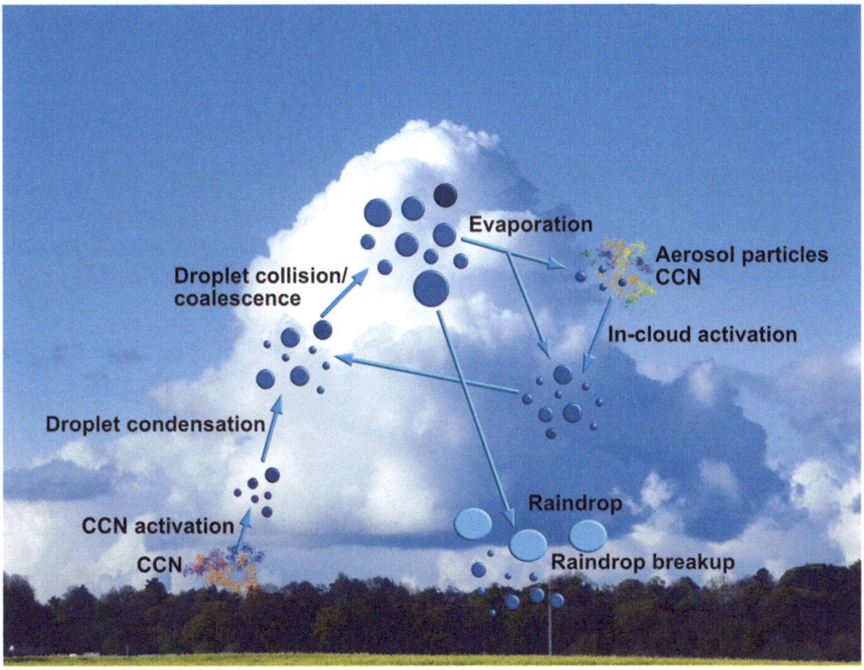


**Fig. 3.27** Scheme of precipitation formation in a warm cloud environment

## 3.5.1 Growth by Diffusion and Condensation

### 3.5.1.1 Single Drop

To become a cloud droplet, a droplet must exceed a certain size $r_c$, the critical radius. Before and after reaching this size, the droplet grows by diffusion of water vapour from ambient supersaturated air.

A simple way to describe the growth of the droplet is shown in Fig. 3.28. The droplet of radius $r$ is surrounded by a vapour field and the concentration of vapour molecules at a distance $R$ from the droplet centre is $n(R)$. The vapour field could be described also in terms of the vapour density or absolute humidity $\rho_v(R)$, where $\rho_v = nm_0$ and $m_0$ is the mass of a water molecule. Isotropy is assumed and thus $n(R)$ and $\rho_v(R)$ do not depend on the direction outwards of the droplet surface. The concentration of molecules in the vapour field is described by the diffusion equation:

$$\frac{\partial n}{\partial t} = D\nabla^2 n \tag{3.23}$$

where $D$ is the molecular diffusion coefficient. If we now work in steady-state or "stationary" conditions, so that $\partial n/\partial t = 0$, Eq. 3.23 becomes.

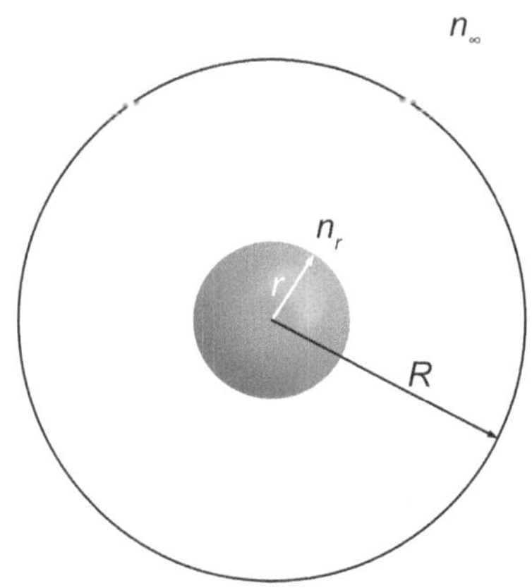


**Fig. 3.28** Scheme of the coordinates for the solution of water vapour diffusion equation toward a growing/evaporating droplet of radius $r$. The vapour field is assumed isotropic

$$\nabla^2 n(R) = 0 = \frac{1}{R^2}\frac{\partial}{\partial R}\left(R^2 \frac{\partial n}{\partial R}\right) \tag{3.24}$$

with the general solution

$$n(R) = C_1 - C_2 / R \tag{3.25}$$

Boundary conditions are as follows:

- as $R \to \infty$, $n \to n_\infty$, the ambient or unperturbed value of the vapour concentration;
- as $R \to$ r, $n \to n_r$, the vapour concentration at the surface of the droplet.

The solution satisfying these conditions is

$$n(R) = n_\infty - \frac{r}{R}(n_\infty - n_r) \tag{3.26}$$

The flux of molecules impinging on the surface of the droplet is $D(\partial n/\partial R)_{R=r}$. The rate of mass increase is thus given by

$$\frac{dm}{dt} = 4\pi r^2 D\left(\frac{\partial n}{\partial R}\right)_{R=r} m_0 \tag{3.27}$$

By combining Eqs. 3.26 and 3.27, we obtain

$$\frac{dm}{dt} = 4\pi r D(n_\infty - n_r)m_0 \tag{3.28}$$

In terms of the vapour density, Eq. 3.28 can be written as

$$\frac{dm}{dt} = 4\pi r D(\rho_v - \rho_{vr}) \tag{3.29}$$

where $\rho_v$ is the ambient vapour density and $\rho_{vr}$ the vapour density at the surface of the droplet.

What we have derived is the diffusion growth equation for a water droplet at rest in a vapour field. If $\rho_v > \rho_{vr}$, the droplet grows, while it evaporates in the case that $\rho_v < \rho_{vr}$. $\rho_v$ is determined from the environmental condition and $\rho_{vr}$ depends on droplet size, composition, and temperature. The temperature is not the same as the ambient temperature and must be determined considering the heat transfer between the droplet and the ambient.

The condensation implies a release of latent heat, which raises the temperature of the droplet with respect to the ambient temperature. The heat diffusion equation produces a result analogous to Eq. 3.29

$$\frac{dQ}{dt} = 4\pi r K(T_r - T) \tag{3.30}$$

where $T$ is the ambient temperature, $T_r$ the temperature at the surface of the droplet, and $K$ the coefficient of thermal conductivity of air.

Equations 3.29 and 3.30 were first derived by James Clerk Maxwell (1890) in an article on the *Encyclopaedia Britannica* when elaborating the theory of the wet bulb thermometer. The theory of a droplet growing at rest in a water vapour field is then called Maxwell's theory.

Using Eqs. 3.29 and 3.30, we derive the rate of change of temperature at the surface of the growing droplet

$$\frac{4}{3}\pi r^3 \rho_L c \frac{dT_r}{dt} = L\frac{dm}{dt} - \frac{dQ}{dt} \tag{3.31}$$

where $\rho_L$ is the density of water, $c$ its specific heat capacity, $L$ the latent heat of condensation and $Q$ the sensible heat. We have assumed a steady-state equilibrium growth so that $dT_r/dt = 0$ and we can derive the balance conditions satisfied by the temperature and vapour fields

$$\frac{\rho_v - \rho_{vr}}{T_r - T} = \frac{K}{LD} \tag{3.32}$$

The ambient conditions are described by $\rho_v$ and $T$ and the ratio $K/LD$ depends weakly on temperature and pressure. The temperature $T_r$ and the density at the surface of the droplet $\rho_{vr}$ are normally unknown. If we recall the equation of Köhler (Eq. 3.20) and the equation of state of water vapour, these two quantities are related by the following expression

$$\rho_{vr} = e_s'(r)/R_v T_r = \left(1 + \frac{a}{r} - \frac{b}{r^3}\right) e_s(T_r)/R_v T_r \tag{3.33}$$

where $e_s(T_r)$ is the equilibrium vapour pressure over a plane water surface at temperature $T_r$ and is derived from the Clausius-Clapeyron equation.

Equations 3.32 and 3.33 form a system of simultaneous equations, which can be solved numerically for $T_r$ and $\rho_{vr}$ to allow for determining the rate of growth of a droplet by condensation.

An analytical approximate solution was proposed by Mason (1971) [see Rogers and Yau 1979 for details]. The rate of growth of a droplet by condensation takes the form suggested by Mason

$$r\frac{dr}{dt} = \frac{(S-1) - \frac{a}{r} + \frac{b}{r^3}}{\left[F_k + F_d\right]} \tag{3.34}$$

where $F_k = \left(\frac{L}{R_v T} - 1\right)\frac{L\rho_L}{KT}$ is the thermodynamic term associated with heat conduction and $F_d = \frac{\rho_L R_v T}{D e_s(T)}$ the term associated with vapour diffusion. For given values of $a$ and $b$, this approximation is shown to be an excellent alternative to the complete numerical solution of the simultaneous Eqs. 3.32 and 3.33.

Note that it is not possible to solve Eq. 3.34 analytically. It can only be solved numerically and graphically for given values of temperature, pressure, saturation ratio, and mass and molecular weight of the CCN.

A list of the rate of growth of droplets growing on sodium chloride CCN at a supersaturation of 0.05%, $p = 900$ hPa and $T = 273$ K is given by Mason (1971). A droplet forming on a large CCN initially grows faster than droplets developing on smaller CCNs, but after reaching a certain radius (in this case about 10 μm) the growth rates are about the same regardless of the nuclear mass. When a droplet becomes sufficiently large, the ratios $a/r$ and $b/r^3$ are negligible with respect to $S - 1$ (see Fig. 3.26) and the approximation suggested by Mason is good.

Then, the droplet radius increases with time as

$$r(t) = \sqrt{r_0^2 + 2\xi t} \tag{3.35}$$

where

$$\xi = (S-1)/\left[F_k + F_d\right] \tag{3.36}$$

We can also write $\xi = (S - 1)\xi_1$ where $\xi_1$ is a normalised condensation growth parameter expressed (in $\mu m^2\ s^{-1}$) as

$$\xi_1 = \frac{1}{F_k + F_d} \tag{3.37}$$

**Table 3.1** Distance covered by a droplet in fall before complete evaporation assuming an isothermal atmosphere with $T = 280$ K and $S = 0.8$

| Initial radius | Distance fallen |
|---|---|
| 1 μm | 2 μm |
| 3 μm | 0.17 mm |
| 10 μm | 2.1 cm |
| 30 μm | 1.69 m |
| 0.1 mm | 208 m |
| 0.15 mm | 1.05 km |

Equation 3.35 has a parabolic form, and this leads to a narrowing of the droplet size distribution as condensation growth proceeds. Let's, for example, consider two droplets growing under the same ambient conditions from initial radii $r_1(0)$ and $r_2(0)$, with $r_2 > r_1$. Equation 3.35 shows that the difference between the squares of the radii remains constant. At any time $t$,

$$r_2(t) - r_1(t) = \frac{r_2^2(0) - r_1^2(0)}{r_2(t) - r_1(t)} \tag{3.38}$$

and the difference of the radii becomes smaller with growth.

It is now important to consider that Eq. 3.34 also describes the rate of evaporation of a droplet when $S < 1$ and $dr/dt < 0$. When knowing the dependence of droplet fall speed on size, Eq. 3.34 can be solved to determine the distance the droplet falls during the time required for its complete evaporation. For droplets smaller than about 50 μm the terminal fall speed increases approximately with the square of the radius and the distance of fall for complete evaporation increases with the fourth power of the radius. This is the concept behind the numbers in Table 3.1.

Note the rapid increase of the distance with droplet radius, which determines the difference between droplets and raindrops (see Fig. 3.1). The latter are the drops that reach the ground before evaporating, while droplets are those that evaporate after leaving the base of the cloud and never reach the ground (e.g., in the virga). The convention accepted in cloud physics is that the boundary between cloud droplets and raindrops is at 0.1 mm. The droplets whose radius is around 0.1 mm are called drizzle, and the meteorological radar starts to detect some scattering.

#### 3.5.1.2 Droplet Populations

In the early stages of cloud development, droplets are still too small to collide with each other or to fall by sedimentation. Their growth is only due to condensation. The ambient saturation ratio then controls the process, and it is consequently necessary to consider the vapour budget of the cloud. The assumption that can be made is that the vapour is made available by saturated air cooled by ascent and lost by condensation on the growing droplets. The rate of change of the saturation ratio can thus be expressed as

$$\frac{dS}{dt} = P - C \tag{3.39}$$

where $P$ is the production term and $C$ the condensation term. In more details

$$\frac{dS}{dt} = Q_1 \frac{dz}{dt} - Q_2 \frac{d\chi}{dt} \tag{3.40}$$

where $dz/dt$ is the vertical air velocity and $d\chi/dt$ the rate of condensation measured in units of mass of condensate per mass of air per unit time. $Q_1$ and $Q_2$ are thermodynamic variables given by

$$Q_1 = \frac{1}{T}\left[\frac{\varepsilon L g}{R' c_p T} - \frac{g}{R'}\right] \tag{3.41}$$

$$Q_2 = \rho\left[\frac{R'T}{\varepsilon e_s} + \frac{\varepsilon L^2}{pTc_p}\right] \tag{3.42}$$

$Q_1 dz/dt$ is the increase of supersaturation due to the cooling by adiabatic ascent and $Q_2 d\chi/dt$ the decrease of supersaturation due to the condensation of vapour on growing droplets. We will not derive explicitly these variables (see Rogers and Yau 1979).

Based on Eq. 3.34 for the growth rate and 3.40 for the saturation ratio, it is possible to start with an initial distribution of CCN, specify the updraft velocity, and calculate the resulting evolution of the droplet spectrum. Calculations show that the supersaturation reaches a peak within about 100 m from the cloud base, and it decreases above this level approaching a constant value. The supersaturation controls the number of CCN that are activated, and this implies that the cloud droplet concentration is determined in the lowest cloud layer. The supersaturation decreases to its steady-state value when a balance is reached between the rate of condensation of the droplets and the rate of increase of supersaturation induced by the updraft. Simple calculations allow for deriving a limiting supersaturation at 0.5% with a time constant of 2 s.

In a growth by condensation regime, it is possible to find the size distribution function at any time, given the same form at an earlier time. Let us consider a sample of cloudy air where the size distribution is described by the function $v_0(r_0)$ where $v_0(r_0)dr_0$ is the number of cloud droplets per unit mass of air with radii in the interval $(r_0, r_0 + dr_0)$. Later, at the time $t$ the droplets will have grown by condensation and the distribution will be narrower as shown schematically in Fig. 3.29. Let us consider the rate at which the number of droplets in the radius interval $\delta r$ changes to derive the expression for the variation of $v(r,t)$ with time. $I(r,t)$ is the droplet current or number of droplets per unit time per unit mass of air passing through the point $r$ in the radius space. The rate of change of the number of droplets in $\delta r$ is given by

$$\frac{\partial}{\partial t}(v\delta r) = -\frac{\partial I}{\partial r}\delta r \tag{3.43}$$

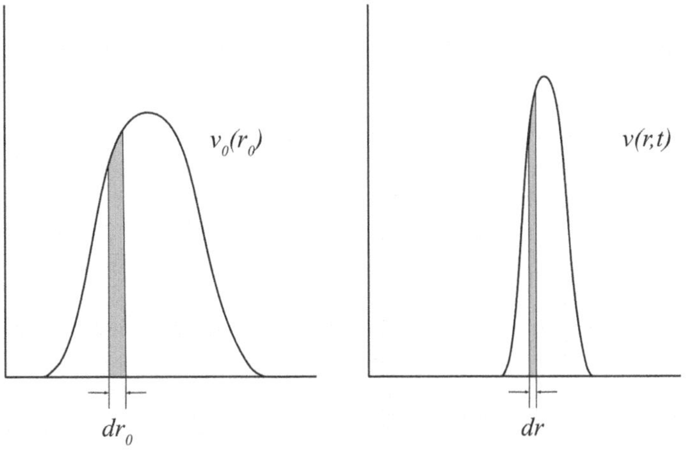


**Fig. 3.29** Schematic of the evolution of a droplet spectrum during growth by condensation

The droplet current is due only to condensation and thus

$$I(r,t) = v\frac{dr}{dt} \tag{3.44}$$

Substituting 3.44 in 3.43 and dividing both sides by $\delta r$ we obtain

$$\frac{\partial v}{\partial t} = -\frac{\partial}{\partial r}\left(v\frac{dr}{dt}\right) \tag{3.45}$$

This is the equation that describes the change in the droplet size distribution or, if we prefer, the continuity equation for droplets. In case of a steady supersaturation, the droplet growth rate is given by the parabolic equation $dr/dt = \xi/r$ where $\xi = (S-1)/(F_k + F_d)$. The solution is

$$v(r,t) = \frac{r}{\sqrt{r^2 - 2\xi t}} v_0\left(\sqrt{r^2 - 2\xi t}\right) \tag{3.46}$$

The result expresses the distribution at any time $t$ in terms of the initial distribution $v_0$ and the growth parameter $\xi$.

Measurements of the droplet size distributions in developing cumulus clouds within a few tens of metres from cloud base indicate narrow distributions centred at radii ranging from about 5 to 10 μm, increasing with distance from cloud base. These measurements are reasonably consistent with theory of growth by condensation. At higher levels and later in the development of the cloud the spectra are broader and extend to larger sizes.

The theory we have discussed above needs some corrections for several effects that we have not included. Such effects are, among others, kinetic effects, non-stationary growth, ventilation effects, the presence of an unsteady updraft, and statistical effects.

In summary, growth by condensation allows droplets to become larger, but it is a somewhat inefficient mechanism to ensure a rapid growth inducing the production of a population of raindrops. In realistic cloud conditions, growth by water-vapour diffusion seldom produces droplets with radii close to 20 μm because of the low magnitude of the supersaturation field and the time available for the growth of the first precipitating drops (~20 min).

At this point the stage needs to be set prior to droplet collision start-up. Several mechanisms have been proposed in time conducive to a droplet spectrum broadening and eventually to the formation of some large droplet in the adiabatic parcel. Two are the theories that were proposed: (1) the homogeneous mixing and (2) the inhomogeneous mixing. In the first mechanism mixing occurs quickly and completely so that all droplets at a given level in the cloud experience the same level of sub saturation and are affected by the same size reduction. This process does explain some spectrum broadening, but not the creation of larger droplets. The second proposed process assumes that mixing occurs in some limited portions of the cloud creating droplet-free volumes of air but saturated. These volumes then mix with the rest of the unaffected cloud. The difference between the two mechanisms is that in the second droplets in some volumes of the cloud may not be affected by size reduction while others do. Details can be found in Telford et al. (1984) and Paluch and Knight (1986). Recently, Lim and Hoffmann (2023) have revisited the processes with small scale models indicating that turbulent mixing is essential in describing a realistic DSD shape at this stage.

### 3.5.2 Growth by Collision and Coalescence

The main problem of cloud physics of warm clouds has been to explain in detail the creation of the first rain about 20 min after the start of the cloud development. During this short time a population of cloud droplets, consisting of around $10^8$ droplets $m^{-3}$ averaging 10 μm in radius, evolves to a population of 1000 drops $m^{-3}$ with a typical diameter of 1 mm. Details are not completely known, but observations show that this 50-fold increase in drop radius is almost entirely due to collision and coalescence between droplets. However, this process does not start significantly until some of the droplets grow to a radius of about 20 μm. The reason is that smaller droplets have small collision cross section and slow settling speed and thus do not collide efficiently with each other (Fig. 3.30).

The rate of collision that a large droplet experiences in time increases rapidly with size (with the fourth power of the radius) and thus coalescence proceeds at a sustained pace once it has started. It is observed that when some of the droplets will have reached a radius of about 30 μm coalescence will be the dominant process for droplet growth. A typical raindrop of 1 mm diameter will have resulted from

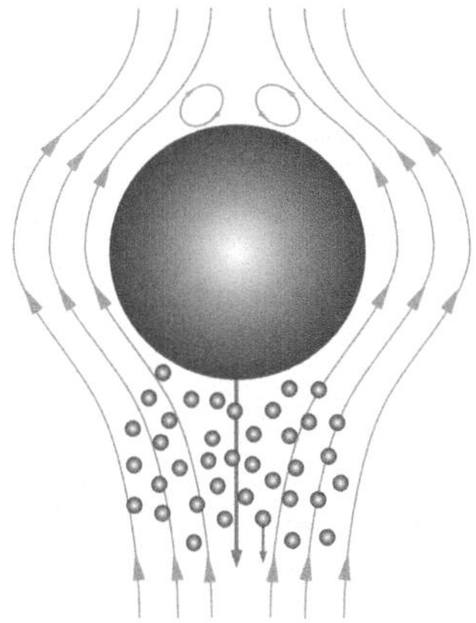

**Fig. 3.30** Scheme of the flow field around a large droplet falling in an ambient full of smaller droplets in a developing cloud. The length of the downward arrows is proportional to the settling speeds of the large and small droplets

something like $10^5$ collisions. However, before the collisions start, some other process would have to take place to produce a few droplets (at least 1 every $10^5$ droplets, or 1 per litre of cloud volume) of 20 μm in radius. We will see later that this is a rather fundamental concept involving stochastic processes within the developing cloud.

Collision, however, does not grant coalescence. When two drops collide, four different things can happen: (1) they may bounce apart; (2) they may coalesce and give birth to a new larger drop; (3) they may coalesce temporarily and then separate maintaining their original identity; (4) they may coalesce temporarily and subsequently break up into several small droplets. The type of interaction depends on the size, trajectory, and speed of the collision, and is also influenced by electrical forces. For sizes <100 μm the relevant interactions are (1) and (2). The ratio between the number of coalescences and the number of collisions is called coalescence efficiency. The growth of a drop by collision and coalescence is governed by the collection efficiency, i.e. the product between collision and coalescence efficiency. Laboratory studies have demonstrated that the coalescence efficiency is close to unity if the droplets are charged or embedded in an electric field. Since at least a weak electric field is always present in a cloud, we can safely assume that collection efficiency equals collision efficiency. Thus, the problem of determining rain development in a warm cloud reduces to determining collision rates among droplets.

#### 3.5.2.1 Droplet Terminal Fall Speed

The first thing we must examine is the speed at which droplets fall in a cloud. The fall geometry is illustrated in Fig. 3.31.

The drag force exerted on a sphere of radius r by a viscous fluid is

$$F_R = \frac{\pi}{2} r^2 u^2 \rho C_D \tag{3.47}$$

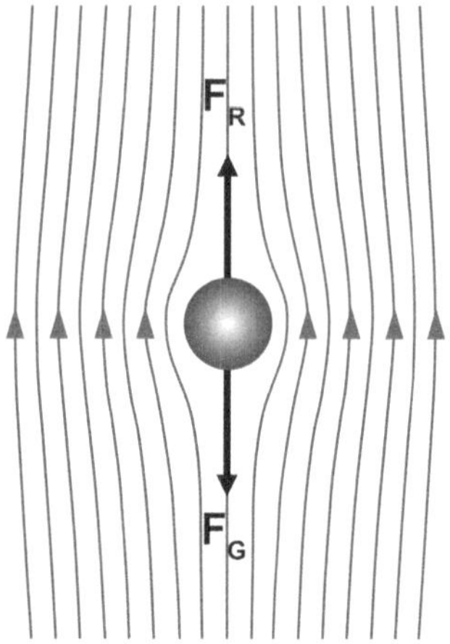


**Fig. 3.31** Fall geometry of a droplet in a flow field. $F_R$ and $F_G$ are the drag and gravitational force, respectively

where $u$ is the velocity of the sphere relative to the fluid, $\rho$ the density of the fluid, and $C_D$ the drag coefficient of the flow. If we introduce the Reynolds number $Re = 2\rho ur/\mu$, with μ the dynamic viscosity of the fluid, eq. 3.47 can be written as

$$F_R = 6\pi\mu ru\left(C_D \,\mathrm{Re}/24\right) \tag{3.48}$$

The gravitational force of the sphere is expressed by

$$F_G = \frac{4}{3}\pi r^3 g\left(\rho_L - \rho\right) \tag{3.49}$$

where $\rho_L$ is the density of the sphere. For a water drop falling in air, $\rho_L \gg \rho$ and a good approximation is

$$F_G = \frac{4}{3}\pi r^3 g\rho_L \tag{3.50}$$

When $F_G = F_R$ the drop falls relative to the air at the so-called terminal fall speed. The equilibrium speed is

$$u^2 = \frac{8}{3}\frac{rg\rho_L}{\rho C_D} \tag{3.51}$$

or

$$u = \frac{2}{9}\frac{r^2 g\rho_L}{\left(C_D \,\mathrm{Re}/24\right)\mu} \tag{3.52}$$

For very small Reynolds numbers we are in the Stokes regime where ($C_D$ $Re$ /24) = 1 and Eq. 3.52 reduces to

$$u = \frac{2}{9}\frac{r^2 g \rho_L}{\mu} = k_1 r^2 \tag{3.53}$$

with $k_1 = 1.19 \times 10^6$ cm$^{-1}$ s$^{-1}$. This is defined as the Stokes law and applies to droplets up to about 30 μm radius.

For high Reynolds number (potential flow) $C_D$ becomes independent of $Re$ and has a value of about 0.45. Equation 3.52 becomes

$$u = k_2 r^{1/2} \tag{3.54}$$

where $k_2 = 2.2 \times 10^3 \left(\frac{\rho_0}{\rho}\right)^{1/2}$ cm$^{1/2}$ s$^{-1}$, with $\rho$ the air density and $\rho_0$ a reference density of 1.2 kg m$^{-3}$, corresponding to the dry air density at 1013 hPa and 20 °C. Raindrops at these high $Re$ are not spherical anymore and the description of their fall speed is just an approximation.

In the intermediate $Re$ range between the Stokes law and the square-root law (40 μm < $r$ < 0.6 mm), the fall speed is approximated by

$$u = k_3 r \tag{3.55}$$

with $k_3 = 8 \times 10^3$ s$^{-1}$.

#### 3.5.2.2 Collision Efficiency

A very simple model can be used to describe the collision efficiency when a droplet of radius r collides with a larger droplet of radius R. The scheme of the model is shown in Fig. 3.32 and is similar to that of any model for the calculation of a cross section in a collision problem.

The method that has been used to determine the collision efficiency of water drops is the *superposition method*, i.e. a method that assumes that each drop moves in the flow field generated by the other drop moving as isolated.

Figure 3.32 shows two spherical drops, one with a larger radius $R$ (the collector drop) and the other with radius $r$ (the impinging drop). The assumption is that the collector drop falls at a higher terminal fall speed and thus catches up with the smaller drop in time. For this pair of drops, the collision efficiency is defined as

$$E = \frac{y_c^2}{(R+r)^2} = \frac{\pi y_c^2}{\pi (R+r)^2} = \frac{A}{A^*} \tag{3.56}$$

where $y_c$ is the critical initial offset, that is the initial horizontal distance of the centre of the impinging drop from the vertical line joining the two drop centres; it identifies the grazing trajectory in the collision with the collector drop. $A^*$ is the area

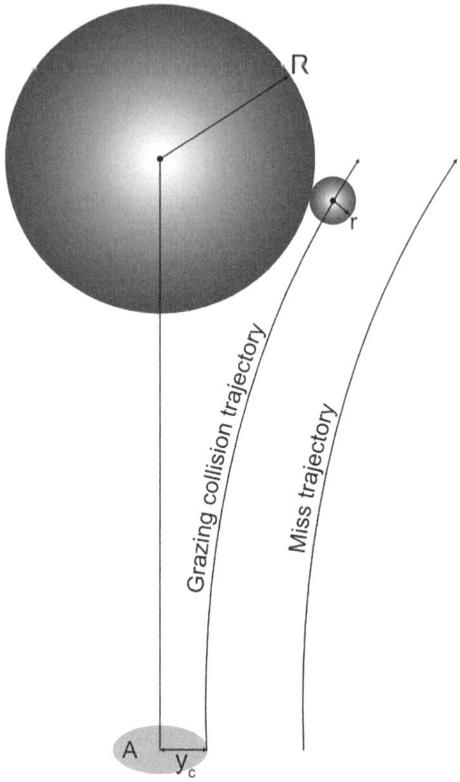


**Fig. 3.32** Geometrical configuration of the collision between two spherical drops. The model assumes that the smaller drop impinges on the larger drop colliding with it or missing it altogether

corresponding to the $R + r$ radius of the two drops aligned at their equator or geometrical collision cross-section.

The calculations of Schlamp et al. (1976) show that the collision efficiency is small for small values of $r/R$ (their Fig. 1). The reason is simple: small droplets have little inertia and are easily deflected by the flow field around the collector drop (see Fig. 3.30). The inertia increases with $r/R$ and the collision efficiency increases alongside up to a radius ratio of about 0.6. Beyond this value the radii are increasingly similar and the relative velocity between the drops is reduced, with an increase of the interaction time. The time of interaction then becomes sufficient for the droplets around the collector drop to get deflected by the flow field. At the same time, however, there is a possibility that the droplets are attracted towards the collector in its wake when the two drops fall nearly at the same speed. This is the "wake capture", which explains why the collision efficiency may exceed unity for $r/R \approx 1$ and is supported by laboratory experiments.

#### 3.5.2.3 The Growth Equation

Let's now suppose that a drop of radius $R$ falls at terminal fall speed through a population of smaller droplets. During unit time it will sweep droplets of radius $r$ from a volume given by

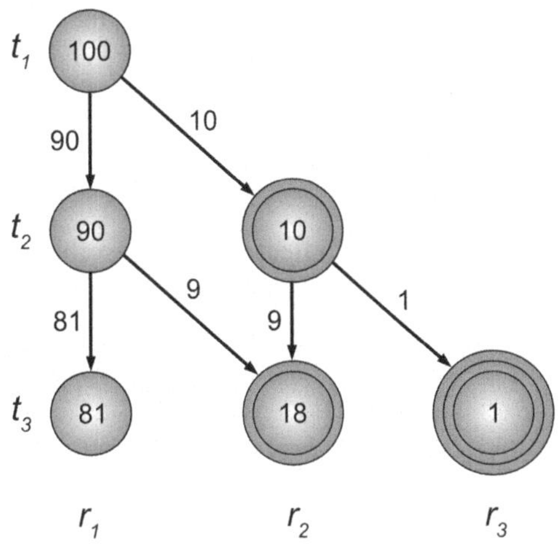


**Fig. 3.33** Scheme of the stochastic growth model. $t_1$, $t_2$ and $t_3$ refer to the time sequence; $r_1$, $r_2$ and $r_3$ are the drop radii. The numbers on the arrows identify the number of droplets transferred to a certain growth category (see text for details)

$$\pi(R+r)^2\left[u(R)-u(r)\right] \tag{3.57}$$

where $u$ is the terminal fall speed. The average number of droplets with radii between $r$ and $r + dr$ collected in the unit time is

$$\pi(R+r)^2\left[u(R)-u(r)\right]n(r)E(R,r)dr \tag{3.58}$$

where $E(R, r)$ is the collection efficiency, i.e. the product of collision efficiency and coalescence efficiency. When the drop radii are smaller than 100 μm the coalescence efficiency is approximated by unity and thus we can use the collision efficiency in 3.58.

The total rate of increase in volume of the collector drop is obtained by integration on all droplet sizes

$$\frac{dV}{dt}=\int_0^R \pi(R+r)^2\frac{4}{3}\pi r^3 E(R,r)\left[u(R)-u(r)\right]dr \tag{3.59}$$

In terms of drop radius

$$\frac{dR}{dt}=\frac{\pi}{3}\int_0^R\left(\frac{R+r}{R}\right)^2\left[u(R)-u(r)\right]n(r)r^3E(R,r)dr \tag{3.60}$$

The change of drop size with altitude is obtained from

$$\frac{dR}{dz}=\frac{dR}{dt}\frac{dt}{dz}=\frac{dR}{dt}\frac{1}{U-u(R)} \tag{3.61}$$

where $U$ is the updraft velocity. If the droplets are much smaller than the collector drop, then $u(r) \approx 0$ and $R + r \approx R$. Thus,

$$\frac{dR}{dt} = \frac{\overline{EM}}{4\rho_L} u(R) \tag{3.62}$$

where $\bar{E}$ is an effective average value of collection efficiency for the droplet population and $M$ the cloud liquid water content (units of mass unit volume). From Eq. 3.61, the change in radius with altitude is approximated by

$$\frac{dR}{dz} = \frac{\overline{EM}}{4\rho_L} \frac{u(R)}{U - u(R)} \tag{3.63}$$

If the updraft is negligible, then

$$\frac{dR}{dz} = -\frac{\overline{EM}}{4\rho_L} \tag{3.64}$$

The equations describe drop growth as a continuous collection process treating the cloud as a continuum. However, growth occurs by discrete events of droplet capture and thus the difference can be quite significant. In the continuous growth model (see accurate descriptions in Rogers and Yau 1979, and Wang 2013) all drops of the same size will grow at the same rate when falling through a cloud of monodisperse drops. The problem with this continuous model is that it produces precipitation-sized drops two or more times too fast. Two more suggestions were advanced to explain quick raindrop formation in warm clouds: electric charges on the drops and the presence of giant CCNs. The first suggestion collides with the fact that the electrification in small clouds is relatively weak, while the second is good for example in Hawaiian trade wind cumulus clouds, but not quite clear for other warm clouds elsewhere. The growth must then follow a more discrete or even stochastic pathway.

### 3.5.3 The Stochastic Growth Model

The stochastic (or statistical) growth model describes unequal growth rates for drops of the same size and explains the observed growth time. Surprisingly enough, some of the drops in the population are somehow "luckier" than the others in collision. This fact has a great impact on the speed of growth of raindrops.

Figure 3.33 illustrates schematically the stochastic growth model. We start with 100 drops of radius $r_l$ at $t_l = 0$. The collision process starts, but only 10% of this initial population has the chance to grow. This is explained by the fact that the small droplets are not uniformly distributed, and a portion of large droplets exists and collide with the others in the time $\Delta t$. So, the scheme says that at time $t_l$ only 10 drops will grow larger to radius $r_2$ while the remaining 90 drops will keep their $r_1$ size. Following the same scheme, at time $t_2$ only one drop out of the 10 $r_2$ drops can grow to $r_3$ and the remaining 9 will stay at $r_2$. At the same time, 9 of the 90 $r_1$ drops will grow to $r_2$ size. Thus, the total number of $r_2$ drops is 18 at this time and 81 drops will remain at the $r_1$ size.

We realise that the stochastic growth model allows for the formation of a broad size spectrum and a quick formation of large drops, while the continuum model does not. In the continuum growth model, the drops share the same average quantity of smaller droplets: the result is that all droplets grow at the same slower pace. As we have discovered before, the collision efficiency becomes significant only when the collector drop size is large enough. This explains why the continuous growth model is so slow in producing large drops.

Before deriving the statistical growth equation, let's define the coagulation coefficient, which describes the likelihood that a drop of radius $R$ will collide and overtake a droplet of radius $r$. The two drops are contained in the unit volume of air. In the unit time, the larger drop sweeps a volume given by $\pi R^2 u(R)$ and the droplet sweeps a volume of $\pi r^2 u(r)$. If the drop will capture the droplet in unit time and neglecting aerodynamic effects, they must be not too far apart initially; in fact, they must be contained in a volume given by $\pi(R + r)^2[u(R) - u(r)]$. If we allow for aerodynamic effects, the volume is reduced to

$$K(R,r) = \pi (R+r)^2 \left| u(R) - u(r) \right| E(R,r) \tag{3.65}$$

where $E(R, r)$ is the collision efficiency. $K(R, r)$ is called the coagulation coefficient. It is interpreted as the probability that a drop of radius $R$ collects a droplet of radius $r$ in unit time, provided they are both present in unit concentration.

The coagulation coefficient is best defined in terms of the drop volumes. If $V$ and $v$ are the volumes of the drops of radius $R$ and $r$, respectively, the coagulation coefficient becomes

$$H(V,v) = K\left[\left(\frac{3V}{4\pi}\right)^{1/3}, \left(\frac{3v}{4\pi}\right)^{1/3}\right] \tag{3.66}$$

Thus, $H(V, v)$ represents the probability that a drop of volume $V$ collects a droplet of volume $v$ in unit time.

Now let's suppose that the drop spectrum is described by $n(v)$ such that $n(v)dv$ is the average number of drops per unit volume whose volumes are between $v$ and $v + dv$. The total number of coalescences experienced by the drops in unit time within this volume range is

$$n(v)dv \int_0^\infty H(V,v) n(V) dV \tag{3.67}$$

The integration accounts for all possible captures of drops in $dv$ by larger drops ($v < V < \infty$). Note that the coalescence reduces the number of drops in $dv$. Meanwhile, the number of drops is increased by the coalescence between all pairs of smaller drops whose volumes sum to $v$. This increase is given by

$$\frac{1}{2}dv\int_{0}^{v} H(\delta,u)n(\delta)n(u)du \tag{3.68}$$

where $\delta = v - u$. The factor 1/2 is necessary to prevent any capture event to be counted twice. Summing up both effects, we derive the rate of change of drop concentration in the volume interval $dv$

$$\frac{\partial}{\partial t}n(v)dv = \frac{1}{2}dv\int_{0}^{v} H(\delta,u)n(\delta)n(u)du - n(v)dv\int_{0}^{\infty} H(V,v)n(V)dV \tag{3.69}$$

This equation is called in various ways: kinetic equation, coagulation equation or stochastic coalescence equation.

Berry and Reinhardt (1974a, b) provided a solution for the Eq. 3.69 by characterising the droplet spectrum by the density function $f(x)$, where $f(x)dx$ is the number of droplets per unit volume of air in the size interval $(x, x + dx)$ with $x$ standing for droplet mass. Mass and volume are related by $x = \rho_{Lv}$ and the density functions $f(x)$ and $n(v)$ are related by $f(x)dx = n(v)dv$. The authors also considered the mass density function $g(x) = xf(x)$. The droplet concentration N and liquid water content L are derived integrating over the density functions

$$N = \int f(x)dx \tag{3.70}$$

$$L = \int g(x)dx \tag{3.71}$$

The mean mass of a droplet is given by

$$x_f = \int xf(x)dx / \int f(x)dx = L / N \tag{3.72}$$

A droplet with size as the one of the mean droplet mass has a radius given by

$$r_f = (3/4\pi\rho_L)^{1/3}\, x_f^{1/3} \tag{3.73}$$

The mean mass of the mass density function is defined by

$$x_g = \int xg(x)/g(x)dx = \int x^2 f(x)dx / \int xf(x)dx \tag{3.74}$$

and its equivalent radius

$$r_g = (3/4\pi\rho_L)^{1/3}\, x_g^{1/3} \tag{3.75}$$

The authors show the evolution of six different droplet spectra by coalescence with $f(x)$ in the form of a gamma distribution with varying $r_f$, a liquid water content of 1 g m$^{-3}$, and a concentration of 166 droplets cm$^{-3}$. The spread of the distributions is characterised by the relative variance $var\ x = (x_g/x_f) - 1$. Their graphs plot the log-increment mass density function $g_l(r)$ defined such that $g_l(r)d(dlnr)$ equals the mass of cloud droplets per unit volume of air with radii in $d(lnr)$. The distribution emphasises the large droplets and is related to the fundamental density function by $g_l(r) = 3x^2f(x)$. The initial unimodal distribution evolves in time by coalescence into a bimodal distribution. Such evolutions agree well with observations.

### 3.5.4 Condensation and Stochastic Coalescence

While drops are growing by collision and coalescence the growth by diffusion continues if the cloud conditions are the right ones. The growth by diffusion is extremely slow with respect to collision and coalescence. However, diffusion has an effect, which stems from the importance of the size sensitivity of the droplet spectrum as we commented before.

Ryan (1974) used Eq. 3.34 to simulate condensation and assumed an updraft speed of 1 m s$^{-1}$. The initial size distribution is described by a gamma function with a concentration of 200 cm$^{-3}$, a liquid water content of 1 g m$^{-3}$, and a dispersion (ratio of standard deviation of radius to mean radius) of 0.2. The temperature lapse rate is assumed moist adiabatic. Pure coalescence produces drops smaller than 60 μm after a time of 1000 s. When condensation occurs the spectrum becomes narrower below 25 μm. The shift of the droplets towards larger sizes by condensation increases their collection efficiency and encourages the growth by collision and coalescence. After 1000 s a considerable number of drizzle-size drops are created indicating the importance of condensation in accelerating coalescence. In other words, the spectrum broadens faster with condensation and coalescence is favoured. The effect is called *condensation broadening*.

Ochs (1978) published another study where he started from the activity spectrum of CCN and calculated the evolution of the distribution spectrum in an ascending air parcel by collision and coalescence. The author considers two different situations. In a maritime cloud condition with an activity spectrum $N_c = 105s^{0.63}$ the supersaturation increases rapidly and CCN are activated. The number concentration of droplets reaches a maximum and the supersaturation falls to a steady value. The droplets deplete the water vapour at the same rate it is made available by the updraft. The droplets then grow and become large enough to start collision and coalescence, which progresses rapidly as indicated by the fast decline of the number of drops. At the same time the supersaturation increases since the drops, which are fewer in number, are not anymore able to consume the excess vapour at the rate it is produced. The increased supersaturation activated new nuclei as indicated by a secondary small peak in the number concentration around 770 s. the effect is only transient since the new droplets are consumed fast by collision and coalescence processes. In the case of a continental cloud with an activity spectrum $N_c = 1450s^{0.84}$ the initial

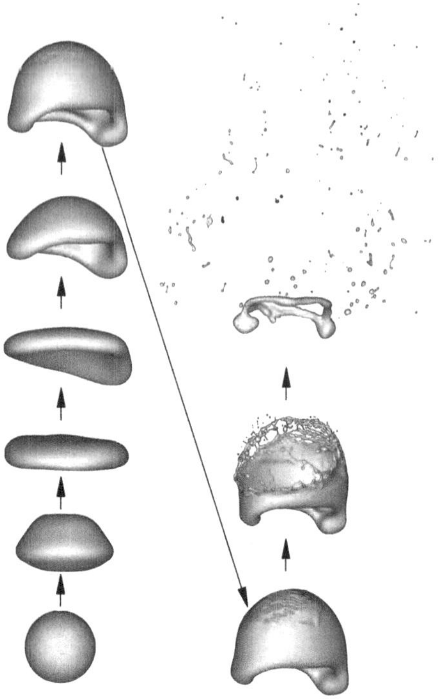

**Fig. 3.34** Schematic of the bag breakup mechanism for large drops

conditions are very similar to those of the previous case except for the fact that the number concentration of the droplets is a factor of 10 larger: the CCN number is obviously much higher. The onset of coalescence is signalled by the reduction of the number concentration of drops as in the previous case. However, the number remains high enough to prevent a rise of the supersaturation and the activation of new nuclei.

### 3.5.5 Drop Breakup

Once raindrops become very large, they are subject to instability and to break into smaller drops. Breakup also happens by effect of collisions between drops. McTaggart-Cowan and List (1975) performed 25,000 collisions between large drops (4.8, 3.6 and 3.0 mm diameter) and small drops (1.8 and 1.0 mm diameter) in a cloud tunnel. They observed 712 collisions finding four distinct types of collision-breakup:

- *Neck or filament*. The small drop hits the large one near its equator, but it is carried away by its own momentum while remaining attached through a tiny filament. The filament eventually breaks into tiny fragments much smaller than the small drop.

- *Sheet*. The small drop hits the large one near its centre and rips away a sheet of water. The sheet becomes unstable and disintegrates into fragments. The small drop loses its identity into the fragments.
- *Disk*. The small drop hits the large one at its centre and forms a disk conglomerate, which disintegrates into fragments.
- *Bag*. The flow field shapes the large disk of the previous category into a bag, which disintegrates into many fragments.

The authors found the following occurrence for the four categories: neck 27%, sheet 55%, disk 18% and bag <0.5%. The various types of breakup depend on the geometry of the collision and on the kinetic energy of the drops as observed by Low and List (1982a, b).

The bag shape of large raindrops, as mentioned before, was also recently observed by Villermaux and Bossa (2009) who observed the disintegration of the drops by effect of the updraft-induced instability. A scheme of the bag breakup is shown in Fig. 3.34.

Clearly breakup influences the shape of the drop size distribution. Low and List (1982a, b) have parameterised the equilibrium size distributions deriving from the three main breakup modes. Several theoretical studies have indicated that the fragments resulting from the filament, sheet, and disk modes together with the solitary large fragments of disk breakup are responsible for the creation of trimodal distributions as, for example, demonstrated by Brown (1988).

## 3.6 Ice Clouds

Clouds at midlatitudes are generally partly or fully glaciated depending on season, temperature vertical profile and height. Also, tropical clouds, especially deep convective clouds, may be partly glaciated. Precipitation formation thus involves ice, which eventually melts forming raindrops. Snow is produced in winter when temperature is around 0 °C down to the ground. Ice hydrometeors form in several ways and the cloud ice physics is only partially known.

A newly formed ice crystal finds itself in a favourable environment for growth since the water vapour is saturated with respect to water and hence supersaturated with respect to ice. The growth by diffusion is thus very rapid and the crystal grows to a size of many tens of microns in a few minutes. The fall speed of the growing crystal is of a few cm $s^{-1}$ and the crystal may reach the ground as an individual crystal. It may also collide with supercooled droplets to form a rimed crystal or with other crystals to form an aggregate of crystals or snowflake.

The growth processes are therefore the same as for a cloud droplet, but the diffusional growth in the case of ice crystals plays a more relevant role than for water droplets due the difference in saturation vapour pressure over water and ice.

### 3.6.1 Ice Nucleation

When a cloud extends vertically at temperatures below 0 °C ice crystals may form, and the cloud starts glaciating. Two phase transitions are involved: freezing of a liquid droplet or direct deposition (sublimation) of vapour to the solid phase. Both homogeneous and heterogeneous nucleation processes are possible (Fig. 3.35).

Homogeneous freezing of a pure water droplet or of a liquid aerosol particle happens when statistical fluctuations of the molecular structure of water produce a stable, icelike structure that can serve as an ice nucleus. There are two conditions that govern the homogeneous freezing process: the size of the stable nucleus and the probability of occurrence of embryonic ice nuclei by random rearrangements of water molecules. These two quantities depend on the surface free energy at the liquid/ice interface (approximately $2.0 \times 10^{-2}$ N $m^{-1}$), analogous to the surface tension at the liquid/vapour interface.

Theory and experiments conclude that droplets smaller than 5 μm freeze spontaneously at a temperature lower than −40 °C and larger droplets freeze at higher temperatures. In natural clouds very few droplets exist as liquid droplets at these temperatures, implying that heterogeneous nucleation has taken place at higher temperatures. However, observations in deep convective clouds (Rosenfeld and Woodley 2000) and aircraft icing at these temperatures have shown that supercooled droplets may still be found in these conditions.

So, homogeneous deposition occurs when vapour molecules hit an ice embryo by random collisions. Theory predicts that homogeneous nucleation happens only in extreme conditions of supersaturation. More than 20 times the supersaturation with respect to ice is needed at a temperature just below 0 °C, and even higher supersaturations are required at lower temperatures. This tells us that homogeneous deposition cannot happen in the atmosphere because these levels of supersaturation never exist.

Significant numbers of ice crystals appear in clouds when the temperature drops below −15 °C and this suggests that heterogeneous nucleation takes place. In fact, water in contact with surfaces freezes at temperatures much higher than −40 °C. Moreover, deposition occurs on surfaces at supersaturations and supercoolings less than the homogeneous values. Thus, ice nucleation in a supercooled and supersaturated environment is favoured by the presence of ice nuclei (INP).

An INP provides a substrate where vapour molecules can deposit, stick and bond together to form aggregates with an icelike structure. This is the reason why not all nuclei can act as INP: they must have a crystal structure very much like the one of ice to start with (e.g., Sosso et al. 2016; Kiselev et al. 2017). When the binding and matching of the lattice structure of the INP are right, the supersaturation or supercooling required to nucleate ice on a substrate may be much lower than those required for homogeneous nucleation.

Figure 3.35 shows that several nucleation mechanisms are possible in a glaciating cloud. Three modes are observed for the activation of INPs: (1) some serve as CCN and then successively as freezing nuclei (freezing by condensation); (2) some cause freezing when they collide with the surface of a supercooled droplet (freezing

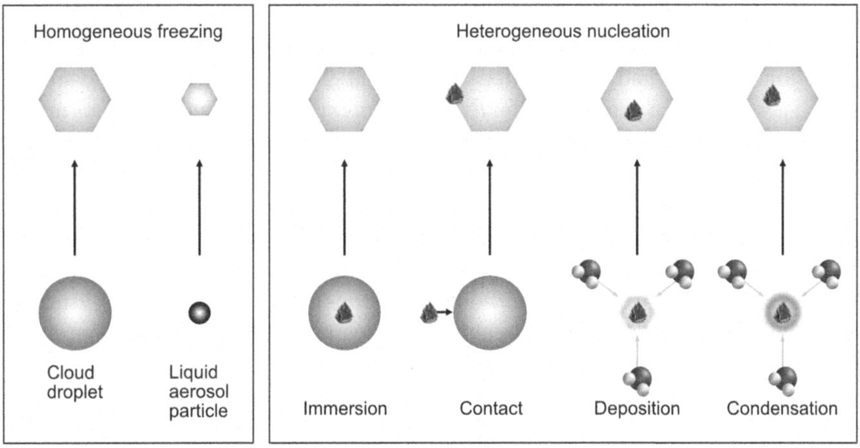


**Fig. 3.35** Ice crystal formation mechanisms: homogeneous freezing and heterogeneous nucleation modes. (Adapted from public domain)

by contact); (3) some others cause freezing when embedded in a droplet (freezing by immersion). Finally, vapour may deposit directly on an INP and form directly the ice structure (freezing by deposition).

All these mechanisms act at the same time, and it is still not clear what is the relative importance of each one of them. It is difficult to distinguish between deposition and freezing nucleation when ice nucleates on an insoluble particle in conditions of supersaturation with respect to liquid water. Even in condition below water supersaturation nucleation does not imply deposition because the INP may contain a soluble fraction. In this case the soluble material may nucleate the liquid phase below water saturation and let the insoluble fraction nucleate ice by freezing. In summary, there is some confusion due to the superposition of different mechanisms. Often "ice nucleation" is used instead of the more specific terms "freezing nucleation" and "deposition nucleation". Other mechanisms are involved including biological material: for a discussion revisiting the terminology of ice nucleation the work of Vali et al. (2015) is a useful reading.

New developments in identifying the heterogeneous freezing mechanisms, their atmospheric relevance, the uncertainties, and the unknowns about INPs are described by Kanji et al. (2017) and Burrows et al. (2022). Note that INPs belong to different types such as soluble, insoluble, and porous: all these types induce different freezing mechanisms that remain to be fully characterised. Recent observations of new particle formation (NPF) from oceanic organic vapours (Chang et al. 2022) suggest that these particles may be contributing to shape the ice formation in clouds as well as bioaerosol (Cornwell et al. 2023).

The temperature of activation of INPs depends on their composition. For example, minerals such as silicates (vaterite and kaolinite) nucleate at temperatures between −7 and −9 °C while volcanic ashes activate around −13 °C (Houghton 1985). Biological aerosol particles are also very effective freezing nuclei. Pollen and sub-pollen particles (deriving from pollen rupture under high humidity conditions) can reach the upper troposphere in convective conditions thus influencing the formation of clouds and precipitation (Zhang et al. 2024). Bacteria nucleate around −9 °C: a well-known example is the *Pseudomonas syringae*, which produces nucleation proteins (INA, Ice Nucleation Active) and is a powerful ice nucleator (Pandey et al. 2016). It is reported to nucleate ice at very high temperatures around −1.8 to −3.8 °C (Maki et al. 1974). Protein aggregates may account for cloud glaciation at moderate supercooling (e.g., apoferritin that starts glaciation at −5 °C, Cascajo-Castresana et al. 2020) as well as viruses. Another biological particle type, cellulose, has been shown to exist as an airborne ubiquitous particle and its microcrystalline structure helps ice nucleation, although the nucleation process becomes significant below about −21 °C (Hiranuma et al. 2015).

### 3.6.2 Ice Crystal Growth by Diffusion

How do ice crystals grow after their initial nucleation stage? We have seen in Fig. 3.2 the various shapes of the crystals in a cloud depending on temperature and vapour saturation. It is important to examine how different the growth of the crystals is with respect to that of droplets.

When the ice crystals nucleate, they are immersed in a cloud environment where the vapour pressure is equal or slightly higher than the equilibrium vapour pressure $e_s$ over liquid water. The saturation ratio relative to ice is written as

$$S_i = \frac{e}{e_i} = \frac{e}{e_s}\frac{e_s}{e_i} = S\left(e_s / e_i\right) \tag{3.76}$$

where $S$ is the saturation ratio with respect to water. The supersaturation $(e_s/e_i) - 1$ is plotted in Fig. 3.36. A water-saturated cloud has a high supersaturation relative to ice and is a favourable environment for crystal growth by diffusion and deposition. Such environment will remain favourable for ice crystal growth as long as liquid droplet are available to evaporate and maintain the vapour pressure at equilibrium relative to water.

The problem in determining the rate of growth of an ice crystal by diffusion is similar to that we have already described for cloud droplets. However, the crystals are not spherical and thus the degree of complexity is higher. The theory of crystal growth was devised by James Clerk Maxwell and is the same as that for droplets. Maxwell solved the equations of heat and mass transfer by using the analogy between the diffusion equation and the equations of electrostatics describing the distribution of potential around a charged conductor. He used the Poisson equation in electrostatics and the Green's theorem showing that the integral over the surface of the conductor of the normal component of $-\nabla\phi$ ($\phi$ is the electrostatic potential) is equal to $4\pi C\phi$, where $C$ is the capacitance of the conductor. Now, if we identify $-D\nabla n$, i.e. the flux of water molecules, with $-\nabla\phi$, we derive that total current of water out of the ice crystal is by analogy $4\pi CD(n_s - n_\infty)$ where $n_s$ is the vapour

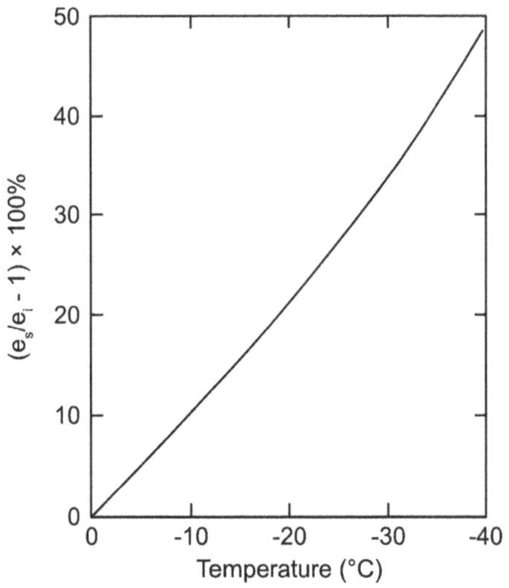


**Fig. 3.36** Supersaturation relative to ice in an atmosphere at equilibrium saturation relative to water

number density at the crystal surface and $n_\infty$ its value far from the surface. The assumptions that $\nabla^2 n = 0$ and that $n_s$ is the same all over the surface are necessary to carry on the analogy. The generalised growth equation thus has the form

$$\frac{dm}{dt} = 4\pi CD\left(\rho_v - \rho_{vr}\right) \tag{3.77}$$

$C$ is the electric capacitance, with length units, which is a function of size and shape of the particle. For a sphere $C = r$ and Eq. 3.77 reduces to the growth equation of a cloud droplet. For a circular disk of radius r, normally used to approximate plate-like crystals, $C = 2r/\pi$. Ice needles are approximated by a prolate spheroid with major and minor semi-axes a and b and $C = A/\ ln\ [(a + A)/b]$ where $A = \sqrt{a^2 - b^2}$. For an oblate spheroid (approximating a disk) $C = \dfrac{a\varepsilon}{arcsin\varepsilon}$ where the ellipticity is $\varepsilon = \sqrt{1 - b^2 / a^2}$.

Real ice crystals have forms much more complex than spheres, disks, and spheroids, but they can be approximated by these forms if we are describing the vapour field at some distance. Irregularities can distort at distances from the surface comparable to the dimensions of the irregularity. In fact, as the vapour approaches the surface, the fine structure of the crystal determines how and where it deposits, but this does not affect the total flux to and from the surface.

During growth the ice crystal surface is heated by the latent heat of sublimation and $\rho_{vr}$ is raised above the value without heating. Under stationary growth $\rho_{vr}$ is determined by the balance between the rates of latent heating and the heat transfer away from the surface:

$$\frac{\rho_v - \rho_{vr}}{T_r - T} = \frac{K}{L_s D} \tag{3.78}$$

Following the same concepts used for water droplets, Eqs. 3.77 and 3.78 can be combined to produce an analytical expression for crystal growth rate. The expression is the same obtained for water droplets replacing *r* by *C*, $e_s(T)$ by $e_i(T)$ and *L* by $L_s$ (the latent heat of sublimation):

$$\frac{dm}{dt} = \frac{4\pi C(S_i - 1)}{\left[\left(\frac{L_s}{R_v T} - 1\right)\frac{L_s}{KT} + \frac{R_v T}{e_s(T)D}\right]} \tag{3.79}$$

As for the droplets, kinetic effects are neglected and in general not well understood. $\rho_{vr}$ may also not be the same all over the surface of the crystal and thus not exactly reflects the equilibrium vapour density over all points of the surface. For these and other reasons the growth rate described by 3.79 may by slower. For small ice crystals times may be up to half those predicted, while for large crystals the ventilation effects tend to increase the growth rate and thus 3.79 may be a better approximation.

The initial growth of ice crystals occurs in a water cloud and $S_i \approx e_s/e_i$ with a strong temperature dependence as shown in Fig. 3.36. The denominator of 3.79 depends on temperature and pressure as for water droplets. Byers (1965) combined the temperature dependence with that of $S_i$ in a water saturated environment determining the dependence of ice crystal growth rate on temperature and pressure (Fig. 3.37). The curves show that the growth rate varies inversely with pressure and that the temperature for maximum growth rate is about −15 °C over a wide range of pressures.

### 3.6.3 Ice Presence in Clouds

Ice nucleation mechanisms in a glaciating cloud might suggest that the number of INPs is the same as that of ice crystals in the cloud. This is not true as the concentrations of ice particles often exceed the measured concentrations of ice nuclei by several orders of magnitude (for a bibliography on the subject see Hobbs and Rangno 1985).

The ice enhancement ratio in a cloud is defined as the ratio of the maximum concentration of ice particles measured in a cloud to the concentration of ice nuclei at cloud top temperature. The latter quantity is given by the empirical relationship of Fletcher (1962) based on standard ice nucleus measuring techniques:

$$n(\Delta T) = n_0 \exp(\beta \Delta T) \tag{3.80}$$

where $n(\Delta T)$ is the concentration of ice nuclei (per litre of air) active at supercoolings $\leq \Delta T$, $n_0 = 10^{-5} L^{-1}$, and $\beta = 0.6$. As noted by Hobbs and Rangno (1985), the

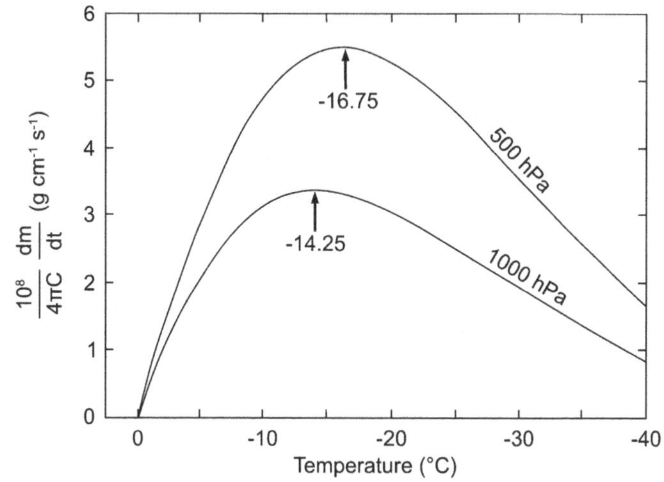


**Fig. 3.37** Normalised ice crystal growth rate as a function of temperature and pressure

relation gives a concentration of ice nuclei of $\approx 1\ L^{-1}$ at −20 °C ($\Delta T = 20\ ^\circ C$) and the concentration changes by a factor of 10 for ≈ 4 °C temperature change.

Hobbs and Rangno (1985) have measured the maximum ice particle concentrations ($I_{max}$) over a wide range of cloud top temperatures (−3 to −32 °C) in mature and aging cumuliform clouds. $I_{max}$ is defined as the peak value of the ice particle concentration averaged over a time interval of 15 s. Ice particles are present in high concentrations in maritime clouds when the cloud top temperature falls below −4 to −7 °C. For the continental clouds the temperature is about −10 °C. Beyond these values $I_{max}$ is not very sensitive to cloud top temperature at least down to −32 °C. Note that the lowest ice particle concentrations to be detected in mature and aging cumuliform clouds with cloud top temperatures between −15 and −25 °C are given approximately by the relation of Fletcher (3.80) for ice nucleus concentrations.

One more consideration to be made is that ice crystals may not be observed in significant quantities until the temperature falls below −10 °C. This means that a supercooled cloud in the atmosphere is a rather common phenomenon. Figure 3.38 summarises results of several measurement campaigns in different parts of the world (see also Houghton 1985). These data of ice likelihood as a function of cloud top temperature may be considered correct, but it must be noted that ice formation in a cloud depends not only on top temperature but also on cloud type, cloud age, and geographical location.

## 3.7 Mixed Phase Clouds

In the previous section we have discovered that finding liquid in clouds that extend well below 0 °C is not uncommon or, better, supercooled water is a most common component of sub-zero clouds. Thus, dividing clouds between warm, cold, and mixed phase is not entirely proper since a certain amount of liquid water is present

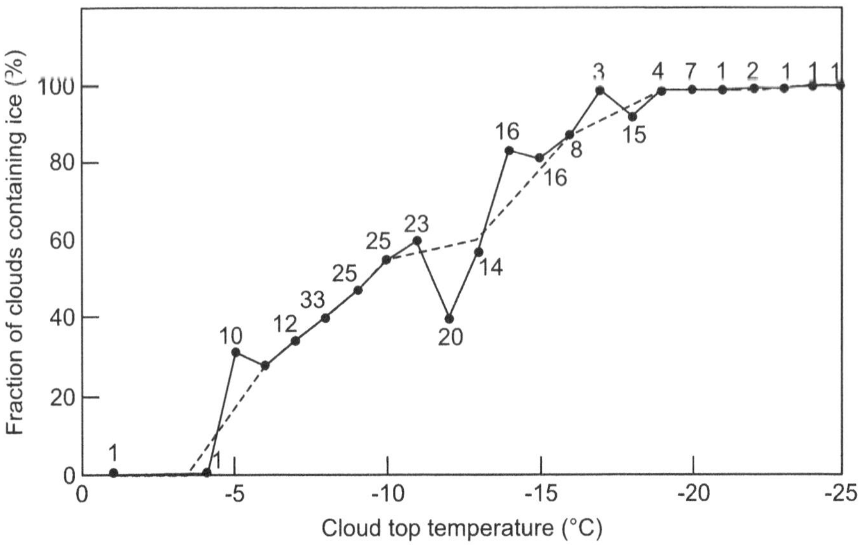


**Fig. 3.38** Fraction of clouds containing ice as a function of cloud top temperature from 258 cloud observations in different regions. Numbers above each point are the number of observations at that temperature; the dashed curve is a three-point running mean

most of the times. However, we will now examine the main principle of the physics of mixed phase clouds.

### 3.7.1 The Wegener-Bergeron-Findeisen Process

In the first half of the last century, several studies were conducted between Germany and Norway (the Bergen school) on the formation of precipitation in cold clouds. The first contribution came from Alfred Wegener (1911) followed by Thor Bergeron (1935) and Walter Findeisen (1938). The three scientists wrote one of the most important pages of cloud physics history identifying the process that is now known as the Wegener-Bergeron-Findeisen (WBF) process (Storelvmo and Tan 2015).

The WBF, also called the cold rain process, is a process of ice crystal growth in mixed phase clouds in conditions where the ambient vapour pressure falls between the saturation vapour pressure over water and the lower saturation vapour pressure over ice. The environment is subsaturated for liquid water but supersaturated for ice and this results in rapid evaporation of liquid water and rapid ice crystal growth through vapour deposition. In other words, the ice crystals grow at the expenses of the evaporation of liquid drops (Fig. 3.39). If the number density of ice is small compared to liquid water, the ice crystals may grow large enough to precipitate, and melting into rain drops while passing through the melting layer, if temperatures at the lower levels are high enough. A revealing photograph in the cloud chamber is

**Fig. 3.39** Scheme of the Wegener-Bergeron-Findeisen process in a mixed phase cloud. The ice crystal is growing in the middle by deposition of water vapour molecules (oxygen with two attached hydrogen atoms) generated by the evaporation of cloud droplets (large spheres)

shown in Fig. 3.40 where an ice crystal is growing at the expenses of the surrounding supercooled droplets that evaporate.

The WBF process is expected to occur under a limited range of conditions and ice particles and liquid droplets in mixed-phase clouds are not always processed in accordance with the WBF mechanism (Korolev 2007). The main pre-condition for the WBF process to take place is that the number of droplets should be much higher than the number of ice crystals. This depends on the fraction of CCN that would later (higher in the cloud) act as INP. Alternatively, the adiabatic updraft must be sufficiently fast so that high supersaturation induces spontaneous nucleation of many more droplets than the number of INPs present in the environment. In both cases, this happens not far below the freezing point to prevent direct nucleation of

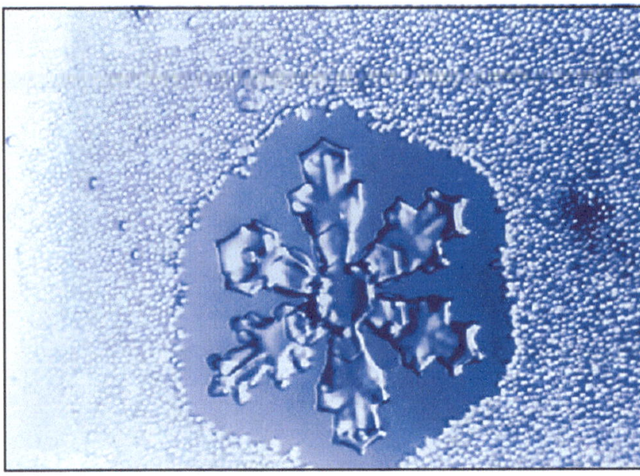

**Fig. 3.40** A branched ice crystal growing in a cloud chamber at the expenses of the evaporation of surrounding supercooled droplets. (Photo kindly provided to first author by Richard L. Pitter)

ice. The growth of the droplets would in fact stop the temperature from rapidly falling below the point of fast nucleation of ice crystals.

The larger supersaturation with respect to ice causes it to grow fast drawing water from the vapour phase. If the vapour pressure drops below the saturation pressure with respect to liquid water, the droplets will cease to grow. This may not happen if the vapour pressure drops rapidly, depending on the slope of the saturation curve, the lapse rate, and the speed of the updraft, or if the drop of the pressure is slow, depending on the number and size of the ice crystals. If the updraft is too fast, all droplets would freeze rather than evaporate.

Similarly, in a downdraft liquid water evaporates causing the vapour pressure to rise, but if the saturation pressure with respect to ice rises too fast in the downdraft, all ice would melt before large ice crystals have formed.

### 3.7.2 Ice Multiplication

Now we have all the ingredients to try to answer one of the fundamental questions of cloud physics: what is the reason why we observe a concentration of ice crystals in clouds that is much higher than the concentration of INPs? In his review article, Mossop (1985) indicated two of fundamental mechanisms: (a) the rime-splintering process, and (b) the mechanical fracture of ice crystals.

More recently, Korolev and Leisner (2020) have critically examined the results of laboratory experiments over the last 70 years that demonstrate the existence of six mechanisms of ice multiplication detailed below.

**Fig. 3.41** Shattering during droplet freezing. (Adapted from Korolev and Leisner 2020. © The Authors under CC-BY-4.0 license)

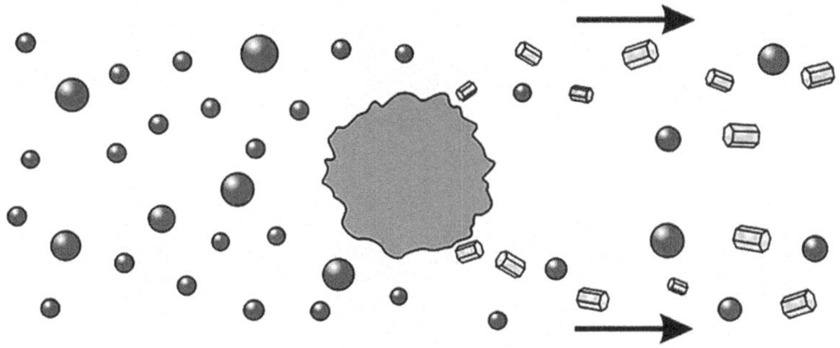

**Fig. 3.42** Splintering during riming. (Adapted from Korolev and Leisner 2020. © The Authors under CC-BY-4.0 license)

***Shattering During Droplet Freezing (Fig. 3.41)*** During the freezing of a cloud droplet, liquid water may be trapped inside a growing ice shell around the droplet. Ice expands during further freezing resulting in a pressure increase inside the ice shell and inducing a shell crack or shatter. This produces secondary ice from the resulting ice fragments. In addition, gases dissolved in the droplet might be released during the process. Gas bubbles may burst upon freezing at the colder droplet surface, resulting in a second source of small ice fragments.

***Splintering During Riming (Fig. 3.42)*** Splintering during ice particle riming was observed by several authors. Hallett and Mossop (1974) found that splinter production is active in the temperature range between −8 and −3 °C. Furthermore, the rate of splinter production had a pronounced maximum at the air temperature of −5 °C and the drop impact velocity of 2.5 m $s^{-1}$. The phenomenon is usually referred to as the Hallett–Mossop mechanism. James et al. (2021) have recently demonstrated experimentally the production of ice during the spreading and retraction phase during the collision of a supercooled drop with an ice particle.

***Fragmentation Due to Ice-Ice Collision (Fig. 3.43)*** The analysis of literature of Korolev and Leisner (2020) suggests that the efficiency of secondary ice production during ice–ice collision depends on several factors: (a) properties of the colliding particles (size, mass, density, shape, surface roughness); (b) air temperature; and (c) relative fall velocity of colliding particles.

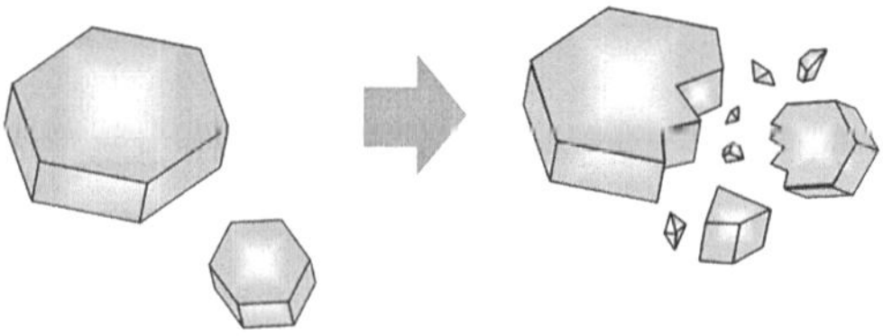

**Fig. 3.43** Fragmentation due to ice-ice collision. (Adapted from Korolev and Leisner 2020. © The Authors under CC-BY-4.0 license)

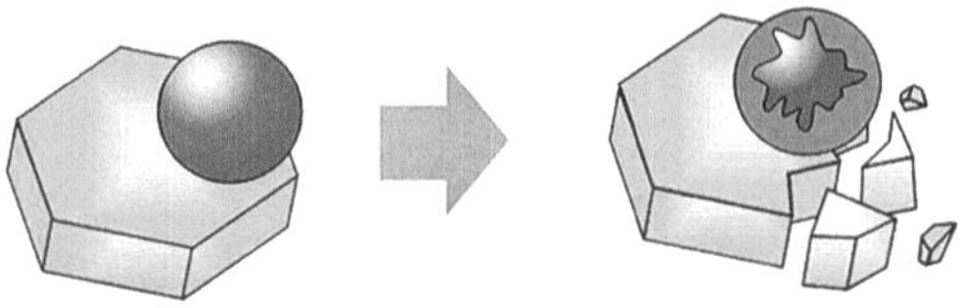

**Fig. 3.44** Fragmentation due to thermal shock. (Adapted from Korolev and Leisner 2020. © The Authors under CC-BY-4.0 license)

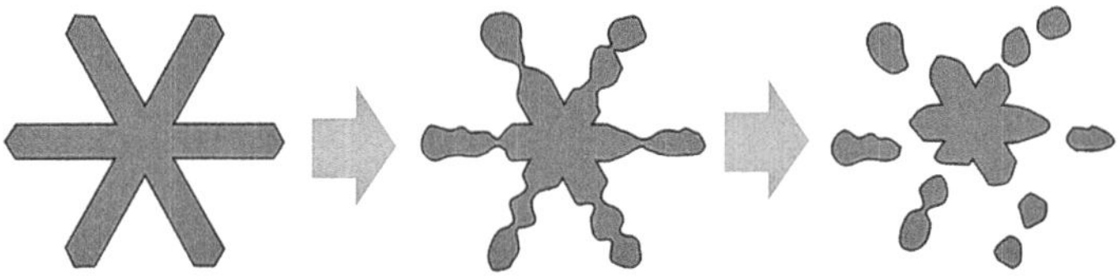

**Fig. 3.45** Fragmentation of a sublimating ice particle. (Adapted from Korolev and Leisner 2020. © The Authors under CC-BY-4.0 license)

***Fragmentation Due to Thermal Shock (Fig. 3.44)*** A supercooled drop riming the surface of an ice crystal freezes, and its temperature rises to the melting point. Some fraction of the latent heat of freezing will be transferred into the ice crystal. A thermal shock may result at the location of the droplet attachments with following ice crystal cracking and splintering due differential expansion of ice (Koenig 1963).

***Fragmentation of Sublimating Ice Particles (Fig. 3.45)*** Ice particle fragmentation and formation of secondary ice may occur during sublimation in subsaturated cloud regions. The first observation is due to Mossop et al. (1974) who observed sublimation of rimed particles under the microscope. They reported detachment of 20 μm size rime.

***Activation of INPs in Transient Supersaturation Around Freezing Drops (Fig. 3.46)*** Several studies provide experimental and theoretical support that activation of supersaturation sensitive INPs in the wake of free-falling freezing drops, wet hailstones, or riming graupel is one of possible mechanisms of secondary ice production in clouds (for details see Korolev and Leisner 2020).

Droplet fragmentation during freezing may be active across a wide range of temperatures. There is an increasing in both convective and stratiform clouds.

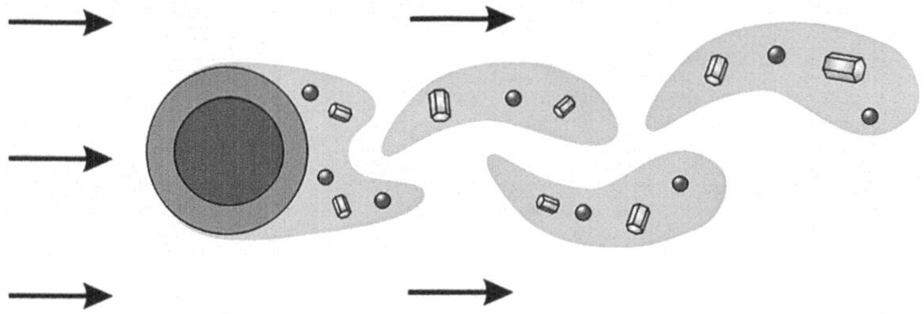

**Fig. 3.46** Activation of INPs in transient supersaturation around freezing drops. (Adapted from Korolev and Leisner 2020. © The Authors under CC-BY-4.0 license)

Rime-splintering mechanisms require the presence of heavily rimed graupel with high fall velocity. Formation in convective mixed-phase cloud regions is in the temperature range between −8 and −3 °C. Ice–ice collisional fragmentation requires a large separation of vertical velocities of ice particles to enhance kinetic energy of their collision. The most likely candidates for this process are lightly and heavily rimed ice particles, once again in mixed-phase convective regions. Precipitation-sized drops and temperatures < −10 °C in mixed-phase convective cloud regions are required for shock fragmentation. However, King and Fletcher (1976) have shown that this mechanism is not likely to be important in natural clouds. The fragmentation of sublimating ice requires spatial proximity of undersaturated and supersaturated cloud regions. Such conditions may occur in cloud regions affected by entrainment and mixing with out-of-cloud dry air. Finally, INP activation in transient supersaturation requires precipitation-sized drops and high supercooling that are typical of convective cloud regions.

Secondary ice production is important to justify the formation of drizzle drops below the melting layer. Korolev et al. (2020) have conducted observations in oceanic tropical mesoscale convective systems (MCS) and midlatitude frontal clouds in the temperature range between 0 and −15 °C and heavily seeded by aged ice particles. They proposed that the initiation of SIPs above the melting layer is stimulated by the recirculation of large liquid drops through the melting layer with convective turbulent updrafts. After re-entering a supercooled environment above the melting layer, they impact with aged ice, freeze, and shatter. The size of the splinters generated during secondary ice production was estimated as 10 μm or less. A scheme of the process is shown in Fig. 3.47. Finally, Korolev et al. (2022) have carried out in situ observations of secondary ice production at temperatures down to −27 °C, well outside the range of the Hallett–Mossop process. More recently, aircraft observations and modelling (Korolev et al. 2024) have led to the introduction of a conceptual model that explains the high ice water content in tropical MCSs as a possible result of secondary ice production induced by fragmentation of freezing drops in the vicinity of the melting layer.

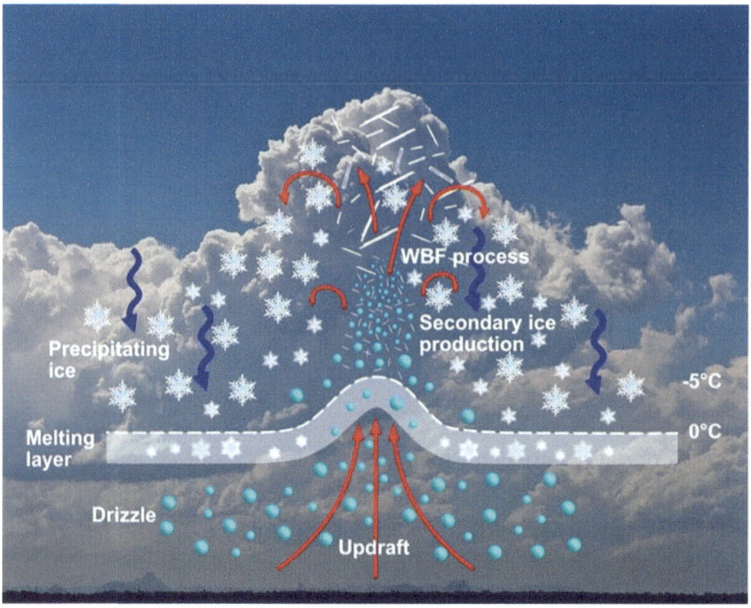


**Fig. 3.47** Conceptual model of the effect of the melting layer on the SIP formation in MCSs and frontal clouds. (Adapted from Korolev et al. 2020. © The Authors under CC-BY-4.0 license)

### 3.7.3 Ice Growth by Accretion

The growth of hydrometeors by accretion takes place when a large precipitation particle overtakes and captures a smaller one. However, the common usage refers to accretion as the capture of supercooled droplets by an ice phase precipitation particle. In the cases where the droplets freeze immediately upon contact, they form a coating of rime, leading to rimed crystals or graupels. If the freezing is delayed, a denser coating of ice is created leading eventually to hail formation, even though not necessarily.

The term coalescence that we have used for small droplet capture by collector cloud drops can be used also in case the collector is a raindrop. The term aggregation normally indicates the sticking of ice crystals to form snowflakes.

The efficiency of collision and capture mechanisms of ice crystals clearly depends on their terminal fall speed, which is shown in Fig. 3.48. The fastest falling objects are graupels (not quite crystals but aggregates of frozen droplets); rimed crystals fall at about 1 m $s^{-1}$, but all the crystal types have a fall velocity smaller than 1 m $s^{-1}$. An empirical formula fits the fall speeds of graupels

$$u = 343D^{0.6} \tag{3.81}$$

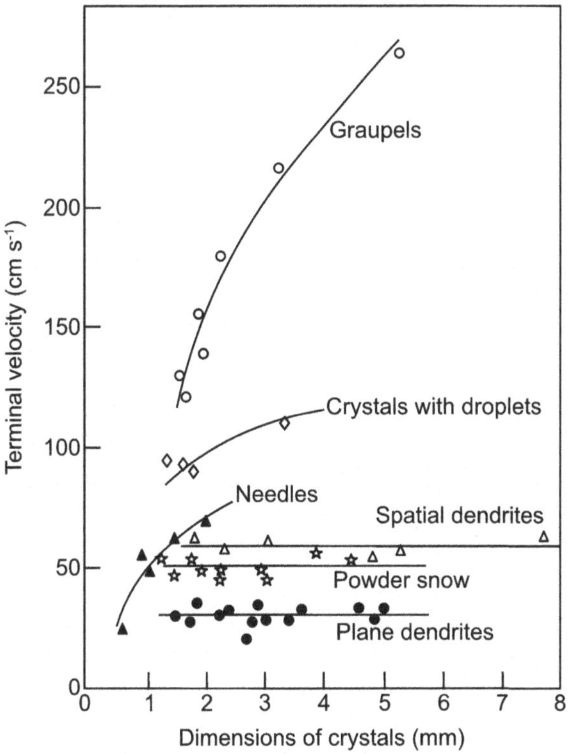


**Fig. 3.48** Measured terminal velocities of ice crystals. (Adapted from Nakaya and Terada 1935. )

with u in cm $s^{-1}$ and $D$, the diameter of the sphere which circumscribes the particle, in cm. Snowflakes are found to fall at about 1 m $s^{-1}$ as long as they keep dry. A general formula for the snowflakes fall speed is

$$u = kD^n \tag{3.82}$$

where D is the melted diameter. For dendrites $k \approx 160$ and $n \approx 0.3$ and columns and plates $k \approx 234$ and $n \approx 0.3$.

Mass and size of the different crystals are related by empirical formulas of the type

$$m = a\mathcal{D}^b \tag{3.83}$$

where $\mathcal{D}$ is the major linear dimension of the crystal.

Collection efficiency is a major problem when studying accretion, especially in case of crystal aggregation. Several studies have been conducted in recent times and an extensive appraisal is given by Wang (2013). Sticking of ice crystals is more frequent for dendritic crystals, but temperature has a strong influence. Temperatures higher than −10 °C were found favourable for an effective aggregation. Considering all the possible uncertainties, an equation for accretional growth, eventually leading to graupel formation can be derived as we did before for water droplets (see 3.62)

$$\frac{dm}{dt} = \bar{E} M \pi R^2 u(R) \tag{3.84}$$

where $m$ is the mass of the particle, $M$ the cloud liquid water content, $R$ the radius of the particle, $u(R)$ the fall speed and $\bar{E}$ the mean collection efficiency.

The same approach can be adopted for the aggregation process considering that snowflakes fall at around 1 m s$^{-1}$ and ice crystals at 0.4–0.5 m s$^{-1}$:

$$\frac{dm}{dt} = \bar{E} M \pi R^2 \Delta u \tag{3.85}$$

where $\Delta u$ is the difference in fall speed of the snowflake and the ice crystals, a constant. The number density N and the density of the frozen water are related by

$$M = N v \rho \tag{3.86}$$

where $v$ is the average volume of the crystals and $\rho$ their density. If the snowflake is assumed to have the same density, then $m = \rho V$ where $V$ is its volume. The growth equation in terms of the volume can be written as

$$\frac{dV}{dt} = B \bar{E} V^{2/3} N v \Delta u \tag{3.87}$$

where $B^3 = \frac{9}{16}\pi$. Naturally, all these equations must be considered as rough approximations, but they provide a reasonable description of the process.

### 3.7.4 Ice Crystal Process Versus Coalescence

From what we have learned so far, cloud particles will grow to precipitation size when aggregation or accretion will make the crystal grow, or coalescence will act effectively for water droplets. In fact, condensation and diffusion are not sufficient to account for an appreciable growth. This latter process is, however, more effective for ice crystals than for water droplets because the vapour is often at equilibrium with respect to water and supersaturated with respect to ice.

Light precipitation can occur in the form of ice crystals of the same size or as drizzle from their melting. These crystals are individual crystals that never experienced aggregation. In warm clouds diffusional growth is too slow and coalescence is always necessary to produce rain.

In cumulus clouds that originally develop at temperatures higher than 0 °C droplets may not freeze even if the cloud reaches heights where the temperature is substantially below 0 °C. In these clouds the ice crystal process and coalescence act together. The question is: which process dominates, and which one is faster? Rogers and Yau (1979) propose an exercise (see also their Fig. 9.8).

The growth condition of the droplet and of the ice crystal are the most favourable to start with. The ice crystal is a stellar dendrite growing in saturated air at −15 °C and 800 hPa from an initial mass of $10^{-8}$ g. The water drop starts with a radius of 25 $\mu$m and grows by continuous collection in a cloud consisting of droplets of 10 $\mu$m radius, with a liquid water content of 1 g m$^{-3}$.

The crystal grows rapidly by diffusion surpassing the size of the drop at 75 s and continues its predominance because the drop is initially limited in its growth by the small collection efficiency. The fractional rate of increase of its mass is initially much higher for the crystal, but it decreases as the crystal grows along a straight line, which means that we have a power law dependence upon the crystal mass. For the drop, the fractional mass increase starts low and increases monotonically to an asymptotic value ($5 \times 10^{-3}$ s). Below 7 min the ice crystals grows more rapidly than the droplet. By this time the collection efficiency has increased enough to make the coalescence process efficient. The droplet reaches the ice crystal in mass at 30 min, when its radius is 160 $\mu$m: this is a drizzle droplet falling at about 1.3 m s$^{-1}$.

So, the answer to the question on the most efficient mechanism for precipitation formation is essentially that both processes can produce precipitation depending on the environmental conditions in the cloud. Precipitation normally starts when particles of 4$\mu$g are formed. This may happen in favourable conditions starting with an ice crystal in about 10 min or by continuous water collection in a time twice as long. The two processes at midlatitudes are quite often intertwined and precipitation may be initiated by both. The development of precipitation particles will then subsequently require either coagulation/coalescence in the water process or accretion/aggregation in the ice process.

### 3.7.5 Graupel and Hail Formation

We have seen that cold clouds contain not only ice crystals and aggregates, but also large amounts of supercooled water, depending on temperature. Riming is thus very frequent because of the impact of water droplets and ice crystals. Rimed crystals grow by deposit on their surface of the droplets that immediately freeze. This process starts from the external rim of the crystal for hydrodynamic reasons and progresses towards the centre when the crystal is subject to turbulence and tumbling. The frozen droplets accumulate and form amorphous hydrometeors named graupels, which are rimed particles less than 5 mm in diameter (Fig. 3.49).

A scheme of graupel formation in a thunderstorm cloud is shown in Fig. 3.50. Graupels fall in thunderstorm showers when they become too heavy to be sustained by the updraft. Note that, however, graupels often feed heavy thunderstorm showers when they melt before reaching the ground thus producing large drops.

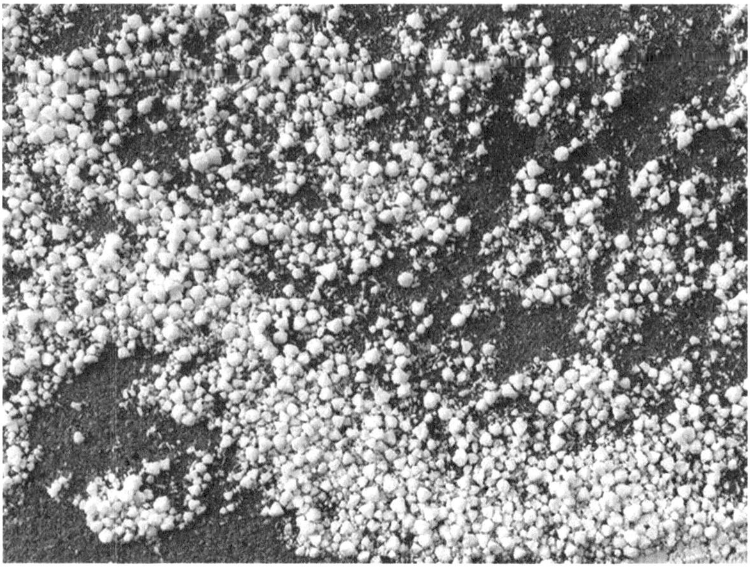

**Fig. 3.49** Graupels (snow pellets) covering the ground after a thunderstorm. (Public domain)

Not all ice crystals act as graupel embryos (Reinking 1975; Harimaya 1976), even those that grow to relatively large sizes over relatively long periods by deposition and accretion. Instead, graupels grow over selected small crystals utilizing such regular crystals as kernels. Their growth mechanisms are influenced by temperature as noted by Harimaya (1990) in Japan and by sublimation and evaporation processes. If we look carefully at Fig. 3.49, we see that most graupel particles are conical and studies have confirmed this observation (Arenberg 1941; Knight and Knight 1973). The reason is due to the lateral motion of the riming particle, following a bell-shaped trajectory. Before reaching the final conical shape the graupel undergoes a series of steps continuously collecting supercooled droplets that freeze inducing a loss of the original crystal embryo shape (Fig. 3.51). The frozen droplets pile up on the growing graupel and thus the surface of the graupel is not smooth, but grainy due to the rime caused by the frozen droplets. This is what distinguishes them from ice pellets. The density of graupels is smaller than that of solid ice because air is trapped in the open spaces between the frozen drops. Density can vary from 0.05 to 0.89 g $cm^{-3}$ depending on the stage of growth. However, note that recent in situ observations have suggested that different graupel formation processes occur in different rain systems (Suzuki et al. 2018) related to water concentration in clouds and rain enhancement. Further studies are needed as the mechanisms are not yet fully understood.

What is the difference between graupels and hail? They are both formed via a riming process in a storm cloud with a high supercooled water content. However, their internal structure is notably different. While graupels have no detectable structure, hailstones are characterised by a layered structure as shown in Fig. 3.52. The

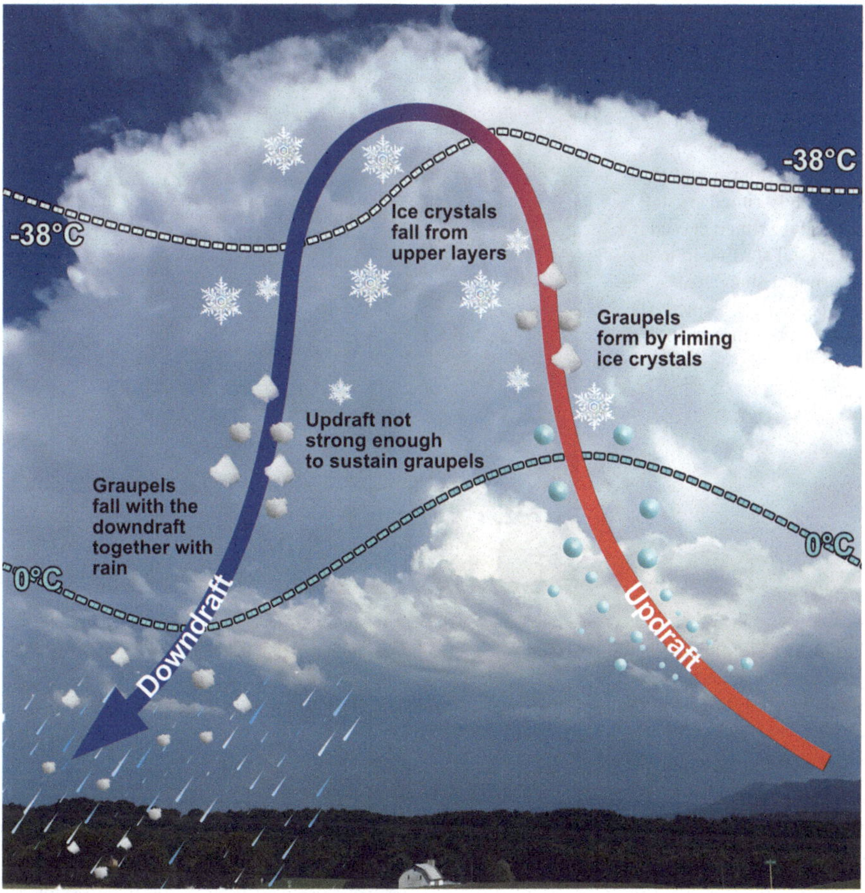


**Fig. 3.50** Scheme of graupel formation in a convective cloud

onion-shaped structure alternates rather transparent and translucid layers whose number is variable depending on the size of the stone.

There are two ingredients of the hailstone growth. First, a small (<1 cm) particle is needed to serve as the nucleus for further growth. Any particle consisting of heavily rimed snow particles of diameters less than 5 mm is defined as graupel, as we have seen. These particles are known as "embryos" and typically take the form of graupel or frozen drops (Knight and Knight 1970a). The growth process happens in the hail growth zone (Fig. 3.53) where the temperature varies approximately between −10 and −25 °C. The reason behind the layered structure is due to the passage of the stone through diverse cloud layers at different temperatures. The more transparent layers indicate that the temperature at the time of the formation of the layer was relatively high. The supercooled drops impinging on the hailstone have thus the time to spread around the impact location and not instantaneously freeze. This implies that the liquid water occupies all the available space, and the ice has a

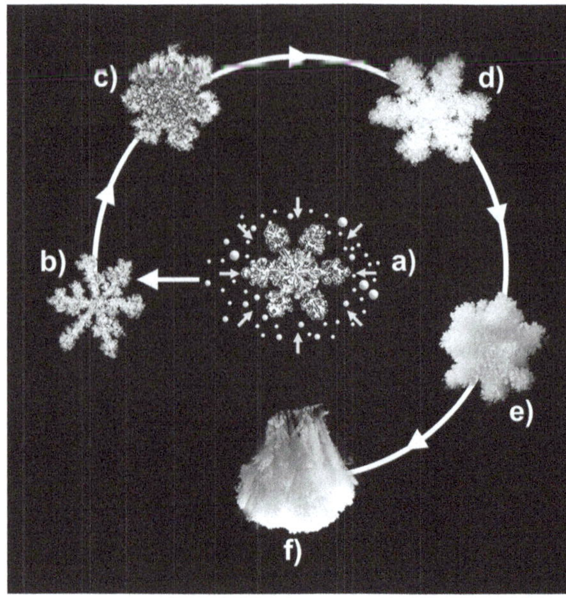


**Fig. 3.51** Graupel formation steps through riming of an ice crystal embryo. From centre, clockwise: (a) supercooled droplets capture by an ice crystal during riming; (b) moderately rimed crystal; (c) heavily rimed crystal; (d) crystal starting to lose its original shape; (e) graupel almost formed; (f) final form of the graupel (conical)

**Fig. 3.52** Small hailstone fallen in Levizzano Rangone (Italy) on 15 June 2022. (Photo by V. Levizzani)

high density and compactness: this type of growth is called wet growth. The contrary is true for the translucid layers that are formed in regions of the cloud at lower temperatures, the so-called dry growth. Here, the supercooled droplets freeze immediately upon impact and the resulting ice structure is consequently full of air bubbles, which lends the ice an opaquer appearance. An intermediate type of growth has also been identified, the spongy growth, and refers to the situations when the growing hailstone contains liquid trapped within the ice structure. For a complete overview of the growth processes the reader is referred to the article by Allen et al. (2020).

When hailstones are collected on the ground, they are mostly spherical or spheroidal. However, natural hailstones exhibit irregular shapes and protuberances called "lobes" (Knight and Knight 1970b). They are classified into cusp lobes and

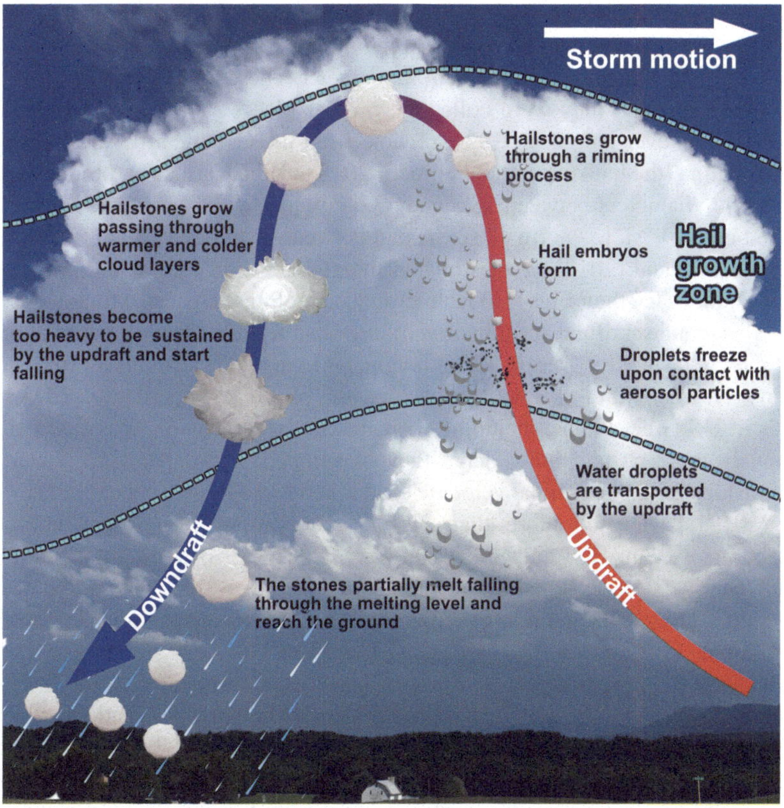


**Fig. 3.53** Hail formation process in a convective storm cloud

icicle lobes, which arise in dry and wet growth, respectively. They can be seen on the ground in those cases when the stone is large enough to be only marginally affected by the melting process. Thus, the internal structure of a hailstone is quite complex and provides insights into the dynamics and thermodynamics of the generating cloud (see for example Fig. 1 of Soderholm and Kumjian 2023). Recent studies have led to further insights into hail formation mechanisms, especially in supercells, suggesting that environments leading to weakened horizontal flow within updrafts may lead to larger maximum hailstone sizes (Kumjian and Lombardo 2020). Shape-specific thermal energy transfer coefficients lead to increasing sizes; the effects of lobes vary depending on the fall speed scheme used. These assumptions can be parameterised efficiently and potentially used in microphysics schemes (Lin et al. 2024).

The largest hailstone on record seems to be the one that fell in Vivian, South Dakota, on 23 July 2010. It had a diameter of 20 cm and weighed an astonishing 0.88 kg. Note that these numbers describe the stone at the very moment of the collection, but they may have been larger due to possible partial melting during the trajectory between the cloud and the ground or during its stay on the ground before collection. Previously, another record-breaking hailstone was collected in Aurora, Nebraska, in 2003, and it was 17.8 cm in diameter. These are the so-called gargantuan hailstones, and they were collected in other places such as Bangladesh (the heaviest stone ever recorded, 1.02 kg, came from the Gopalganj District in 1986) and Argentina (the hailstorm on 8 February 2018 in Villa Carlos Paz produced hailstones with diameters between 18.8 and 23.6 cm, probably beating the Vivian recognised record). More "normal" sizes span between 5 mm and 15 cm in diameter. Prein and Holland (2018) have estimated the global average large hail probability over the globe (see their Fig. 11a) from ground samplings. They also found that four predictors enable skilful discrimination of large hail frequencies on a regional scale: atmospheric instability, freezing level height, 0–3 km wind shear and storm relative helicity. Heymsfield et al. (2014) measured the physical properties of 2295 hailstones in US Great Plains, including their maximum dimension, mass, and cross-sectional area. Using these data, they developed size-dependent power-law relationships for their terminal velocities and kinetic energies. Recently, Grieser and Hill (2019) have used more than 37,000 observations from the Community Collaborative Rain, Hail and Snow Network (CoCoRaHS) finding that total hail kinetic energy is proportional to the diameter of the largest hailstone to a power of less than 2. These data and relationships are used by insurance companies to calculate hail risk.

The need for more comprehensive approaches combining information from different sources is necessary to improve our understanding of where and when severe hail most often occurs. In this direction, the use of high resolution, convection-permitting reanalysis products together with geostationary satellite observations is instrumental to identify hail-favourable environmental conditions. Giordani et al. (2024) have identified a proxy based on a filter to identify convective updraughts potentially linked to the formation of hailstones in a thunderstorm by considering the surrounding environment.

### 3.7.6 Aggregates and Snowflakes

Ice crystals may grow large enough to be sufficiently heavy and generate precipitation. However, snowfall consisting of single ice crystals is not usually the norm. Single, monodisperse ice crystals normally melt passing through the melting layer and reach the ground as inconspicuous rain or even drizzle. Snowfall is composed by snowflakes that are aggregates of several snow crystals. The crystals collide with each other and remain attached to form a lumped structure.

Hobbs et al. (1974) collected ice crystals and aggregates in clouds and at ground stations for two winter seasons. They found that the maximum dimensions of aggregates and their probability of occurrence decrease with decreasing temperatures: this is because at higher temperatures the crystals may partially melt during collision and thus stick together more easily. Both characteristics show a local maximum near the dendritic growth region, and this is understandable since dendrites are characterised by branches that favour the process of aggregate formation. The authors claim that below −15 °C and for particle concentrations less than 0.1 $cm^{-3}$ aggregates are unlikely to form; above −5 °C and for particle concentrations higher than 1 $cm^{-3}$ the chance of aggregate formation is above 50%.

Fujiyoshi and Wakahama (1985) investigated the type, degree of riming and size of snow particles from snow bands. They found three types of snow particles: non rimed or slightly rimed snow crystals, rimed or heavily rimed snow crystals, and heavily rimed or graupel-like snow particles. Non rimed or lightly rimed snow provided up to about 30% of the total number of snow particles when precipitation was very intense (>5 mm $h^{-1}$), and regardless of the intensity of precipitation, there appear not to be differences in the size distribution of snow particles.

The forms of the snowflakes are highly variable and depend on temperature and falling angles of the individual snow crystals. Only recently instruments were developed to photograph snowflakes in a cloud. One of such instruments, the Multi-Angle Snowflake Camera (MASC) (Garrett et al. 2012), captures high-resolution photographs of the flakes from three angles while simultaneously measuring their fall speed. An example of photographs of flakes and graupels obtained with the MASC is visible in Fig. 3.54. The snowflakes captured by the MASC do not look like those we are normally used to see in photos. The reason is that for the first time

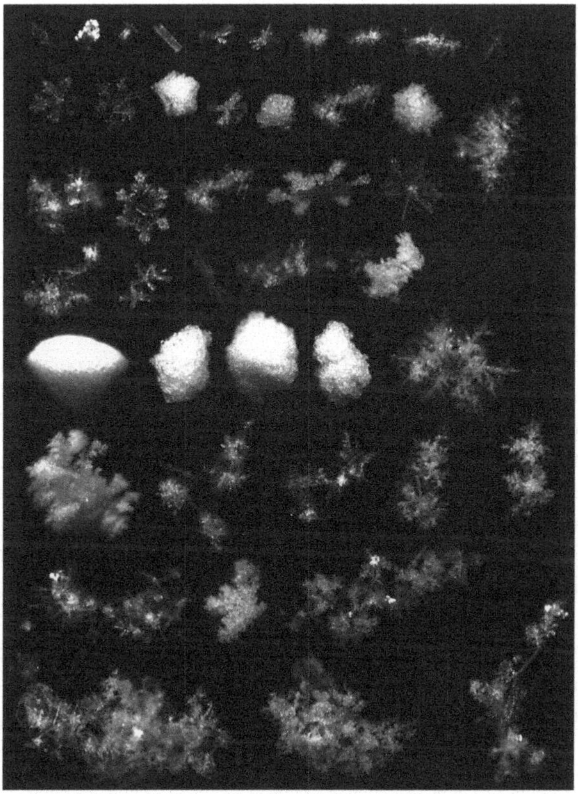

**Fig. 3.54** Snowflakes, rimed flakes and graupels captured by the Multi-Angle Snowflake Camera (MASC) of the University of Utah (Garrett et al. 2012). (Source: T. J. Garrett, Department of Atmospheric Sciences, University of Utah, Salt Lake City, reproduced with permission)

the flakes are photographed untouched as they fall in air to the ground. Garrett et al. (2015) examined the photographs of nearly 73,000 hydrometeors in free fall to determine the aspect ratio and orientation of aggregates, moderately rimed particles, and graupel. These observations indicate that there can be a much broader range of orientation angles, with a larger median value than indicated by previous observations. The authors found that more vertical orientations for frozen particles become increasingly common with higher turbulence.

Jiang et al. (2019) have demonstrated that the maximum dimension of the flakes is not in the horizontal plane, and that the rotational angles of the maximum dimensions in the horizontal plane are not uniform but instead are correlated with the wind direction. Moreover, Dunnavan et al. (2019) have argued that simulations and observations prove that aggregates are rarely oblate and exhibit a triaxial ellipsoid shape. These studies are increasingly important for the optimisation of precipitation retrievals from radar and satellite PMW sensors (Liao et al. 2013; Nowell et al. 2013; Ori et al. 2014, 2021; Honeyager et al. 2016).

### 3.7.7 Melting

Mixed clouds are characterised by vertical temperature gradients and a melting level. Ice hydrometeors melt, partially or totally, depending on temperature and their melting influences precipitation formation mechanisms. Approximately 60% of the liquid precipitation reaching the Earth surface results from melting of ice particles such as ice crystals, snowflakes, graupel, and hailstones, not only in polar regions and midlatitudes but even in the tropics (e.g., Heymsfield et al. 2020; Mülmenstädt et al. 2015).

However, melting processes are far from being completely understood. What we know is based on laboratory experiments in cloud tunnels in controlled conditions and theoretical modelling due to the difficulty of in situ measurements. An accurate overview of the studies from the 1950s to the present is provided by Theis et al. (2022).

Oraltay and Hallett (2005) have conducted experiments on melting, freezing, and evaporation of individual and aggregates of snow crystals in the laboratory under controlled temperature, relative humidity, and air velocity. The authors argue that the fallout of ice aggregates to the melting level, the changing depth of any isothermal layer, and the capability of forming or not forming a radar bright band all need to consider the particle habit and melting evolution.

Ice particles melt differently depending on their shape, density, fall speed and air temperature. Parameters that are essential to describe the melting process physically are the melting rate (ice mass decrease as a function of time), the total melting time of the ice particle, the shape variation of melting ice particles, the melting behaviour describing changes of the fall mode during melting, and, in the case of droplet shedding, the mass and size distribution of the shed droplets.

Laboratory experiments in the vertical cloud tunnel were conducted by Rasmussen and Pruppacher (1982) on the rate of melting of ice spheres of radii smaller than 500 microns. The experimental results were discussed producing a comprehensive theory of the melting of ice spheres with the fluid layer eccentric with respect to the inner ice core (Rasmussen et al. 1982, 1984a). Successively, Rasmussen et al. (1984b) investigated the melting behaviour of spherical ice particles up to radii of several centimetres. These laboratory and theoretical results were applied by Rasmussen and Heymsfield (1987a, b, c) to describe the melting behaviour of graupel and hailstones.

The melting of aggregates of snowflakes is different in that water penetrates the ice mesh. Mitra et al. (1990) conducted experiments in the vertical cloud tunnel and devised a theory computing distances needed by snowflakes to completely melt finding a good agreement with radar observations.

Faraday (1859) discovered the mode of ice surface melting and the presence of a thin water layer, called quasi-liquid layer (QLL), that homogeneously wets ice surfaces. Murata et al. (2016) have demonstrated theoretically and experimentally that there are more than two wetting states. Furthermore, QLLs show up not only in supersaturated conditions, but also in undersaturated conditions. Finally, they are totally absent at the equilibrium. The authors argue that the QLLs are a metastable transient state formed through water vapour growth and sublimation of ice. Their model shows that a unique interfacial potential controls both wetting and thermodynamic behaviour of QLLs.

## 3.8 Cloud Electricity

The Earth atmosphere is electrically charged at any time, and the presence of electrically charged particles is not necessarily linked to the presence of clouds. This was discovered by Benjamin Franklin and investigated by others, inspired by Franklin, in the summer of 1753. Fair weather conditions are in fact characterised by the constant polarisation of charges that maintains a vertical electric field. Thus, clouds develop in this field and are subject to charging that induces separation of charge between cloud hydrometeors. The electric charging has a strong influence on the hydrometeor formation and on their speed of growth. Precipitation formation is strongly influenced by the presence of the electric field, especially in thunderstorm clouds.

The static magnetic field of the Earth is generated by electric currents in the melted iron core of the Earth's core. Its shape can be approximated by a magnetic dipole. These natural atmospheric (and cosmic) electromagnetic fields are also an important driver of Earth currents (or telluric currents), and their dynamics, in both soil and water. Static electric fields are present throughout the Earth's atmosphere as part of the global electric circuit that extends from lower ionospheric layers to the surface of the Earth (Fig. 3.55).

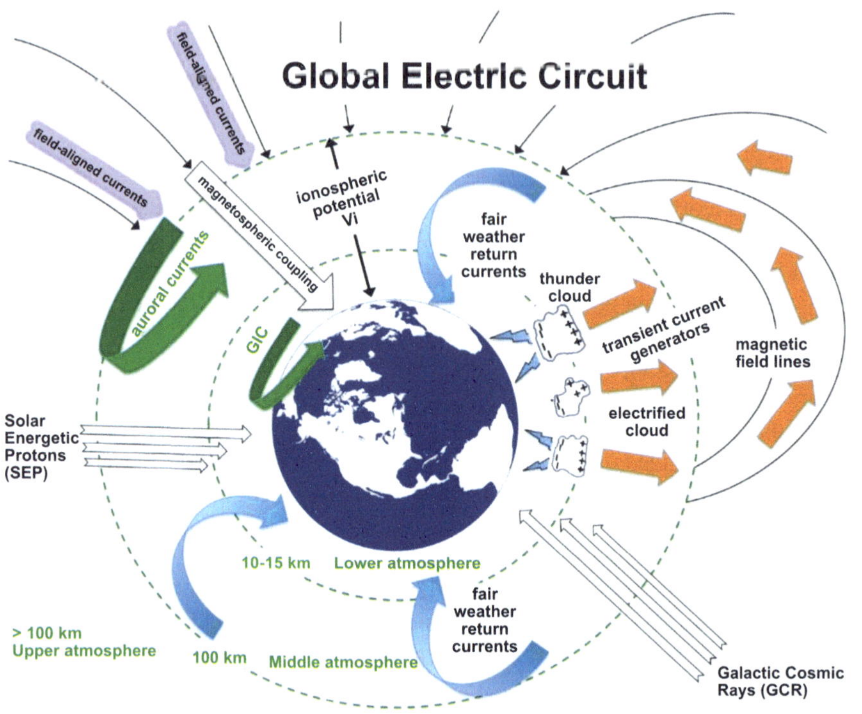


**Fig. 3.55** Phenomena collectively driving the global electric circuit. (Source: J. Forbes, University of Colorado, Boulder, reproduced with permission)

### 3.8.1 Fair Weather Electric Field

In the wake of Franklin's studies an electric field was discovered in regions of the Earth away from thunderstorms, the fair-weather electric field (Harrison 2004, 2011). Lord Kelvin investigated the matter through the development of electrostatic instrumentation. The electrostatic field is a consequence of a potential difference (~250 kV) between the electrically conducting ionosphere and the Earth's surface.

Ions in the higher atmosphere are almost entirely produced by cosmic rays. They consist of positive and negative charges equally distributed in number. Apart from recombination that takes place, enough free ions are present at any time and the ensemble of ions behaves like a fluid, called plasma, which responds to electric and magnetic fields. The layer starting at the altitude of ~60 km is called electrosphere and overlaps with the ionosphere. The Earth's surface, land and water, is also conducting and can be approximated by an equipotential surface. The atmosphere between the two charged surfaces acts as an insulating layer. The resistance is not uniform and depends on local conditions: clear air, cloud, and polluted regions have all different resistances.

However, the atmosphere is not a perfect insulator, and a leakage current of 1500 A exists in the cloudless regions. Such current brings positive charges to the surface that neutralise the negative charges present thereon. The presence of the leakage current, if not counteracted by an opposite charging phenomenon, would totally discharge the global charge of the circuit. A simple calculation tells us that the global charge of $5.1 \times 10^5$ C would be discharged in ~385 s (~6.5 min). Thus, without a charging mechanism of some sort supplying the negative charges the positive ions coming from extraterrestrial source would bring the Earth surface to electrical neutrality and the fair-weather electric field would disappear in about 6.5 min. The fair-weather field, however, exists and thus there must be some mechanism that supplies the negative charges to maintain the field.

### 3.8.2 The Global Electric Circuit and Thunderstorms

The structure of the field was explained through the global circuit model of Wilson (1903, 1920, 1929), the inventor of the cloud chamber. The atmosphere is sandwiched between the ionosphere and the surface and the whole system behaves like a capacitor, as we have seen above (Fig. 3.56). What is the charging mechanism that prevents the capacitor to completely discharge in a short time?

All clouds, even the small ones, are normally electrified, but the electric field they host is usually small. Thunderclouds are different in that they produce a large number of electric charges that are eventually transported to the ground,

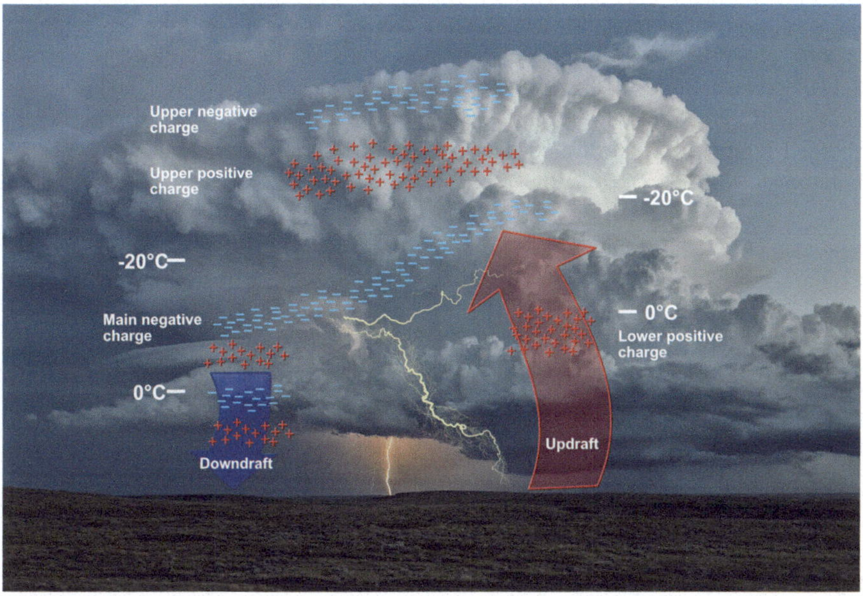


**Fig. 3.56** Charge distribution in a storm cloud

maintaining the global electric circuit. This happens through lightning strokes. Studies have demonstrated that thunderstorms have a tripole structure of charge distribution in their updraft core (Fig. 3.56). The electrical structure of a storm, though, remains a bit speculative (e.g., Soula 2012).

A quick calculation can be done to support the notion of lightning from thunderstorms as an effective charging mechanism of the global circuit. The global frequency of lightning strokes is >100 $s^{-1}$, up to 250 $s^{-1}$, of which 20% (50 $s^{-1}$) are cloud to ground flashes. Now, if ~30 C are transported to the surface by each lightning, we obtain a current of just 1500 A, which can counteract the fair-weather current of the opposite sign. If we assume that each storm cell lasts about 25 min, each storm cell generates a current of ~1 A.

The particles in the upper layer, where the temperature is lower than −20 °C, are mainly cloud ice or snow aggregates: they mostly carry positive charges. In the middle level, raindrops and graupel/hail carry negative charges. A region of concentrated negative charges exists between −10 and −20 °C: here is where ice particles grow more effectively, as we have previously seen. Near the cloud base there is a small positive charge region. The structure is a bit more complex than this simple model can suggest. The polar structure may consist of more than three poles or, in certain storms, inverted with a main negative charge centre above a positive charge centre at midlevel (~ −10 °C).

The charge centres are responsible for the buildup of high electric fields. Such fields induce a breakdown, which is however produced also by other factors. One of such factors is the resistance of air which needs to be drastically reduced, thus increasing the charge transfer, the electron avalanche. The typical environmental field during the breakdown is of the order of 100–300 V $m^{-1}$, but the critical field at the very moment of breakdown is much higher. At this point in time, the lightning occurs. There are several types of lightning flashes (Fig. 3.57): cloud to ground (CG), between clouds (CC), within a single cloud or intracloud (IC), and cloud to clear air (CA). Note that, because of the different distribution of polarity of the charged regions in a storm and the differential accumulation of charges, not all lightning strokes start from the region of predominant negative polarity. Some

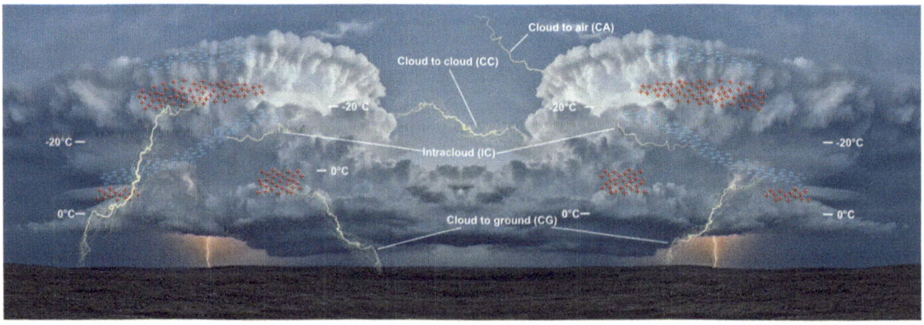


**Fig. 3.57** The types of lightning flashes in thunderstorm clouds: cloud to ground (CTG), between two nearby clouds (cloud to cloud, CC), within a single cloud (intracloud, IC), and cloud to air (CA)

lightning conveys positive charges, and this is reflected in measurements and theory (Nag and Rakov 2012) and in recent satellite observations (Leal and Rakov 2024). Negative lightning makes up to 90–95% of all strikes while the positive lightning accounts for about 5% and is mostly confined to winter months and to the dissipating stage of a thunderstorm. Positive lightning is usually composed of one single stroke while negative lightning normally consists of two or more. The damaging power of positive lightning is greater than that of the negative ones. This is because such strokes originate normally in the upper part of the storm and thus the amount of air they must burn to hit the ground is much greater than in the case of negatives strokes. The flash duration is longer, and the peak charge and potential can be ten times greater than in the case of a negative stroke, as much as $3 \times 10^5$ A.

The global density of lightning flashes is very variable from place to place. Satellite observations have documented recently the global distribution (Christian et al. 2003), the flash rate density (Albrecht et al. 2016), and the location of the most intense flashes (Zipser et al. 2006). A lightning climatology from satellite sensors has been proposed by Cecil et al. (2014) identifying a mean global flash rate from the merged climatology is 46 flashes $s^{-1}$. This varies from around 35 flashes $s^{-1}$ in February (austral summer) to 60 flashes $s^{-1}$ in August (boreal summer). The peak annual flash rate at 0.5° scale is 160 flashes $km^{-2}$ $year^{-1}$ in eastern Congo. The peak monthly average flash rate at 2.5° scale is 18 flashes $km^{-2}$ $month^{-1}$ from early April to early May in the Brahmaputra Valley of far eastern India. Recent studies focus on the significant increase of lightning in some high altitude or high latitude regions, such as the Tibetan Plateau and Arctic regions (e.g., Qie et al. 2024) to identify the role of aerosol particles in modulating lightning variations under certain dynamical and thermodynamic conditions, even on a global scale.

Peterson (2024) has recently identified a new class of relatively compact convective storms with concentrated lightning activity. These storms are mostly found throughout the Gulf of Mexico and east of South Africa (among other coastal and oceanic regions) and have been associated with convective intensity that rival the top MCSs. However, their horizontal and vertical dimensions are small by comparison, and this is the reason why the hydrometeors involved in the electrification processes are more concentrated, enabling the observed extreme lightning densities.

### 3.8.3 Cloud Electrification Mechanisms

A fundamental question, very much in line with what we need to know on precipitation formation mechanisms concerns the processes of cloud electrification. How does a cloud electric field build up? The details of cloud charging processes are still under investigation and thus no conclusive word can be spent on which is the dominating one. However, there is a general agreement on the basic concepts.

Studies of cloud microphysics have postulated three types of mechanisms for cloud electrification: convective charging, inductive charging, and non-inductive charging.

**Convective Charging** Positive ions are the predominant type near the Earth's surface, the ratio of positive to negative ions being approximately 1.2. The net positive charge near the surface of the Earth is swept up into the developing cumulus cloud in the early development stages. The updraft carries the positive charge into the forming cloud and, eventually, lifts the positive ions to the very top of the cloud. The positive area within the updraft causes a negative ion "curtain" to form on the top and side boundaries of the cumulus cloud. Entrainment of the net negatively charged environmental parcels form much of the cloud's downdraft, therefore creating the negatively charged lower portion of the cloud.

**Inductive Charging** Two sub-categories belong to this type. Selective ion capture and rebound of particles. The first mechanism is constructed around the fact that the Earth electric field induces a charge separation within the cloud constituents, negative on the top and positive on the bottom. As the hydrometeor falls through the cloud, it attracts free negative ions to the bottom hemisphere, but positive ions cannot "catch up" to the particle to attach to the top. The net negatively charged cloud constituents migrate to the lower, warmer region of the cloud. By removing the attached negative ions from the higher, colder region of the cloud a net positive charge remains near the top of the cloud. Such theory produces the correct polarity but is found to be effective only under weak electric fields, and thus quantitatively unrealistic. The second mechanism starts from the charge separation by charge migration within individual hydrometeors due to the Earth's electric field. Negative charges stay on the top, and positive charges on the bottom. Larger, faster falling particle and a smaller particle may collide, and a net positive charge is transferred to the smaller particle because the larger particle has greater surface area, net negative to the larger, warmer particle. The heavier, negative hydrometeors fall to the bottom of the cloud, while positive, lighter particle are lifted to the top. There are several problems with this theory, but an evident one is the assumption that the collision angle is such that positive and negative surface regions of the respective hydrometeors collide.

**Non-inductive Charging** Several mechanisms belong to this category. The thermoelectric effect that describes the separation of charges due to the difference of temperature between ice hydrometeors (ice crystals and hail pellets). The most important mechanism is the electrification during riming. The idea is that a substantial electric charge can be deposited on a riming graupel in a collision with an ice crystal in a cloud with supercooled water vapour present. This is considered in general as the most promising mechanism, though not completely known in all the details.

We have seen that in the central part of the storm air moves rapidly upwards (updraft) and the temperature is between −15 and −25 °C. In this cloud region, the combination of temperature and rapid upward air movement produces a mixture of supercooled droplets, small ice crystals, and graupel. The updraft carries the supercooled droplets and very small ice crystals upward. At the same time, the graupel tends to fall or be suspended in the rising air. The differences in the movement of the precipitation cause collisions to occur. When the rising ice crystals collide with graupel, the ice crystals become positively charged and the graupel becomes negatively charged (Fig. 3.58).

The updraft carries the positively charged ice crystals toward the cloud top. The larger and denser graupel is either suspended in the middle of the thunderstorm cloud or falls toward the lower part of the storm. The result is that the upper part of the thunderstorm cloud becomes positively charged while the middle to lower part of the thunderstorm cloud becomes negatively charged.

Some of these charges can be redistributed by updrafts and downdrafts. In addition, there is a small but important positive charge buildup near the bottom of the thunderstorm cloud due to precipitation and higher temperatures.

### 3.8.4 Effects of the Electric Field on Microphysics

We have discussed in the previous sections of this chapter the various mechanisms of hydrometeor formation and the formation of precipitation. Electric fields present in a cloud obviously play a role in modifying microphysical processes. Electrofreezing (possibly induced by cavitation) is one of these phenomena. Then, electric fields induce changes in terminal velocity of the hydrometeors because of the deformation of drops in the electric field. The electric field or the charges exert an influence on the vibrational frequencies of drops. Condensation nuclei are produced by the Rayleigh bursting of solution droplets. Finally, and most important of all, electric forces influence collision, coalescence and aggregation efficiencies.

However, while these processes are important for the formation of cloud hydrometeors and may enhance their growth, they are not totally significant in the development of precipitation. Note that electric fields influence the DSD of droplets and drops in a cloud and thus may have some influence on precipitation rates (e.g., Mudiar et al. 2022). Studies are conducted such as those of Ambaum et al. (2022) who show that the attractive force between charged conductive spheres increases

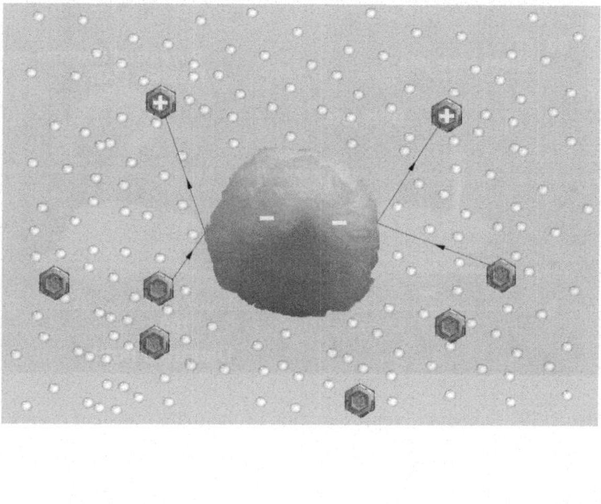

**Fig. 3.58** Schematic of the non-inductive charging mechanism in cold regions of a thundercloud (below −15 °C) by collision of ice crystals and a graupel

with increasing charge variance. The importance of this electrostatic effect between water drops is evaluated for its potential to enhance rain formation.

One more important issue regards the emerging evidence of coupling between large- and small-scale atmospheric electrical phenomena and various biological processes in terrestrial environments (Hunting et al. 2021). The current understanding of this connectivity is not sufficient, and a more direct appraisal on how atmospheric electricity can affect various levels of biological organisation across multiple ecosystems is crucial.

### 3.8.5 Transient Events

Lightning from convective activity is not the only phenomenon associated with intense thunderstorms. Other large-scale phenomena happen above the thunderstorms cloud top (Fig. 3.59). Although there is no proof that they are linked to precipitation formation, we will mention them for completeness.

Wilson (1924) predicted that electrical breakdown should occur in the atmosphere high above large thunderstorms. In the following decades, high altitude electrical discharges directed towards the outer space were reported by aircraft pilots and observed from the Space Shuttle until the first direct visual evidence was documented in 1989 (a history of the early speculations and findings can be found in

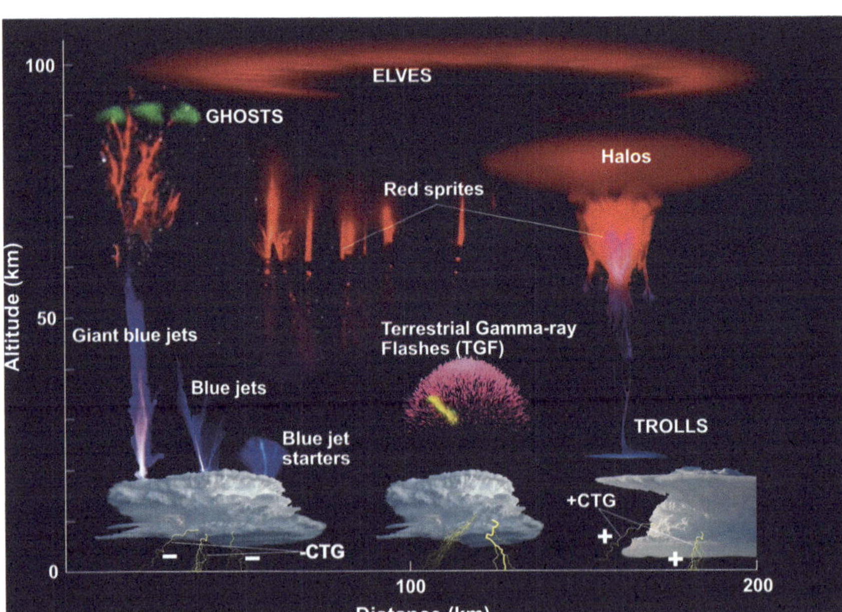


**Fig. 3.59** Schematic representation of the main transient events associated with thunderclouds

Williams 2001). These phenomena fall under the category of Transient Luminous Events (TLEs).

Two classes of vertically extensive optical flashes above thunderclouds have been identified (e.g., Pasko et al. 2002): red sprites and blue jets. Sprites initiate near the base of the ionosphere, develop very rapidly downwards at speeds which can exceed $10^7$ m s$^{-1}$, and assume different geometrical forms. In contrast, blue jets develop upwards from cloud tops at speeds of the order of $10^5$ m s$^{-1}$ and have a blue conical shape. Sprites are triggered by positive lightning discharges between the cloud and the ground, while blue jets seem to be initiated between the upper positive charge region of a thundercloud and the negative screening layer above this region. Blue starters are shorter and brighter than blue jets, reaching a height of about 20 km. Gigantic jets are another category of luminous events that appear to initiate as an intracloud flash between the middle negative and the upper positive regions in a thundercloud. The negatively charged leader escapes upward from the cloud toward the ionosphere reaching higher altitudes than normal blue jets up to 90 km. Recently, observations of sprites and jets became more organised, and a climatology of their occurrence is becoming possible as shown by Arnone et al. (2020). Impacts on atmospheric chemistry are also investigated (Gordillo-Vázquez and Pérez-Invernón 2021).

A few other TLEs were identified such as the ELVES (Emissions of Light and Very low frequency perturbations due to Electromagnetic pulse Sources), which appear as a dim, flattened, expanding glow around 400 km in diameter that lasts for about 1 ms. TROLLs (Transient Red Optical Luminous Lineaments) occur after strong sprites and appear as red spots with faint tails. They show up as a rapid series of events, starting as a red glow that forms after a sprite tendril, that later produces a red streak downward. GHOSTs (GreenisH Optical emission from Sprite Tops) are faint, green glows that appear within the footprint of a red sprite, remaining after the red has dissipated, and fading away in ms.

Other observations have suggested that there may be additional phenomena to be uncovered, such as true "cloud to stratosphere" upward lightning discharges (Lyons et al. 2003). These phenomena appear to be neither sprites nor elves and depart from our current understanding of the phenomenology of blue jets. More in depth studies are obviously needed.

Finally, Terrestrial Gamma-ray Flashes (TGFs) are bursts of gamma rays produced in the Earth's atmosphere. Although their formation mechanism is still being investigated, it is presumed that TGF photons are emitted by electrons traveling at speeds very close to the speed of light that collide with the nuclei of atoms in the air and release their energy in the form of gamma rays (bremsstrahlung). Many energetic electrons form by avalanche growth driven by electric fields. This electric field is most probably provided by lightning, as most TGFs occur within a few ms of a lightning event. Marisaldi et al. (2024) have shown that TGFs are relatively common in tropical thunderclouds. TGFs are characterised by high intensities and have characteristic durations of tens to hundreds of μs. They are not to be confused with gamma-ray glows that last for one to hundreds of seconds, have moderate intensities and originate from quasi-stationary thundercloud fields. Recent airborne observations have identified one more class of these phenomena, the Flickering Gamma-Ray Flashes, which have more pulses, and each pulse has a longer duration than ordinary TGFs (Østgaard et al. 2024). They seem to be the missing link between pulses and TGFs.

## 3.9 Microphysical Parameterisation in Models

The representation of microphysical processes in models continues to pose a major challenge leading to uncertainty in numerical weather forecasts and climate simulations. The problem is not solved even if considerable progress has been recently made. Such a problem, according to Morrison et al. (2020), needs to be divided in two parts: (i) how to represent the population of cloud and precipitation particles, given the impossibility of simulating all particles individually within a cloud, and (ii) uncertainties in the microphysical process rates owing to fundamental gaps in knowledge of cloud physics. Probabilistic models are being proposed that combine elements from statistical and physical modelling. These models, besides providing rigorous constraint of schemes, add value to quantifying uncertainty systematically.

Liu et al. (2023) in their review paper concentrate on several key topics that hold great potential for advancing bulk microphysics parameterisations: multi-moment bulk microphysics parameterisations and the role of the spectral shape of hydrometeor size distributions; discrete vs "continuous" representation of hydrometeor types; turbulence-microphysics interactions including turbulent entrainment-mixing processes and stochastic condensation; theoretical foundations for the mathematical expressions used to describe hydrometeor size distributions and hydrometeor morphology; and approaches for developing bulk microphysics parameterisations. They also present the spectral bin scheme and particle-based scheme (especially, super-droplet method) for representing explicit microphysics. Finally, they indicate particle-resolved direct numerical simulation (DNS) models as an emerging technique to investigate turbulence-microphysics interactions at the most fundamental level by tracking individual particles and resolving the smallest turbulent eddies in turbulent clouds.

Further studies are needed to understand how more refined microphysical parameterisations can improve surface precipitation forecast in all weather conditions. An example is given by Dedekind et al. (2024) who have shown that enhanced IN concentrations from collisional breakup reduce localised regions of higher precipitation and, thereby, improve the model performance in terms of surface precipitation over the domain for winter orographic precipitation (Fig. 3.60).

## 3.10 What We Know and What We Need to Understand

After more than six decades of laboratory experiments, in situ samplings and theoretical investigations, we have reached a substantial level of knowledge of cloud microphysical processes leading to precipitation formation. However, there are still

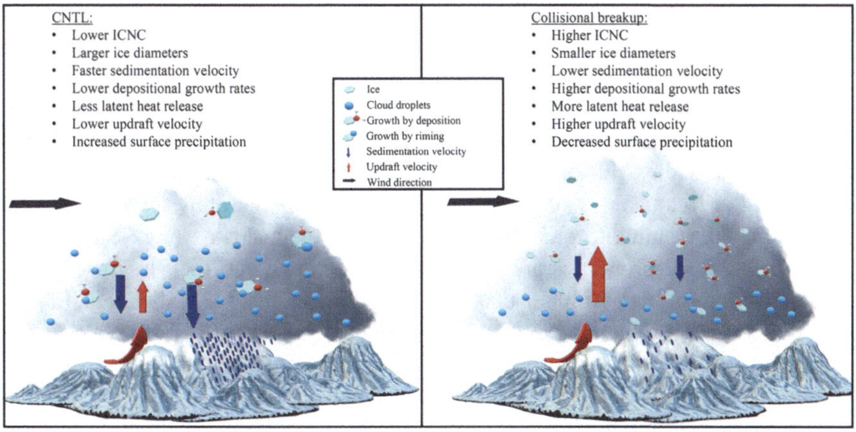


**Fig. 3.60** Schematic of the impact of collisional breakup of ice particles on the cloud structure and surface precipitation: control run (left) and parameterised model (right). (From Dedekind et al. 2024. © The Authors under CC-BY-4.0 license)

gaps both in empirical and theoretical description of cloud processes. For example, gaps concern ice nucleation, aggregation and splintering of ice particles, collision rates in turbulent flows, and breakup of drops. Moreover, the natural variability of clouds, cloud particles and aerosol are very large: the different particle habits (including degrees of riming), the time-spatial structures in clouds, as well as the particle size distributions. Finally, the strong nonlinearity and high complexity of cloud processes hinders rigorous analytic and theoretical approaches.

After an entire chapter on describing the numerous findings of cloud physics both in the experimental and theoretical areas, one must ask the question whether we do know enough of the fall mechanism of hydrometeors and the changes in a population of cloud droplets in their transition to precipitation. Köhler's theory explains the initial droplet growth but struggles with large hydrometeors. More studies are being conducted using stochastic approaches and the topic is all but fully investigated (e.g., Santos Gutiérrez et al. 2024).

More is needed on ice hydrometeors and their characteristics. Recent works suggest that we still need to work on these aspects. Wang (2021) argues that only the most recent advances of computer technologies allow for in depth studies on the flow fields around ice hydrometeors. This will obviously have an influence in cloud modelling and radar meteorology. Lin et al. (2024) have in turn shown that it is time to abandon the paradigm of spherical hailstones. They have demonstrated that the choice of hailstone size-mass relation and fall speed scheme have the strongest influence on hail sizes. Using non-spherical, tumbling hailstones reduces the number of large hailstones produced. Thus, more studies are on their way on this emerging section of cloud physics (see also Wang 2002). One more topic that needs attention is the exact calculation of the scattering properties of ice hydrometeors. This does not refer only to ice crystals or aggregates, but also to frozen droplets and frozen-droplet aggregates that are observed within deep convective clouds and influence cloud electrification and precipitation formation (Kim et al. 2024). Finally, freezing drizzle in variable and complex cloud environments is worth exploring (Faber et al. 2024).

Further studies are necessary especially in cloud-aerosol interactions and the formation of atmospheric ice (Burrows et al. 2022). This is especially true for Arctic conditions where the cycle of aerosol capable of ice cloud formation is still scarcely known (e.g., Creamean et al. 2022). Comprehensive microphysics datasets are needed to better simulate precipitation formation mechanisms (e.g., King et al. 2024a, b). Aerosol properties are not completely known and their hygroscopicity needs to be determined (Kulmala et al. 2023; Pöhlker et al. 2023). Stier et al. (2024) have argued that the extent of aerosol effects on precipitation at smaller scales is not entirely clear. While there is broad consensus and strong evidence that aerosol perturbations microphysically increase cloud droplet numbers and decrease droplet sizes, slowing precipitation droplet formation, the overall aerosol effect on precipitation across scales remains highly uncertain.

Cloud electricity has been extensively studied, but we are far from a full understanding of cloud electrification mechanisms. For example, the role of ice multiplication in winter clouds is not sufficiently explored (Yang et al. 2024). Another important aspect is the mechanisms of charge transfer involving the ice phase. While the collisional mechanism of thundercloud charging is well established, the nature of the charge transfer between the colliding ice particles is not very well understood at the atomic level, and no present theory can explain it in full detail; theories are proposed but the problem is not clearly solved (Jungwirth et al. 2005). Winter storm electric fields are scarcely studied and understood as regards the charge of snowflakes and ice pellets together with the location of flashes with respect to the snow showers (Harkema et al. 2024). Studies should concentrate on the influence of lightning discharges on atmospheric chemistry, for example the $NO_x$ production and the tropospheric ozone chemistry (Pickering et al. 2024). Finally, the fluctuations of the Earth's electric field is not completely understood and recent studies investigate its variations with high relative humidity values and low wind speed, leading to foggy conditions (Yair and Yaniv 2023).

Global cloud-resolving models provide unique opportunities to investigate mechanisms that are currently not well represented in global climate models and to robustly connect local effects with larger scales. However, as noted by Morrison et al. (2020), the way to represent cloud microphysical processes in models is far from being resolved. As noted by Seifert (2011), there are still gaps in the empirical and theoretical description of cloud processes like ice nucleation, aggregation and splintering of ice particles, collision rates in turbulent flows, breakup of drops, etc. The natural variability of clouds, cloud particles and aerosol is overwhelmingly large, e.g., the different particle habits (including degrees of riming), the time-spatial structures in clouds, as well as the particle size distributions, etc. The strong nonlinearity and high complexity of cloud processes hinders any rigorous analytic and theoretical approaches. Mixed-phase clouds that contain ice and supercooled liquid particles are especially hard to constrain (Hofer et al. 2024). Moreover, it has recently been shown that supercooled-liquid water in mixed-phase clouds is not homogeneously mixed, neither vertically nor horizontally. For addressing critical gaps in cloud physics knowledge, sustained investment for observational advances from laboratory experiments, new probe development, and next-generation instruments in space are needed.

## References


Albrecht, R.I., Goodman, S.J., Buechler, D.E., Blakeslee, R.J., Christian, H.J.: Where are the lightning hotspots on Earth? Bull. Am. Meteorol. Soc. **97**(11), 2051–2068 (2016). https://doi.org/10.1175/BAMS-D-14-00193.1

Allen, J.T., Giammanco, I.M., Kumjian, M.R., Punge, H.J., Zhang, Q., Groenemeijer, P., Kunz, M., Ortega, K.: Understanding hail in the Earth system. Rev. Geophys. **58**, e2019RG000665 (2020). https://doi.org/10.1029/2019RG000665

Ambaum, M.H.P., Auerswald, T., Eaves, R., Harrison, R.G.: Enhanced attraction between drops carrying fluctuating charge distributions. Proc. R. Soc. A. **478**, 20210714 (2022). https://doi.org/10.1098/rspa.2021.0714

AMS: Hydrometeor. Glossary of Meteorology. https://glossary.ametsoc.org/wiki/Hydrometeor. Last accessed 10 Jan 2024 (2020)

Andronache, C. (ed.): Mixed Phase Clouds—Observations and Modeling. Elsevier (2018)

Arenberg, D.L.: The formation of small hail and snow pellets. Bull. Am. Meteorol. Soc. **22**(3), 113–116 (1941). https://doi.org/10.1175/1520-0477-22.3.113

Arnone, E., Bór, J., Chanrion, O., Barta, V., Dietrich, S., Enell, C.-F., Farges, T., Füllekrug, M., Kero, A., Labanti, R., Mäkelä, A., Mezuman, K., Odzimek, A., Popek, M., Prevedelli, M., Ridolfi, M., Soula, S., Valeri, D., van der Velde, O., Yair, Y., Zanotti, F., Zoladek, P., Neubert, T.: Climatology of transient luminous events and lightning observed above Europe and the Mediterranean Sea. Surv. Geophys. **41**, 167–199 (2020). https://doi.org/10.1007/s10712-019-09573-5

Baumgardner, D., McFarquhar, G.M., Heymsfield, A.J. (eds.): Ice Formation and Evolution in Clouds and Precipitation - Measurement and Modelling Challenges. American Meteorological Society (2018)

Bergeron, T.: On the physics of clouds and precipitation. In: Proceedings of the Fifth Assembly of UGGI, Lisbon, vol. 2, pp. 156–160 (1935)

Berry, E.X., Reinhardt, R.L.: An analysis of cloud drop growth by collection: part I. Double distributions. J. Atmos. Sci. **31**(7), 1814–1824 (1974a). https://doi.org/10.1175/1520-0469(1974)031<1814:AAOCDG>2.0.CO;2

Berry, E.X., Reinhardt, R.L.: An analysis of cloud drop growth by collection: part II. Single initial distributions. J. Atmos. Sci. **31**(7), 1825–1831 (1974b). https://doi.org/10.1175/1520-0469(1974)031<1825:AAOCDG>2.0.CO;2

Block, K., Haghighatnasab, M., Partridge, D.G., Stier, P., Quaas, J.: Cloud condensation nuclei concentrations derived from the CAMS reanalysis. Earth Syst. Sci. Data. **16**, 443–470 (2024). https://doi.org/10.5194/essd-16-443-2024

Boucher, O.: Atmospheric Aerosols—Properties and Climate Impacts. Springer Nature (2015)

Brasseur, G.P., Jacob, D.J.: Modeling of Atmospheric Chemistry. Cambridge University Press (2017)

Brown Jr., P.S.: The effects of filament, sheet, and disk breakup upon the drop spectrum. J. Atmos. Sci. **45**(4), 712–718 (1988). https://doi.org/10.1175/1520-0469(1988)045<0712:TEOFSA>2.0.CO;2

Burrows, S.M., McCluskey, C.S., Cornwell, G., Steinke, I., Zhang, K., Zhao, B., Zawadowicz, M., Raman, A., Kulkarni, G., China, S., Zelenyuk, A., DeMott, P.J.: Ice-nucleating particles that impact clouds and climate: observational and modeling research needs. Rev. Geophys. **60**, e2021RG000745 (2022). https://doi.org/10.1029/2021RG000745

Byers, H.R.: Elements of Cloud Physics. University of Chicago Press (1965)
Carslaw, K.S. (ed.): Aerosols and Climate. Elsevier (2022)
Cascajo-Castresana, M., David, R.O., Iriarte-Alonso, M.A., Bittner, A.M., Marcolli, C.: Protein aggregates nucleate ice: the example of apoferritin. Atmos. Chem. Phys. **20**, 3291–3315 (2020). https://doi.org/10.5194/acp-20-3291-2020
Cecil, D.J., Buechler, D.E., Blackslee, R.J.: Gridded lightning climatology from TRMM-LIS and OTD: dataset description. Atmos. Res. **135–136**, 404–414 (2014). https://doi.org/10.1016/j.atmosres.2012.06.028
Cha, J.W., Koo, H.J., Kim, B.Y., Belorid, M., Hwang, H.J., Kim, M.H., Chang, K.-H., Lee, Y.H.: Analysis of rain drop size distribution to elucidate the precipitation process using a cloud microphysics conceptual model and in situ measurement. Asia-Pac. J. Atmos. Sci. **59**, 257–269 (2023). https://doi.org/10.1007/s13143-022-00299-w
Chang, R.Y.-W., Abbatt, J.P.D., Boyer, M.C., Prakash Chaubey, J., Collins, D.B.: Characterizing the hygroscopicity of growing particles in the Canadian Arctic summer. Atmos. Chem. Phys. **22**, 8059–8071 (2022). https://doi.org/10.5194/acp-22-8059-2022
Charlson, R.J., Seinfeld, J.H., Nenes, A., Kulmala, M., Laaksonen, A., Facchini, M.C.: Reshaping the theory of cloud formation. Science. **292**(5524), 2025–2026 (2001). https://doi.org/10.1126/science.1060096
Christian, H.J., Blakeslee, R.J., Boccippio, D.J., Boeck, W.L., Buechler, D.E., Driscoll, K.T., Goodman, S.J., Hall, J.M., Koshak, W.J., Mach, D.M., Stewart, M.F.: Global frequency and distribution of lightning as observed from space by the Optical Transient Detector. J. Geophys. Res. Atmos. **108**(D1), ACL 4-1–ACL 4-15 (2003). https://doi.org/10.1029/2002JD002347
Cornwell, G.C., McCluskey, C.S., Hill, T.C.J., Levin, E.T., Rothfuss, N.E., Tai, S.-L., Petters, M.D., DeMott, P.J., Kreidenweis, S., Prather, K.A., Burrows, S.M.: Bioaerosols are the dominant source of warm temperature immersion-mode INPs and drive uncertainties in INP predictability. Sci. Adv. **9**, eadg3715 (2023). https://doi.org/10.1126/sciadv.adg3715
Cotton, W.R., Anthes, R.A.: Storm and Cloud Dynamics. Elsevier (1992)
Cotton, W.R., Bryan, G., van den Heever, S.: Storm and Cloud Dynamics—The Dynamics of Clouds and Precipitating Mesoscale Systems. Elsevier (2011)
Creamean, J.M., Barry, K., Hill, T.C.J., Hume, C., DeMott, P.J., Shupe, M.D., Dahlke, S., Willmes, S., Schmale, J., Beck, I., Hoppe, C.J.M., Fong, A., Chamberlain, E., Bowman, J., Scharien, R., Persson, O.: Annual cycle observations of aerosols capable of ice formation in Central Arctic clouds. Nat. Commun. **13**, 3537 (2022). https://doi.org/10.1038/s41467-022-31182-x
Crutzen, P.J., Ramanathan, V. (eds.): Clouds, Chemistry and Climate. Springer Nature (1996)
Curry, J.A., Webster, P.J.: Thermodynamics of Atmospheres & Oceans. Academic Press (1999)
Dedekind, Z., Lauber, A., Ferrachat, S., Lohmann, U.: Sensitivity of precipitation formation to secondary ice production in winter orographic mixed-phase clouds. Atmos. Chem. Phys. **21**, 15115–15134 (2024). https://doi.org/10.5194/acp-21-15115-2021
Deirmendjian, D.: Scattering and polarization properties of polydispersed suspensions with partial absorption. In: Kerker, M. (ed.) Electromagnetic Scattering, pp. 171–189. Pergamon Press (1963)
Dolan, B., Fuchs, B., Rutledge, S.A., Barnes, E.A.: Primary modes of global drop size distributions. J. Atmos. Sci. **75**, 1453–1476 (2018). https://doi.org/10.1175/JAS-D-17-0242.1
Dufour, L., Defay, R.: Thermodynamics of Clouds. Academic Press (1963)
Dunnavan, E.L., Jiang, Z., Harrington, J.Y., Verlinde, J., Fitch, K., Garrett, T.J.: The shape and density evolution of snow aggregates. J. Atmos. Sci. **76**, 3919–3940 (2019). https://doi.org/10.1175/JAS-D-19-0066.1
Dusek, U., Franck, G.P., Hildebrandt, L., Curtius, J., Schneider, J., Walter, S., Chand, D., Drewnick, F., Hings, S., Jung, D., Borrmann, S., Andreae, M.O.: Size matters more than chemistry for cloud-nucleating ability of aerosol particles. Science. **312**(5778), 1375–1378 (2006). https://doi.org/10.1126/science.1125261
Faber, S., Bernstein, B., Landolt, S., Jacobson, D., DiVito, S., Wolde, M., Korolev, A., Heckman, I., Nichman, L., Nguyen, C.: Airborne observations of highly variable and complex freezing drizzle and mixed-phase environments. J. Appl. Meteorol. Climatol. **63**, 1343–1361 (2024). https://doi.org/10.1175/JAMC-D-24-0056.1

Faraday, M.: Experimental Researches in Chemistry and Physics. Taylor & Francis (1859)

Findeisen, W.: Die Kolloidmeteorologischen Vorgange beider Niederschlagsbildung. Meteorol. Z. **55**, 121–133 (1938). https://doi.org/10.1127/metz/2015/0675

Fletcher, N.H.: The Physics of Rainclouds. Cambridge University Press (1962)

Friedlander, S.K.: Smoke, Dust and Haze—Fundamentals of Aerosol Behavior. Wiley (1977)

Fujiyoshi, Y., Wakahama, G.: On snow particles comprising an aggregate. J. Atmos. Sci. **42**(15), 1667–1673 (1985). https://doi.org/10.1175/1520-0469(1985)042<1667:OSPCAA>2.0.CO;2

Garrett, T.J., Fallgatter, C., Shkurko, K., Howlett, D.: Fall speed measurement and high-resolution multi-angle photography of hydrometeors in free fall. Atmos. Meas. Tech. **5**, 2625–2633 (2012). https://doi.org/10.5194/amt-5-2625-2012

Garrett, T.J., Yuter, S.E., Fallgatter, C., Shkurko, K., Rhodes, S.R., Endries, J.L.: Orientations and aspects ratios of falling snow. Geophys. Res. Lett. **42**(11), 4617–4622 (2015). https://doi.org/10.1002/2015GL064040

Giordani, A., Kunz, M., Bedka, K.M., Punge, H.J., Paccagnella, T., Pavan, V., Cerenzia, I.M.L., Di Sabatino, S.: Characterizing hail-prone environments using convection-permitting reanalysis and overshooting top detections over south-central Europe. Nat. Hazards Earth Syst. Sci. **24**, 2331–2357 (2024). https://doi.org/10.5194/nhess-24-2331-2024

Gordillo-Vázquez, F.J., Pérez-Invernón, F.J.: A review of the impact of transient luminous events on the atmospheric chemistry: past, present, and future. Atmos. Res. **252**, 105432 (2021). https://doi.org/10.1016/j.atmosres.2020.105432

Grieser, J., Hill, M.: How to express hail intensity—modeling the hailstone size distribution. J. Appl. Meteorol. Climatol. **58**, 2329–2345 (2019). https://doi.org/10.1175/JAMC-D-18-0334.1

Gunn, K.L.S., Marshall, J.S.: The distribution with size of aggregate snowflakes. J. Meteorol. **15**, 452–461 (1958). https://doi.org/10.1175/1520-0469(1958)015<0452:TDWSOA>2.0.CO;2

Hallett, J., Mossop, S.C.: Production of secondary ice particles during the riming process. Nature. **249**, 26–28 (1974). https://doi.org/10.1038/249026a0

Harimaya, T.: The embryo and formation of graupel. J. Meteorol. Soc. Jpn. **54**(1), 42–51 (1976). https://doi.org/10.2151/jmsj1965.54.1_42

Harimaya, T.: A climatological study on the mechanism of graupel formation. J. Fac. Sci. Hokkaido Univ. Ser. VII. **8**(5), 437–447 (1990) Available at https://eprints.lib.hokudai.ac.jp/dspace/bitstream/2115/8774/1/8(5)_p437-447.pdf. Last accessed 22 Apr 2024

Harimaya, T., Ishida, H., Muramoto, K.: Characteristics of snowflake size distributions connected with the difference of formation mechanisms. J. Meteorol. Soc. Jpn. **78**(3), 233–240 (2000). https://doi.org/10.2151/jmsj1965.78.3_233

Harimaya, T., Kodama, H., Muramoto, K.: Regional differences in snowflake size distributions. J. Meteorol. Soc. Jpn. **82**(3), 895–903 (2004). https://doi.org/10.2151/jmsj.2004.895

Harkema, S.S., Mansell, E.R., Fierro, A.O., Carey, L.D., Schultz, C.J., Matsui, T., Berndt, E.B.: Explicitly resolving lightning and electrification processes from the 10–12 April 2019 thundersnow outbreak. J. Geophys. Res. Atmos. **129**, e2023JD039987 (2024). https://doi.org/10.1029/2023JD039987

Harrison, R.G.: The global atmospheric electrical circuit and climate. Surv. Geophys. **25**(5–6), 441–484 (2004). https://doi.org/10.1007/s10712-004-5439-8

Harrison, R.G.: Fair weather atmospheric electricity. J. Phys. Conf. Ser. **301**, 012001 (2011). https://doi.org/10.1088/1742-6596/301/1/012001

Heymsfield, A.J., Giammanco, I.M., Wright, R.: Terminal velocities and kinetic energies of natural hailstones. Geophys. Res. Lett. **41**, 8666–8672 (2014). https://doi.org/10.1002/2014GL062324

Heymsfield, A.J., Schmitt, C., Chen, C.C.J., Bansemer, A., Gettelman, A., Field, P.R., Liu, C.: Contributions of the liquid and ice phases to global surface precipitation: observations and global climate modeling. J. Atmos. Sci. **77**(8), 2629–2648 (2020). https://doi.org/10.1175/JAS-D-19-0352.1

Hiranuma, N., Möhler, O., Yamashita, K., Tajiri, T., Saito, A., Kiselev, A., Hoffmann, N., Hoose, C., Jantsch, E., Koop, T., Murakami, M.: Ice nucleation by cellulose and its potential contribution to ice formation in clouds. Nat. Geosci. **8**, 273–277 (2015). https://doi.org/10.1038/ngeo2374

Hobbs, P.V.: Ice Physics. Oxford University Press (1974)
Hobbs, P.V. (ed.): Aerosol-Cloud-Climate Interactions. Academic Press (1993)
Hobbs, P.V., Deepak, A. (eds.): Clouds - Their Formation, Optical Properties, and Effects. Academic Press (2012)
Hobbs, P.V., Rangno, A.L.: Ice particle concentrations in clouds. J. Atmos. Sci. **42**(23), 2523–2549 (1985). https://doi.org/10.1175/1520-0469(1985)042<2523:IPCIC>2.0.CO;2
Hobbs, P.V., Chang, S., Locatelli, J.D.: The dimensions and aggregation of ice crystals in natural clouds. J. Geophys. Res. Atmos. **79**(15), 2199–2206 (1974). https://doi.org/10.1029/JC079i015p02199
Hofer, S., Hahn, L.C., Shaw, J.K., McGraw, Z.S., Bruno, O., Hellmuth, F., Pietschnig, M., Mostue, I.A., David, R.O., Carlsen, T., Storelvmo, T.: Realistic representation of mixed-phase clouds increases projected climate warming. Commun. Earth Environ. **5**, 390 (2024). https://doi.org/10.1038/s43247-024-01524-2
Honeyager, R., Liu, G., Nowell, H.: Voronoi diagram-based spheroid model for microwave scattering of complex snow aggregates. J. Quant. Spectrosc. Radiat. Transf. **170**, 28–44 (2016). https://doi.org/10.1016/j.jqsrt.2015.10.025
Houghton, H.G.: Physical Meteorology. MIT Press (1985)
Houze Jr., R.A.: Cloud Dynamics, 2nd edn. Academic Press (2014)
Hunting, E.R., Matthews, J., de Arróyabe, F., Hernáez, P., England, S.J., Kourtidis, K., Koh, K., Nicoll, K., Harrison, R.G., Manser, K., Price, C., Dragovic, S., Cifra, M., Odzimek, A., Robert, D.: Challenges in coupling atmospheric electricity with biological systems. Int. J. Biometeorol. **65**, 45–58 (2021). https://doi.org/10.1007/s00484-020-01960-7
Ichimura, I., Fujiwara, M., Yanase, T.: The size distribution of cloud droplets measured in small maritime cumulus clouds. J. Meteorol. Soc. Jpn. **58**(5), 403–415 (1980). https://doi.org/10.2151/jmsj1965.58.5_403
James, R.L., Phillips, V.T.J., Connolly, P.J.: Secondary ice production during the break-up of freezing water drops on impact with ice particles. Atmos. Chem. Phys. **21**, 18519–18530 (2021). https://doi.org/10.5194/acp-21-18519-2021
Jayachandran, V.N., Varghese, M., Murugavel, P., Todekar, K.S., Bankar, S.P., Malap, N., Dinesh, G., Safai, P.D., Rao, J., Konwar, M., Dixit, S., Prabha, T.V.: Cloud condensation nuclei characteristics during the Indian summer monsoon over a rain-shadow region. Atmos. Chem. Phys. **20**(12), 7307–7334 (2020). https://doi.org/10.5194/acp-20-7307-2020
Jiang, Z., Verlinde, J., Clothiaux, E.E., Aydin, K., Schmitt, C.: Shapes and fall orientations of ice particle aggregates. J. Atmos. Sci. **76**, 1903–1916 (2019). https://doi.org/10.1175/JAS-D-18-0251.1
Joss, J., Thams, J.C., Waldvogel, A.: The variation of raindrop size distributions at Locarno. In: Proceedings of the International Conference on Cloud Physics, Toronto, pp. 369–373 (1968)
Junge, C.E.: The size distribution and aging of natural aerosols as determined from electrical and optical data on the atmosphere. J. Meteorol. **12**, 13–25 (1955). https://doi.org/10.1175/1520-0469(1955)012<0013:TSDAAO>2.0.CO;2
Junge, C.E.: Vertical profiles of condensation nuclei in the stratosphere. J. Meteorol. **18**, 501–509 (1961). https://doi.org/10.1175/1520-0469(1961)018<0501:VPOCNI>2.0.CO;2
Junge, C.E.: Air Chemistry and Radioactivity. Academic Press (1963)
Jungwirth, P., Rosenfeld, D., Buch, V.: A possible new molecular mechanism of thundercloud electrification. Atmos. Res. **76**(1–4), 190–205 (2005). https://doi.org/10.1016/j.atmosres.2004.11.016
Kanji, Z.A., Ladino, L.A., Wex, H., Boose, Y., Burkert-Kohn, M., Cziczo, D.J., Krämer, M.: Overview of ice nucleating particles. Meteorol. Monogr. **58**(1), 1.1-1.33 (2017). https://doi.org/10.1175/AMSMONOGRAPHS-D-16-0006.1
Khadir, T., Riipinen, I., Talvinen, S., Heslin-Rees, D., Pöhlker, C., Rizzo, L., Machado, L.A.T., Franco, M.A., Kremper, L.A., Artaxo, P., Petäjä, T., Kulmala, M., Tunved, P., Ekman, A.M.L., Krejci, R., Virtanen, A.: Sink, source or something in-between? Net effects of precipitation on aerosol particle populations. Geophys. Res. Lett. **50**, e2023GL104325 (2023). https://doi.org/10.1029/2023GL104325

Khrgian, A., Mazin, I.P.: The size distribution of droplets in clouds. Trudy Ts. A. O. **7**, 56 (1952)

Khvorostyanov, V.L., Curry, J.A.: Aerosol size spectra and CCN activity spectra: reconciling the lognormal, algebraic, and power laws. J. Geophys. Res. Atmos. **111**, D12202 (2006). https://doi.org/10.1029/2005JD006532

Khvorostyanov, V.L., Curry, J.A.: Thermodynamics, Kinetics, and Microphysics of Clouds. Cambridge University Press (2014)

Kikuchi, K., Komeda, T., Higuchi, K., Yamashita, A.: A global classification of snow crystals, ice crystals, and solid precipitation based on observations from middle latitudes to polar regions. Atmos. Res. **132-133**, 460–472 (2013). https://doi.org/10.1016/j.atmosres.2013.06.006

Kim, J., Park, S., McFarquhar, G.M., Baran, A.J., Cha, J.-W., Lee, K., Lee, S.S., Jung, C.H., Lim, K.-S., Um, J.: Improved calculation of single-scattering properties of frozen droplets and frozen-droplet aggregates observed in deep convective clouds. Atmos. Chem. Phys. **24**, 12707–12726 (2024). https://doi.org/10.5194/acp-24-12707-2024

King, W.D., Fletcher, N.H.: Thermal shock as an ice multiplication mechanism. Part II: experimental. J. Atmos. Sci. **33**, 97–102 (1976). https://doi.org/10.1175/1520-0469(1976)033<0097:TSAAIM>2.0.CO;2

King, F., Pettersen, C., Dolan, B., Shates, J., Posselt, D.: Primary modes of Northern Hemisphere snowfall particle size distributions. J. Atmos. Sci. (2024a). https://doi.org/10.1175/JAS-D-24-0076.1

King, F., Pettersen, C., Bliven, L.F., Cerrai, D., Chibisov, A., Cooper, S.J., L'Ecuyer, T., Kulie, M.S., Leskinen, M., Mateling, M., McMurdie, L., Moisseev, D., Nesbitt, S.W., Petersen, W.A., Rodriguez, P., Schirtzinger, C., Stuefer, M., von Lerber, A., Wingo, M.T., Wolff, D.B., Wong, T., Wood, N.: A comprehensive northern hemisphere particle microphysics data set from the Precipitation Imaging Package. Earth Space Sci. **11**, e2024EA003538 (2024b). https://doi.org/10.1029/2024EA003538

Kiselev, A., Bachmann, F., Pedevilla, P., Cox, S.J., Michaelides, A., Gerthsen, D., Leisner, T.: Active sites in heterogeneous ice nucleation—the example of K-rich feldspars. Science. **355**, 367–371 (2017). https://doi.org/10.1126/science.aai8034

Knight, C.A., Knight, N.C.: Hailstone embryos. J. Atmos. Sci. **27**(4), 659–666 (1970a). https://doi.org/10.1175/1520-0469(1970)027<0659:HE>2.0.CO;2

Knight, C.A., Knight, N.C.: Lobe structures of hailstones. J. Atmos. Sci. **27**(4), 667–671 (1970b). https://doi.org/10.1175/1520-0469(1970)027<0667:LSOH>2.0.CO;2

Knight, C.A., Knight, N.C.: Conical graupel. J. Atmos. Sci. **30**, 118–124 (1973). https://doi.org/10.1175/1520-0469(1973)030<0118:CG>2.0.CO;2

Koenig, L.R.: The glaciating behavior of small cumulonimbus clouds. J. Atmos. Sci. **20**, 29–47 (1963). https://doi.org/10.1175/1520-0469(1963)020<0029:TGBOSC>2.0.CO;2

Köhler, H.: The nucleus in and the growth of hygroscopic droplets. Trans. Faraday Soc. **32**, 1152–1161 (1936). https://doi.org/10.1039/TF9363201152

Kokhanovsky, A.A.: Cloud Optics. Springer Nature (2010)

Kondratyev, K.Y., Ivlev, L.S., Krapivin, V.F., Varotsos, C.A.: Atmospheric Aerosol Properties—Formation, Processes and Impacts. Springer Nature (2006)

Korolev, A.: Limitations of the Wegener–Bergeron–Findeisen mechanism in the evolution of mixed-phase clouds. J. Atmos. Sci. **64**(9), 3372–3375 (2007). https://doi.org/10.1175/JAS4035.1

Korolev, A., Leisner, T.: Review of experimental studies of secondary ice production. Atmos. Chem. Phys. **20**, 11767–11797 (2020). https://doi.org/10.5194/acp-20-11767-2020

Korolev, A., Heckman, I., Wolde, M., Ackerman, A.S., Fridlind, A.M., Ladino, L.A., Lawson, R.P., Milbrandt, J., Williams, E.: A new look at the environmental conditions favorable to secondary ice production. Atmos. Chem. Phys. **20**, 1391–1429 (2020). https://doi.org/10.5194/acp-20-1391-2020

Korolev, A., DeMott, P.J., Heckman, I., Wolde, M., Williams, E., Smalley, D.J., Donovan, M.F.: Observation of secondary ice production in clouds at low temperatures. Atmos. Chem. Phys. **22**, 13103–13113 (2022). https://doi.org/10.5194/acp-22-13103-2022

Korolev, A., Qu, Z., Milbrandt, J., Heckman, I., Cholette, M., Wolde, M., Nguyen, C., McFarquhar, G.M., Lawson, P., Fridlind, A.M.: High ice water content in tropical mesoscale convective systems (a conceptual model). Atmos. Chem. Phys. **24**, 11849–11881 (2024). https://doi.org/10.5194/acp-24-11849-2024

Kulmala, M., Lintunen, A., Lappalainen, H., Virtanen, A., Yan, C., Ezhova, E., Nieminen, T., Riipinen, I., Makkonen, R., Tamminen, J., Sundström, A.-M., Arola, A., Hansel, A., Lehtinen, K., Vesala, T., Petäjä, T., Bäck, J., Kokkonen, T., Kerminen, V.-M.: Opinion: the strength of long-term comprehensive observations to meet multiple grand challenges in different environments and in the atmosphere. Atmos. Chem. Phys. **23**, 14949–14971 (2023). https://doi.org/10.5194/acp-23-14949-2023

Kumjian, M.R., Lombardo, K.: A hail growth trajectory model for exploring the environmental controls on hail size: model physics and idealized tests. J. Atmos. Sci. **77**(8), 2765–2791 (2020). https://doi.org/10.1175/JAS-D-20-0016.1

Laaksonen, A., Korhonen, P., Kulmala, M., Charlson, R.J.: Modification of the Köhler equation to include soluble trace gases and slightly soluble substances. J. Atmos. Sci. **55**(5), 853–862 (1998). https://doi.org/10.1175/1520-0469(1998)055<0853:MOTKHE>2.0.CO;2

Lamb, D., Verlinde, J.: Physics and Chemistry of Clouds. Cambridge University Press (2011)

Laws, J.O., Parsons, D.A.: The relation of raindrop-size to intensity. Trans. Am. Geophys. Union. **24**(2), 452–460 (1943). https://doi.org/10.1029/TR024i002p00452

Lazaridis, M.: Bacteria as Cloud Condensation Nuclei (CCN) in the atmosphere. Atmosphere. **10**(12), 786 (2019). https://doi.org/10.3390/atmos10120786

Leal, A.F.R., Rakov, V.A.: Processes in negative and positive CG lightning flashes detected from space by GLM. Electr. Power Syst. Res. **229**, 110183 (2024). https://doi.org/10.1016/j.epsr.2024.110183

Lenoble, J., Remer, L., Tanré, D.: Aerosol Remote Sensing. Springer (2013)

Levin, L.M., Litvinov, I.V. (eds.): Cloud Physics and Weather Modification. Amerind Publishing Co. Pvt. Ltd (1977)

Liao, L., Meneghini, R., Kreutzer Nowell, H., Liu, G.: Scattering computations of snow aggregates from simple geometrical particle models. J. Sel. Top. Appl. Earth Obs. Remote Sens. **6**(3), 1409–1417 (2013). https://doi.org/10.1109/JSTARS.2013.2255262

Lim, J.-S., Hoffmann, F.: Between broadening and narrowing: how mixing affects the width of the droplet size distribution. J. Geophys. Res. Atmos. **128**(8), e2022JD037900 (2023). https://doi.org/10.1029/2022JD037900

Lin, Y., Kumjian, M.R., Soderholm, J., Giammanco, I.: Modeling non-spherical hailstones. J. Atmos. Sci. (2024). https://doi.org/10.1175/JAS-D-23-0231.1

Liu, Y.G., Yau, M.-K., Shima, S.-I., Lu, C.S., Chen, S.S.: Parameterization and explicit modeling of cloud microphysics: approaches, challenges, and future directions. Adv. Atmos. Sci. **40**(5), 747–790 (2023). https://doi.org/10.1007/s00376-022-2077-3

Lohmann, U., Lüönd, F., Mahrt, F.: An Introduction to Clouds. From the Microscale to Climate. Cambridge University Press (2016)

Low, T.B., List, R.: Collision, coalescence and breakup of raindrops. Part I: experimentally established coalescence efficiencies and fragment size distributions in breakup. J. Atmos. Sci. **39**(7), 1591–1606 (1982a). https://doi.org/10.1175/1520-0469(1982)039<1591:CCABOR>2.0.CO;2

Low, T.B., List, R.: Collision, coalescence and breakup of raindrops. Part II: parameterization of fragment size distributions. J. Atmos. Sci. **39**(7), 1591–1606 (1982b). https://doi.org/10.1175/1520-0469(1982)039<1607:CCABOR>2.0.CO;2

Lyons, W.A., Nelson, T.E., Armstrong, R.A., Pasko, V.P., Stanley, M.A.: Upward electrical discharges from thunderstorm tops. Bull. Am. Meteorol. Soc. **84**, 445–454 (2003). https://doi.org/10.1175/BAMS-84-4-445

Magono, C., Lee, C.W.: Meteorological classification of natural snow crystals. J. Fac. Sci. Hokkaido Univ. Ser. VII. **2**, 321–335 (1966)

Maki, L.R., Galyan, E.L., Chang-Chien, M.M., Caldwell, D.R.: Ice nucleation induced by Pseudomonas syringae. Environ. Microbiol. Ecol. **28**(3), 456–459 (1974). https://doi.org/10.1128/am.28.3.456-459.1974

Marisaldi, M., Østgaard, N., Mezentsev, A., Lang, T., Grove, J.E., Shy, D., Heymsfield, G.M., Krehbiel, P., Thomas, R.J., Stanley, M., Sarria, D., Schultz, C., Blakeslee, R., Quick, M.G., Christian, H., Adams, I., Kroodsma, R., Lehtinen, N., Ullaland, K., Yang, S., Hasan Qureshi, B., Søndergaard, J., Husa, B., Walker, D., Bateman, M., Mach, D., Cummer, S., Pazos, M., Pu, Y., Bitzer, P., Fullekrug, M., Cohen, M., Montanya, J., Younes, C., van der Velde, O., Roncancio, J.A., Lopez, J.A., Urbani, M., Santos, A.: Highly dynamic gamma-ray emissions are common in tropical thunderclouds. Nature. **634**, 57–60 (2024). https://doi.org/10.1038/s41586-024-07936-6

Marshall, J.S., Palmer, W.M.K.: The distribution of raindrops with size. J. Meteorol. **5**, 165–166 (1948). https://doi.org/10.1175/1520-0469(1948)005<0165:TDORWS>2.0.CO;2

Mason, B.J.: The Physics of Clouds, 2nd edn. Clarendon Press, Oxford (1971)

Maxwell, J. C.: The Scientific Papers of James Clerk Maxwell, vol. 2, pp. 636–640 (Niven, W.D., Ed.) (1890)

McTaggart-Cowan, J.D., List, R.: Collision and breakup of water drops at terminal velocity. J. Atmos. Sci. **32**(7), 1401–1411 (1975). https://doi.org/10.1175/1520-0469(1975)032<1401:CABOWD>2.0.CO;2

Miles, N.L., Verlinde, J., Clothiaux, E.E.: Cloud droplet size distributions in low-level stratiform clouds. J. Atmos. Sci. **57**, 295–311 (2000). https://doi.org/10.1175/1520-0469(2000)057<0295:CDSDIL>2.0.CO;2

Miller, R.C., Anderson, R.J., Kassen, J.L., Hagen, D.E.: Homogeneous nucleation rate measurements for water over a wide range of temperature and nucleation rate. J. Chem. Phys. **78**, 3204–3211 (1983). https://doi.org/10.1063/1.445236

Mitra, S.K., Vohl, O., Ahr, M., Pruppacher, H.R.: A wind tunnel and theoretical study of the melting behavior of atmospheric ice particles. IV: experiment and theory for snow flakes. J. Atmos. Sci. **47**(5), 584–591 (1990). https://doi.org/10.1175/1520-0469(1990)047<0584:AWTATS>2.0.CO;2

Morawska, L., Thomas, S., Jamriska, M., Johnson, G.: The modality of particle size distributions of environmental aerosols. Atmos. Environ. **33**, 4401–4411 (1999). https://doi.org/10.1016/S1352-2310(99)00217-4

Morrison, H., van Lier-Walqui, M., Fridlind, A.M., Grabowski, W.W., Harrington, J.Y., Hoose, C., Korolev, A., Kumjian, M.R., Milbrandt, J.A., Pawlowska, H., Posselt, D.J., Prat, O.P., Reimel, K.J., Shima, S.-I., van Diedenhoven, B., Xue, L.: Confronting the challenge of modeling cloud and precipitation microphysics. J. Adv. Model. Earth Syst. **12**, e2019MS001689 (2020). https://doi.org/10.1029/2019MS001689

Mossop, S.C.: The origin and concentration of ice crystals in clouds. Bull. Am. Meteorol. Soc. **66**(3), 264–273 (1985). https://doi.org/10.1175/1520-0477(1985)066<0264:TOACOI>2.0.CO;2

Mossop, S.C., Brownscombe, J.L., Collins, G.J.: The production of secondary ice particles during riming. Q. J. R. Meteorol. Soc. **100**, 427–436 (1974). https://doi.org/10.1002/qj.49710042514

Mudiar, D., Hazra, A., Pawar, S.D., Karumuri, R.K., Konwar, M., Mukherjee, S., Srivastava, M.K., Goswami, B.N., Williams, E.: Role of electrical effects in intensifying rainfall rates in the tropics. Geophys. Res. Lett. **49**, e2021GL096276 (2022). https://doi.org/10.1029/2021GL096276

Mülmenstädt, J., Sourdeval, O., Delanoë, J., Quaas, J.: Frequency of occurrence of rain from liquid-, mixed-, and ice-phase clouds derived from A-Train satellite retrievals. Geophys. Res. Lett. **42**(15), 6502–6509 (2015). https://doi.org/10.1002/2015GL064604

Murata, K., Asakawa, H., Nagashima, K., Furukawa, Y., Sazaki, G.: Thermodynamic origin of surface melting on ice crystals. PNAS. **113**(44), E6741–E6748 (2016). https://doi.org/10.1073/pnas.1608888113

Nag, A., Rakov, V.A.: Positive lightning: an overview, new observations, and inferences. J. Geophys. Res. Atmos. **117**, D08109 (2012). https://doi.org/10.1029/2012JD017545

Nakaya, U.: Snow Crystals: Natural and Artificial. Harvard University Press (1954)

Nakaya, U., Terada Jr., T.: Simultaneous observations of the mass, falling velocity and form of individual snow crystals. J. Fac. Sci. Hokkaido Univ. Ser. II. **1**, 191–201 (1935) Available at https://eprints.lib.hokudai.ac.jp/dspace/bitstream/2115/34452/1/1_P191-200.pdf. Last accessed 15 Apr 2024

North, G.R., Pyle, J.A., Zhang, F. (eds.): Encyclopedia of Atmospheric Sciences, 2nd edn. Elsevier (2014)

Nowell, H., Liu, G., Honeyager, R.: Modeling the microwave single-scattering properties of aggregate snowflakes. J. Geophys. Res. Atmos. **118**, 7873–7885 (2013). https://doi.org/10.1002/jgrd.50620

O'Donnell, S.E., Akherati, A., He, Y., Hodshire, A.L., Shilling, J.E., Kuang, C., Fast, J.D., Mei, F., Schobesberger, S., Thornton, J.A., Smith, J.N., Jathar, S.H., Pierce, J.R.: Look up: probing the vertical profile of new particle formation and growth in the planetary boundary layer with models and observations. J. Geophys. Res. Atmos. **128**, e2022JD037525 (2023). https://doi.org/10.1029/2022JD037525

Ochs III, H.T.: Moment-conserving techniques for warm cloud microphysical computation. Part II. Model testing and results. J. Atmos. Sci. **33**(10), 1959–1973 (1978). https://doi.org/10.1175/1520-0469(1978)035<1959:MCTFWC>2.0.CO;2

Okazaki, M., Oishi, S., Awata, Y., Yanase, T., Takemi, T.: An analytical representation of raindrop size distribution in a mixed convective and stratiform precipitating system as revealed by field observations. Atmos. Sci. Lett. **24**, e1155 (2023). https://doi.org/10.1002/asl.1155

Oke, T.R., Mills, G., Christen, A., Vogt, J.A. (eds.): Urban Climates. Cambridge University Press (2017)

Oraltay, R.G., Hallett, J.: The melting layer: a laboratory investigation of ice particle melt and evaporation near 0°C. J. Appl. Meteorol. Climatol. **44**(2), 206–220 (2005). https://doi.org/10.1175/JAM2194.1

Ori, D., Maestri, T., Rizzi, R., Cimini, D., Montopoli, M., Marzano, F.S.: Scattering properties of modeled complex snowflakes and mixed-phase particles at microwave and millimeter frequencies. J. Geophys. Res. Atmos. **119**, 9931–9947 (2014). https://doi.org/10.1002/2014JD021616

Ori, D., von Terzi, L., Karrer, M., Kneifel, S.: snowScatt 1.0: Consistent model of microphysical and scattering properties of rimed and unrimed snowflakes based on the self-similar Rayleigh–Gans approximation. Geosci. Model Dev. **14**, 1511–1531 (2021). https://doi.org/10.5194/gmd-14-1511-2021

Østgaard, N., Mezentsev, A., Marisaldi, M., Grove, J.E., Quick, M., Christian, H., Cummer, S., Pazos, M., Pu, Y., Stanley, M., Sarria, D., Lang, T., Schultz, C., Blakeslee, R., Adams, I., Kroodsma, R., Heymsfield, G., Lehtinen, N., Ullaland, K., Yang, S., Hasan Qureshi, B., Søndergaard, J., Husa, B., Walker, D., Shy, D., Bateman, M., Bitzer, P., Fullekrug, M., Cohen, M., Montanya, J., Younes, C., van der Velde, O., Krehbiel, P., Roncancio, J.A., Lopez, J.A., Urbani, M., Santos, A., Mach, D.: Flickering gamma-ray flashes, the missing link between gamma glows and TGFs. Nature. **634**, 53–56 (2024). https://doi.org/10.1038/s41586-024-07893-0

Paluch, I.R., Knight, C.A.: Does mixing promote cloud droplet growth? J. Atmos. Sci. **43**(18), 1994–1998 (1986). https://doi.org/10.1175/1520-0469(1986)043<1994:DMPCDG>2.0.CO;2

Pandey, R., Usui, K., Livingstone, R.A., Fischer, S.A., Pfaendtner, J., Backus, E.H.G., Nagata, Y., Fröhlich-Nowoisky, J., Schmüser, L., Mauri, S., Scheel, J.F., Knopf, D.A., Pöschl, U., Bonn, M., Weidner, T.: Ice-nucleating bacteria control the order and dynamics of interfacial water. Sci. Adv. **2**, e1501630 (2016). https://doi.org/10.1126/sciadv.1501630

Pasko, V.P., Stanley, M.A., Mathews, J.D., Inan, U.S., Wood, T.G.: Electrical discharge from a thundercloud top to the lower ionosphere. Nature. **416**, 152–154 (2002). https://doi.org/10.1038/416152a

Paul, S.K.: Cloud drop spectra at different levels and with respect to cloud thickness and rain. Atmos. Res. **52**, 303–314 (2000). https://doi.org/10.1016/S0169-8095(99)00037-X

Peterson, M.: The thunderstorms with the greatest lightning densities on Earth. Earth Space Sci. **11**, e2023EA003304 (2024). https://doi.org/10.1029/2023EA003304

Petters, M.D., Kreidenweis, S.M.: A single parameter representation of hygroscopic growth and cloud condensation nucleus activity. Atmos. Chem. Phys. **7**, 1961–1971 (2007). https://doi.org/10.5194/acp-7-1961-2007

Petters, M.D., Kreidenweis, S.M.: A single parameter representation of hygroscopic growth and cloud condensation nucleus activity—part 2: including solubility. Atmos. Chem. Phys. **8**, 6273–6279 (2008). https://doi.org/10.5194/acp-8-6273-2008

Petty, G.W., Huang, W.: The modified gamma size distribution applied to inhomogeneous and non-spherical particles: key relationships and conversions. J. Atmos. Sci. **68**(7), 1460–1473 (2011). https://doi.org/10.1175/2011JAS3645.1

Pickering, K.E., Li, Y., Cummings, K.A., Barth, M.C., Allen, D.J., Bruning, E.C., Pollack, I.B.: Lightning $NO_x$ in the 29–30 May 2012 Deep Convective Clouds and Chemistry (DC3) severe storm and its downwind chemical consequences. J. Geophys. Res. Atmos. **129**, e2023JD039439 (2024). https://doi.org/10.1029/2023JD039439

Pincus, R., Winker, D., Bony, S., Stevens, B. (eds.): Shallow Clouds, Water Vapor, Circulation, and Climate Sensitivity. Springer Nature (2018)

Pöhlker, M.L., Pöhlker, C., Quaas, J., Mülmenstädt, J., Pozzer, A., Andreae, M.O., Artaxo, P., Block, K., Coe, H., Ervens, B., Gallimore, P., Gaston, C.J., Gunthe, S.S., Henning, S., Herrmann, H., Krüger, O.O., McFiggans, G., Poulain, L., Raj, S.S., Reyes-Villegas, E., Royer, H.M., Walter, D., Wang, Y., Pöschl, U.: Global organic and inorganic aerosol hygroscopicity and its effect on radiative forcing. Nat. Commun. **14**, 6139 (2023). https://doi.org/10.1038/s41467-023-41695-8

Prabha, T.V., Khain, A., Maheshkumar, R.S., Pandithurai, G., Kulkarni, J.R., Konwar, M., Goswami, B.N.: Microphysics of premonsoon and monsoon clouds as seen from in situ measurements during the Cloud Aerosol Interaction and Precipitation Enhancement Experiment (CAIPEEX). J. Atmos. Sci. **68**, 1882–1901 (2011). https://doi.org/10.1175/2011JAS3707.1

Prein, A.F., Holland, G.J.: Global estimates of damaging hail hazard. Weather Clim. Extrem. **22**, 10–23 (2018). https://doi.org/10.1016/j.wace.2018.10.004

Pringle, K.J., Tost, H., Pozzer, A., Pöschl, U., Lelieveld, J.: Global distribution of the effective aerosol hygroscopicity parameter for CCN activation. Atmos. Chem. Phys. **10**, 5241–5255 (2010). https://doi.org/10.5194/acp-10-5241-2010

Pruppacher, H.R., Klett, J.D.: Microphysics of Clouds and Precipitation, 2nd edn. Springer Nature (2010)

Qie, X., Yair, Y., Di, S., Huang, Z., Jiang, R.: Lightning response to temperature and aerosols. Environ. Res. Lett. **19**, 083003 (2024). https://doi.org/10.1088/1748-9326/ad63bf

Rakov, V.A., Uman, M.A.: Lightning: Physics and Effects. Cambridge University Press (2007)

Randall, D.: Atmosphere, Clouds, and Climate. Princeton University Press (2012)

Rasmussen, R.M., Heymsfield, A.J.: Melting and shedding of graupel and hail. Part I: model physics. J. Atmos. Sci. **44**, 2754–2763 (1987a). https://doi.org/10.1175/1520-0469(1987)044<2754:MASOGA>2.0.CO;2

Rasmussen, R.M., Heymsfield, A.J.: Melting and shedding of graupel and hail. Part II: sensitivity study. J. Atmos. Sci. **44**, 2764–2782 (1987b). https://doi.org/10.1175/1520-0469(1987)044<2764:MASOGA>2.0.CO;2

Rasmussen, R.M., Heymsfield, A.J.: Melting and shedding of graupel and hail. Part III: investigation of the role of shed drops as hail embryos in the 1 August CCOPE severe storm. J. Atmos. Sci. **44**, 2783–2803 (1987c). https://doi.org/10.1175/1520-0469(1987)044<2783:MASOGA>2.0.CO;2

Rasmussen, R.M., Pruppacher, H.R.: A wind tunnel and theoretical study of the melting behavior of atmospheric ice particles. I: a wind tunnel study of frozen drops of radius <500 μm. J. Atmos. Sci. **9**, 152–158 (1982). https://doi.org/10.1175/1520-0469(1982)039<0152:AWTATS>2.0.CO;2

Rasmussen, R.M., Levizzani, V., Pruppacher, H.R.: A numerical study of the heat transfer through a fluid layer with recirculating flow between concentric and eccentric spheres. Pure Appl. Geophys. **120**, 702–720 (1982). https://doi.org/10.1007/BF00876656

Rasmussen, R.M., Levizzani, V., Pruppacher, H.R.: A wind tunnel and theoretical study of the melting behavior of atmospheric ice particles. II: a theoretical study for frozen drops of radius >500 micrometers. J. Atmos. Sci. **41**(3), 374–380 (1984a). https://doi.org/10.1175/1520-0469(1984)041<0374:AWTATS>2.0.CO;2

Rasmussen, R.M., Levizzani, V., Pruppacher, H.R.: A wind tunnel and theoretical study of the melting behavior of atmospheric ice particles. III: Experiment and theory for spherical ice particles of radius <500 micrometers. J. Atmos. Sci. 381–388 (1984b). https://doi.org/10.1175/1520-0469(1984)041<0381:AWTATS>2.0.CO;2

Reinking, R.F.: Formation of graupel. J. Appl. Meteorol. Climatol. **14**(5), 745–754 (1975). https://doi.org/10.1175/1520-0450(1975)014<0745:FOG>2.0.CO;2

Rogers, R.R., Yau, M.K.: A Short Course in Cloud Physics. Butterworth-Heinemann (1979)

Ropelewski, C.F., Arkin, P.A.: Climate Analysis. Cambridge University Press (2019)

Rosenfeld, D., Woodley, W.L.: Deep convective clouds with sustained supercooled liquid water down to $-37.5\,^{\circ}$C. Nature. **405**, 440–442 (2000). https://doi.org/10.1038/35013030

Ruehl, C.R., Davies, J.F., Wilson, K.R.: An interfacial mechanism for cloud droplet formation on organic aerosols. Science. **351**(6280), 1447–1450 (2016). https://doi.org/10.1126/science.aad4889

Ryan, B.F.: Growth of drops by coalescence: the effect of different collection kernels and of additional growth by condensation. J. Atmos. Sci. **31**(7), 1942–1948 (1974). https://doi.org/10.1175/1520-0469(1974)031<1942:GODBCT>2.0.CO;2

Santos Gutiérrez, M., Chekroun, M.D., Koren, I.: Gibbs states and Brownian models for coexisting haze and cloud droplets. Sci. Adv. **10**, eadq7518 (2024). https://doi.org/10.1126/sciadv.adq7518

Schlamp, R.J., Grover, S.N., Pruppacher, H.R., Hamielec, A.E.: A numerical investigation of the effect of electric charges and vertical external electric fields on the collision efficiency of cloud drops. J. Atmos. Sci. **33**(9), 1747–1755 (1976). https://doi.org/10.1175/1520-0469(1976)033<1747:ANIOTE>2.0.CO;2

Seifert, A.: Uncertainty and complexity in cloud microphysics. In: Proceedings ECMWF Workshop on Model Uncertainty, 20–24 June, pp. 77–86. Available at https://www.ecmwf.int/sites/default/files/elibrary/2011/12166-uncertainty-and-complexity-cloud-microphysics.pdf (2011)

Seinfeld, J.H., Pandis, S.N.: Atmospheric Chemistry and Physics: from Air Pollution to Climate Change, 3rd edn. Wiley (2016)

Shulman, M.L., Jacobson, M.C., Charlson, R.J., Synovec, R.E., Young, T.E.: Dissolution behavior and surface tension effects of organic compounds in nucleating cloud droplets. Geophys. Res. Lett. **23**(3), 277–280 (1996). https://doi.org/10.1029/95GL03810

Siebesma, P., Bony, S., Jakob, C., Stevens, B. (eds.): Clouds and Climate. Cambridge University Press (2020)

Soderholm, J.S., Kumjian, M.R.: Automating the analysis of hailstone layers. Atmos. Meas. Tech. **16**(3), 695–706 (2023). https://doi.org/10.5194/amt-16-695-2023

Sosso, G.C., Chen, J., Cox, S.J., Fitzner, M., Pedevilla, P., Zen, A., Michaelides, A.: Crystal nucleation in liquids: open questions and future challenges in molecular dynamics simulations. Chem. Rev. **116**, 7078–7116 (2016). https://doi.org/10.1021/acs.chemrev.5b00744

Soula, S.: Electrical environment in a storm cloud. J. Aerospace Lab. **5**, 1–10 (2012) Available at https://aerospacelab.onera.fr/sites/default/files/2024-01/AL05-01.pdf

Stier, P., van den Heever, S.C., Christensen, M.W., Gryspeerdt, E., Dagan, G., Saleeby, S.M., Bollasina, M., Donner, L., Emanuel, K., Ekman, A.M.L., Feingold, G., Field, P., Forster, P., Haywood, J., Kahn, R., Koren, I., Kummerow, C., L'Ecuyer, T., Lohmann, U., Ming, Y., Myhre, G., Quaas, J., Rosenfeld, D., Samset, B., Seifert, A., Stephens, G., Tao, W.-K.: Multifaceted aerosol effects on precipitation. Nat. Geosci. **17**, 719–732 (2024). https://doi.org/10.1038/s41561-024-01482-6

Storelvmo, T., Tan, I.: The Wegener-Bergeron-Findeisen process—its discovery and vital importance for weather and climate. Meteorol. Z. **24**(4), 455–461 (2015). https://doi.org/10.1127/metz/2015/0626

Stull, R.: Practical meteorology: An algebra-based survey of atmospheric science. Version 1.02b. University of British Columbia. Available at https://www.eoas.ubc.ca/books/Practical_Meteorology/. Last accessed 10 Jan 2024 (2017)

Sullivan, S.C., Hoose, C. (eds.): Clouds and their Climatic Impact: Radiation, Circulation, and Precipitation. AGU-Wiley (2023)

Suzuki, K., Nakagawa, K., Kawano, T., Mori, S., Katsumata, M., Syamsudin, F., Yoneyama, K.: Videosonde-observed graupel in different rain systems during Pre-YMC Project. SOLA. **14**, 148–152 (2018). https://doi.org/10.2151/sola.2018-026

Telford, J.W., Keck, T.S., Chai, S.K.: Entrainment at cloud tops and the droplet spectra. J. Atmos. Sci. **41**(21), 3170–3179 (1984). https://doi.org/10.1175/1520-0469(1984)041<3170:EACTAT>2.0.CO;2

Theis, A., Diehl, K., Mitra, S.K., Borrmann, S., Szakáll, M.: Melting of atmospheric ice particles. In: Michaelides, S. (ed.) Precipitation Science—Measurement, Remote Sensing, Microphysics and Modeling, pp. 423–471. Elsevier (2022). https://doi.org/10.1016/B978-0-12-822973-6.00003-2

Thomson, S.W.: On the equilibrium of vapour at a curved surface of liquid. Phil. Mag., Ser. 4. **42**(282), 448–452 (1871). https://doi.org/10.1080/14786447108640606

Tomasi, C., Fuzzi, S., Kokhanovsky, A. (eds.): Atmospheric Aerosols: Life Cycles and Effects on Air Quality and Climate. Wiley-VCH Verlag GmbH & Co. (2016)

Twomey, S., Wojciechowski, T.A.: Observations of the geographical variation of cloud nuclei. J. Atmos. Sci. **26**(4), 648–651 (1969). https://doi.org/10.1175/1520-0469(1969)26<648:OOTGVO>2.0.CO;2

Um, J., McFarquhar, G.M., Hong, Y.P., Lee, S.-S., Jung, C.H., Lawson, R.P., Mo, Q.: Dimensions and aspect ratios of natural ice crystals. Atmos. Chem. Phys. **15**, 3933–3956 (2015). https://doi.org/10.5194/acp-15-3933-2015

Uman, M.A.: Lightning. Dover Publications (2012)

Vali, G., DeMott, P.J., Möhler, O., Whale, T.F.: Technical note: a proposal for ice nucleation terminology. Atmos. Chem. Phys. **15**, 10263–10270 (2015). https://doi.org/10.5194/acp-15-10263-2015

Villermaux, E., Bossa, B.: Single-drop fragmentation determines size distribution of raindrops. Nat. Phys. **5**, 697–702 (2009). https://doi.org/10.1038/nphys1340

Wallace, J.M., Hobbs, P.V.: Atmospheric Science: an Introductory Survey, 2nd edn. Academic Press (2006)

Wang, P.K.: Ice Microdynamics. Academic Press (2002)

Wang, P.K.: Physics and Dynamics of Clouds and Precipitation. Cambridge University Press (2013)

Wang, P.K.: Motions of Ice Hydrometeors in the Atmosphere: Numerical Studies and Implications. Springer Nature (2021)

Wegener, A.: Thermodynamik der Atmosphäre. Barth (1911)

Wex, H., Stratmann, F., Topping, D., McFiggans, G.: The Kelvin versus the Raoult term in the Köhler equation. J. Atmos. Sci. **65**(12), 4004–4016 (2008). https://doi.org/10.1175/2008JAS2720.1

Whitby, K.: The physical characteristics of sulfur aerosols. Atmos. Environ. **12**, 135–159 (1978). https://doi.org/10.1016/0004-6981(78)90196-8

Williams, E.R.: Sprites, elves, and glow discharge tubes. Phys. Today. **54**(11), 41–47 (2001). https://doi.org/10.1063/1.1428435

Wilson, C.T.R.: Atmospheric electricity. Nature. **68**, 102–104 (1903). https://doi.org/10.1038/068102d0

Wilson, C.T.R.: Investigations on lightning discharges and on the electric field of thunderstorms. Philos. Trans. R. Soc. Lond. A. **221**(582–593), 73–115 (1920). https://doi.org/10.1098/rsta.1921.0003

Wilson, C.T.R.: The electric field of a thundercloud and some of its effects. Proc. Phys. Soc. Lond. **37**, 32D (1924). https://doi.org/10.1088/1478-7814/37/1/314

Wilson, C.T.R.: Some thundercloud problems. J. Frank. Inst. **208**(1), 1–12 (1929). https://doi.org/10.1016/S0016-0032(29)90935-2

WMO: International cloud atlas. Manual on the observation of clouds and other meteors. WMO-No. 407. https://cloudatlas.wmo.int/en/home.html. Last accessed 12 Apr 2024 (2017)

Yair, Y., Yaniv, R.: The effects of fog on the atmospheric electrical field close to the surface. Atmosphere. **14**, 549 (2023). https://doi.org/10.3390/atmos14030549

Yang, J., Huang, S., Yang, T., Zhang, Q., Deng, Y., Liu, Y.: Impact of ice multiplication on the cloud electrification of a cold-season thunderstorm: a numerical case study. Atmos. Chem. Phys. **24**, 5989–6010 (2024). https://doi.org/10.5194/acp-24-5989-2024

Yangang, L., Laiguang, Y., Weinong, Y., Feng, L.: On the size distribution of cloud droplets. Atmos. Res. **35**, 201–216 (1995). https://doi.org/10.1016/0169-8095(94)00019-A

You, C.-H., Kim, H.-J., Suh, S.-H., Jung, W., Kang, M.-Y.: The characteristics of raindrop size distributions in different climatological regions in South Korea. Remote Sens. **14**(20), 5137 (2022). https://doi.org/10.3390/rs14205137

Zhang, Y., Subba, T., Matthews, B.H., Pettersen, C., Brooks, S.D., Steiner, A.L.: Effects of pollen on hydrometeors and precipitation in a convective system. J. Geophys. Res. Atmos. **129**, e2023JD039891 (2024). https://doi.org/10.1029/2023JD039891

Zipser, E.J., Cecil, D.J., Liu, C., Nesbitt, S.W., Yorty, D.P.: Where are the most intense thunderstorms on Earth? Bull. Am. Meteorol. Soc. **87**, 1057–1072 (2006). https://doi.org/10.1175/BAMS-87-8-1057

# Precipitation

# 4

If we search for the term "precipitation" in the Glossary of Meteorology (AMS 2024a) we find that it is defined as "all hydrometeors formed in the atmosphere, including liquid, solid, or a combination of the two (e.g., resulting from incomplete melting or freezing or from accretion), that are large enough to fall as a result of gravity". The Glossary continues telling us that "the more common term rainfall is also used in this total sense to include not only amounts of rain, but also the water equivalents of frozen precipitation. For obvious reasons, precipitation is the preferred general term". It may seem a most trivial definition, but it is not since *rainfall* is often used interchangeably for precipitation and this creates misunderstandings when trying to appreciate weather and climate processes. Very few books exist specifically devoted to precipitation (Sumner 1988; Strangeways 2007; Michaelides 2008; Dohring and Dixon 2012; Martín and Roberts 2013; Testik and Gebremichael 2013; Comino 2021; Morbidelli 2022). Precipitation is the subject of one or more chapters in more general books on weather and climate (Randall 2012; Hartmann 2016; Baumgardner et al. 2018; Ropelewski and Arkin 2019).

## 4.1 Global Annual Precipitation

Figure 4.1 shows the average global precipitation (liquid and solid) in each country. Note that it does not say anything about the fact that there are inhomogeneities within the single country and thus the map is not an indication of the places where it rains most on Earth with, of course, no measures over the oceans.

The amount of water that falls as precipitation each year is approximately 505,000 $km^3$: 398,000 $km^3$ over oceans and 107,000 $km^3$ over land. This means that the globally averaged annual precipitation is 990 mm, but over land it is only 715 mm.

The average annual precipitation rate is used in climate classification systems such as the Köppen classification proposed for the first time in 1918 and



V. Levizzani, C. Kidd, *Precipitation*, Geophysics and Environmental Physics,
https://doi.org/10.1007/978-3-031-97096-2_4

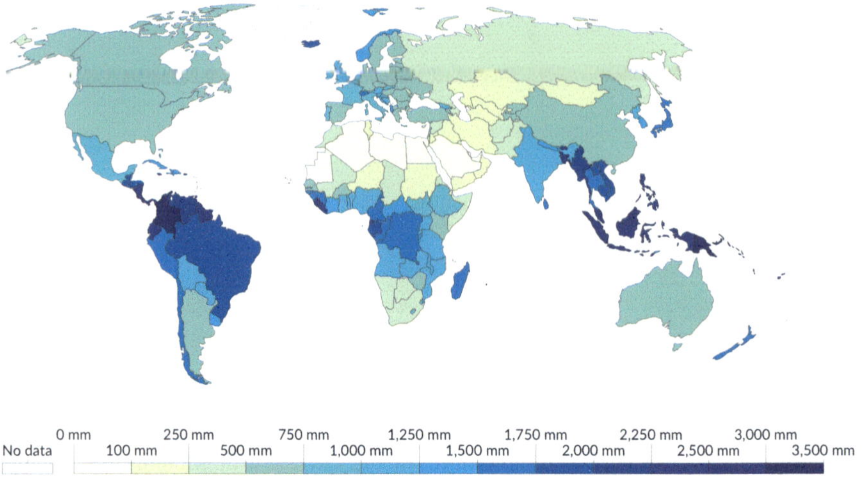


**Fig. 4.1** Average global precipitation (liquid and solid) in millimetres per year in each country. Data sources are multiple and compiled by the World Bank in 2024. (https://ourworldindata.org/grapher/average-precipitation-per-year, © The Authors under CC-BY-4.0 license)

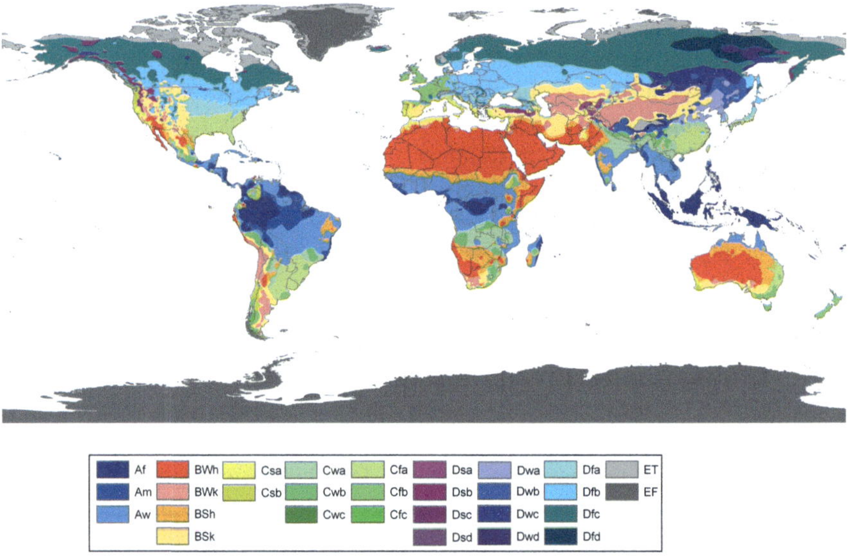


**Fig. 4.2** Köppen classification of climate regimes. Symbols are detailed in Table 4.1. (Adapted from Peel et al. 2007. © The Authors under CC-BY-4.0 license)

successively modified till 1936 (Köppen 1936), shown in Fig. 4.2. The Köppen classification is not based on atmospheric pressure, and latitudinal belts of winds, air masses, fronts or perturbations: locations are assigned to a specific climate subgroup based only on temperature and precipitation rate, provided the available observations are long enough to provide statistically significant means. The

**Table 4.1** Köppen climate classification symbols description table (see Fig. 4.2)

| 1st | 2nd | 3rd |
|---|---|---|
| A (Tropical) | f (Rainforest)<br>m (Monsoon)<br>w (Savanna, dry winter)<br>s (Savanna, dry summer) | |
| B (Dry) | W (Arid desert)<br>S (Semi-arid steppe) | h (Hot)<br>k (cold) |
| C (Temperate) | w (Dry winter)<br>f (No dry season)<br>s (Dry summer) | a (Hot summer)<br>b (Warm summer)<br>c (Cold summer) |
| D (Continental) | w (Dry winter)<br>f (No dry season)<br>s (Dry summer) | a (Hot summer)<br>b (Warm summer)<br>c (Cold summer)<br>d (Very cold winter) |
| E (Polar) | T (Tundra)<br>F (Ice cap) | |

classification consists of (Table 4.1) five main groups in capital letters and eight subgroups assigned to them. A further subdivision is added to differentiate variations of temperature. Eleven subtypes result from the combination of the first two groups. Note that this classification is highly variable with precipitation regimes that are subject to severe local modifications due to climatic change. Moreover, the classification is associated with a degree of subjectivity, and it has been criticised by several authors leading to modifications (Peel et al. 2007; Beck et al. 2018).

## 4.2 Precipitation Types

Precipitation is normally divided in three categories based on its phase when it falls and eventually when it reaches the ground: (1) liquid/falling as liquid water, (2) freezing/mixed/falling as liquid water and freezing on contact with the surface, and (3) frozen. The formation of precipitation hydrometeors is described in detail in Chap. 3. The different types of precipitation are univocally identified in the Meteorological Aerodrome Reports (METAR) used by meteorologists and aircraft pilots. Such reports are standardised by the International Civil Aviation Organization (ICAO) and mainly come from airport observations or other permanent weather observations at the surface; their frequency is hourly or even half-hourly, but they may be even more frequent in presence of special meteorological conditions. In the following we will provide the METAR code for each precipitation type.

### 4.2.1 Liquid Precipitation

Intensity, size, fall speed, and density of particles (number of drops per square metre of air) for various types of liquid precipitation, from fog to cloudburst, are reported in Table 4.2. Note that not all the categories reported in the first column correspond

**Table 4.2** Intensity, size, fall speed, and density of particles (number of drops per square meter of air) for various types of liquid precipitation. (Adapted from Lull 1959, public domain)

| | Precipitation intensity (mm h$^{-1}$) | Drop median diameter (mm) | Drop fall speed (m s$^{-1}$) | Drop number (s$^{-1}$ m$^{-2}$) |
|---|---|---|---|---|
| Fog | 0.13 | 0.01 | 0.003 | $7 \times 10^6$ |
| Mist | 0.05 | 0.1 | 0.21 | $27 \times 10^3$ |
| Drizzle | 0.1 | 0.96 | 4.10 | 151 |
| Light rain | 1.0 | 1.24 | 4.80 | 280 |
| Moderate rain | 1.5 | 1.60 | 5.70 | 495 |
| Heavy rain | 6.0 | 2.05 | 6.70 | 495 |
| Excessive rain | 16.0 | 2.40 | 7.30 | 818 |
| Cloudburst | 40.0 | 2.85 | 7.90 | 1220 |

to official definitions of the WMO, but they rather serve the purpose of identifying the phenomenon in general terms.

Apart from fog, which is normally not identified as associated with precipitation, the first form of liquid precipitation is drizzle (DZ). Drizzle drops are generally smaller than 0.5 mm in diameter and are much more numerous than rain drops. They are formed in stratiform clouds or even stratocumulus. The amount of precipitation associated with this precipitation type is around 1 mm per day, i.e. very light. Drizzle wets all surfaces and may considerably reduce the visibility (see Fig. 4.3). The detection of drizzle is not easy since the droplets may evaporate before reaching the ground making it difficult to be detected at the surface. However, meteorological radars can detect drizzle as the first sign of falling hydrometeors from stratiform clouds. Modern high-frequency radars are particularly suited for observing drizzle. Drizzle conditions tend to be frequent over large areas in the cooler regions over the subtropical oceans dominated by shallow marine stratocumulus and trade wind cumulus clouds, which form entirely within the marine boundary layer. Several studies are under way to explore the mechanisms of drizzle formation and their influence on the thickness of marine stratocumulus clouds, which is very relevant for the warming of the surface of the world's oceans. Drizzle formation seems to be greatly influenced by aerosols in coastal areas. Large amounts of drizzle are found especially in clean air masses where the number of CCNs is lower. Aerosols from human activities tends to suppress drizzle formation thus removing less moisture from forming clouds, which tend to become thicker.

The most common precipitation type at low and mid latitudes is rain (RA) although it is often subdivided in categories depending on its intensity. The Glossary of Meteorology of the AMS identifies rain in the form of falling liquid water drops with diameters greater than 0.5 mm. If they are widely scattered, the drops may be smaller (AMS 2024b).

For observing purposes, the intensity of rainfall at any given time and place may be classified as (1) "light" with intensity varying between a trace and 2.5 mm h$^{-1}$, the maximum rate of fall being no more than 0.25 mm in 6 min; (2) "moderate",

**Fig. 4.3** Drizzle conditions in New York's Kips Bay with the UN Headquarters and the Empire State Building in the background. (Photo by J. Henderson, public domain)

from 2.6 to 7.6 mm $h^{-1}$, the maximum rate of fall being no more than 0.76 mm in 6 min; (3) "heavy", over 7.6 mm $h^{-1}$ or more than 0.76 mm in 6 min; (4) "violent", when the rate is >50 mm $h^{-1}$. However, intensity is not the only category to classify precipitation since the amount of water is related also to the size distribution of falling drops: this is the reason why measuring precipitation intensity by means of rain gauges is not sufficient and disdrometers are very important to determine the effects at the ground, including erosivity. We normally feel rain when we are under it and are forced to open the umbrella, but rain has a distinctive pattern in the atmosphere. When we observe rainfall from far away, it appears to us as bands of water falling from the clouds above, which reduce the visibility and appears as moving streaks of water (see Fig. 4.4).

Rain production is mostly generated when moisture moves along weather fronts. If enough moisture and upward motion is present, precipitation falls from convective clouds such as cumulonimbus, which can organize into narrow rainbands. In mountainous areas, heavy precipitation is associated with upslope flow on the windward sides of the terrain, which forces moist air to condense and fall out as rainfall along the mountain ranges. The movement of the monsoon trough, or ITCZ, brings rainy seasons that are crucial for the water cycle in these areas. The organisation of precipitation is more thoroughly described in Sect. 4.3.

The last category that describes rainfall is the cloudburst (METAR code CB), where a great amount of precipitation concentrated in a short period of time. Such events can dump large amounts of water (Fig. 4.5). For example, 25 mm of water corresponds to 25,000 t $km^{-2}$. The term is not exactly approved in the

**Fig. 4.4** Fall streaks over Beinn Eich, Luss Hills, Scotland. (Photo by M. Klajban, public domain)

**Fig. 4.5** Cloudburst during an intense convective storm over Phoenix, AZ. (Photo by NWS Phoenix, public domain)

meteorological practice, but it is widely used and corresponds to rainfall >100 mm $h^{-1}$, although in certain situations 1 mm $min^{-1}$ and 50 mm $h^{-1}$ are also accepted by meteorological services. The consequences on the ground can be disastrous and cloudburst are often responsible for flash flood conditions. Cloudburst are associated normally with processes in clouds when large droplets grow rapidly by coagulation with smaller droplets. However, while cloudbursts often happen by orographic lifting, they can also occur when water vapour mixes with cold air producing sudden condensation.

### 4.2.2 Freezing/Mixed Precipitation

Rain or drizzle that freezes on contact with a surface within a subfreezing air mass is classified with the adjective "freezing" and becomes freezing rain or freezing drizzle. Typically, freezing conditions happen in winter when warm air lies under cold air and the depth of the warm layer determines the type of precipitation falling below the cloud (see Fig. 4.6).

Freezing drizzle (METAR code FZDZ) is drizzle that freezes on contact with the ground. It is like freezing rain (FZRA) since they both develop from liquid precipitation falling in subfreezing air and freeze on the surface. However, they are different since the development mechanisms are different. Freezing rain forms when frozen precipitation falls through a melting layer (column 2 of Fig. 4.6) and turns into liquid. Freezing drizzle instead forms via the supercooled warm-rain process, i.e. droplets coalesce until becoming heavy enough to fall out of the cloud in subfreezing conditions. The liquid water will not freeze if the ambient temperature is above −8 °C, via supercooling. When ice crystals are present, the liquid droplets will freeze on them and will be removed before they can grow large enough to fall out of the cloud. Freezing drizzle thus develops in shallow low-level stratus clouds when air saturation occurs below the layer where ice crystals can develop and grow.

The effects on the ground of freezing drizzle are the creation of a layer of ice or glaze. The amount of ice is generally not significant, but road conditions can nevertheless be slippery and cause hazardous conditions. Moreover, freezing drizzle is

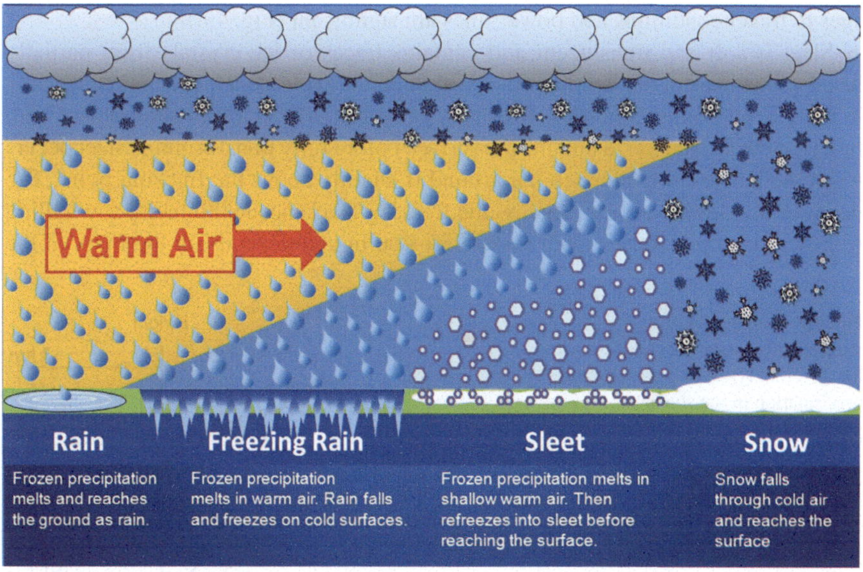


**Fig. 4.6** Winter precipitation types when a cuneus of warm air lies below the cloud. (Source: Northwest Indiana Weather Forecasting Office of the NWS, https://www.weather.gov/bmx/outreach_wwaw, public domain)

extremely dangerous for aircraft icing because the supercooled droplets freeze on the airframe, considerably modifying the flight performance.

Freezing rain is composed entirely of liquid drops that become supercooled when passing through a subfreezing layer a few hundred metres above the ground. It develops when falling snow encounters a layer of warm air aloft, typically around the 800 hPa level, and the snow melts and becomes rain. As the rain continues to fall, it passes through a layer of subfreezing air just above the surface and cools to a temperature below freezing. If this layer of subfreezing air is sufficiently deep, the raindrops may have time to freeze into ice pellets (sleet, METAR code RASN, third column of Fig. 4.6) before reaching the ground. However, if the subfreezing layer of air at the surface is shallow, the raindrops will not have time to freeze and will hit the ground as supercooled rain. It is not straightforward to identify the conditions for freezing rain from observations. One indirect way is to identify the bright band on the radar imagery, which indicates the presence of a warm layer above ground where snow can melt and possibly produce freezing rain. A close look at the radar vertical cross section may be necessary in these cases, although not conclusive. Additional information from weather stations and numerical models is required. Differences between freezing rain and sleet are evident in the upper air analysis: freezing rain events have higher 850 hPa temperatures and lower surface temperatures (Gay and Davis 1993; see also Fig. 4.7).

Upon hitting the ground, the supercooled rain freezes on the surfaces, such as ground, tree branches and electrical wires producing glaze, which is thicker than that caused by freezing drizzle. A storm producing freezing rain is often classified as an ice storm. The amount of precipitation is not particularly high, but the damages to travel conditions and power lines are significant (Fig. 4.8).

Sleet is soft and translucent and contains traces of ice crystals from partially fused snowflakes in the US definition, although may refer to a mix of rain and snow in the UK and other countries. It usually occurs briefly as a transition phase from rain to snow or vice-versa but hits the surface before fully transforming. Sleet occurs when the temperature close to the ground is slightly above the freezing point. The depth of low-level warm air (below the freezing level) needed to melt snow falling from above to rain varies from about 230–460 m and depends on the mass of the

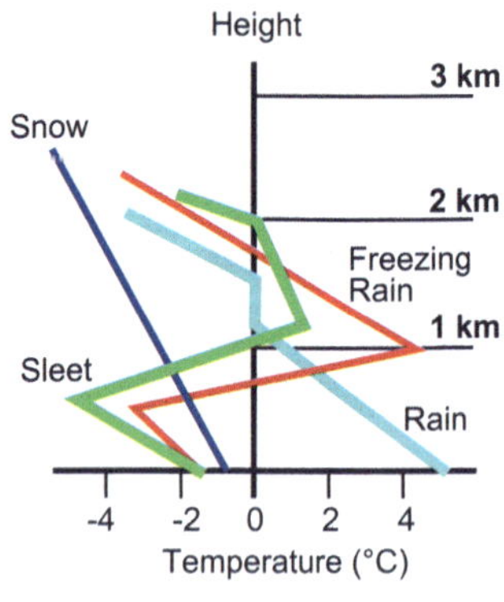


**Fig. 4.7** Schematic of temperature versus height for different types of precipitation in ice storms

**Fig. 4.8** Crews at work to restore power after an ice storm in Flint, MI. (Photo by Consumers Energy, public domain)

flakes and the lapse rate of the melting layer. Rain and snow typically mix when the melting layer depth falls between these values as rain starts forming when in that range (green line in Fig. 4.7).

Drizzle and snow may also mix under certain conditions and the slush created from them is called drizzle and snow mixed (METAR code DZSN).

### 4.2.3 Frozen Precipitation

The types of frozen precipitation are numerous and differ in their formation mechanisms and cloud environment.

As we have seen in Chap. 3, snow (METAR code SN) consists of individual crystals growing on ice nuclei and aggregates of crystals that form from the collision of individual crystals. The crystal formation is quite complex and involves a variety of shapes that we have examined previously.

Snow-producing clouds occur within larger weather systems, mostly low-pressure areas, which typically incorporate warm and cold fronts (Fig. 4.9). Two additional and locally productive sources of snow are lake-effect (also sea-effect) storms and orographic effects. Snowsqualls are generated as intense convective lines full of moisture in cold fronts when the temperature on the ground is near freezing. Such snowsqualls are moderately heavy snowfall events with blowing snow and strong, gusty surface winds. They are commonly referred to as a *whiteout*,

**Fig. 4.9** Multispectral image acquired by the GOES imager of a very large snowstorm with snow falling from Tennessee Valley into the Mid-Atlantic and New England on December 10, 2013, 1445 UTC. The fast-moving system created a swath of 5 to 20 cm of snow from southwest Virginia to the New England coast. (Source: NOAA, public domain)

localised in time and location: the snow accumulation may or may not be significant. Snowsqualls last less than 30 min locally, but the line can cover large distances. Frontal squalls may develop shortly ahead of the frontal line or immediately behind it. when there is significant vertical growth and mixing, the squall may embed cumulonimbus clouds with lightning and thunder, called thundersnow. In warm fronts snow is produced when warm and moist air passes over below-freezing air creating precipitation at their boundary.

When cold air (e.g., Arctic air) moves over large areas of open waters in winter, convective clouds develop causing heavy snow showers caused by the large amount of moisture evaporating from the water body. A typical example are the snowstorms over the Great Lakes of North America, but also those developing on the Adriatic coast of Italy when the *burian* wind coming from the Eastern European plains blows across the Adriatic Sea. Whiteout conditions characterise narrow corridors from shores to inland areas aligned along the prevailing wind direction. The temperature difference between the water and the 850 hPa surface should be at least 13 °C, the surface temperature be around freezing, the water unfrozen, the distance over the lake at least 100 km and the wind shear with height should be less than 30° from the surface to 850 hPa for lake-effect snow to develop. Many centimetres of snow may be deposited in such conditions along lines called snowbelts (see Fig. 4.10).

Snowstorm intensity may be categorised by visibility and depth of accumulation. Snowfall's intensity is determined by visibility, as follows (AMS 2024c):

- Light: visibility greater than 1 km
- Moderate: visibility restrictions between 0.5 and 1 km
- Heavy: visibility is less than 0.5 km

The *International Classification for Seasonal Snow on the Ground* defines "height of new snow" as the depth of freshly fallen snow, in centimetres as measured with a ruler, that accumulated on a snowboard during an observation period of 24 h, or other observation interval. However, melting, compacting, blowing and drifting contribute to the difficulty of measuring snowfall.

One more form of falling snow are snow grains (METAR code SG) that are very small (<1 mm) white, opaque grains of ice, flat or elongated. They do not bounce upon impact like snow pellets and fall in small amounts from stratus clouds or fog, never in a shower.

The next category of frozen precipitation is ice crystals (IC) also called diamond dust. They are ground-level clouds composed of very small ice crystals, which form in clear or nearly clear skies commonly observed in Antarctica or in the Arctic, but occurring anywhere with temperatures well below the freezing level. These ice crystals form with a temperature inversion at the surface and warmer air above the ground mixes with the colder air near the surface. Warmer air contains more water vapour and thus the mixing will transport water vapour

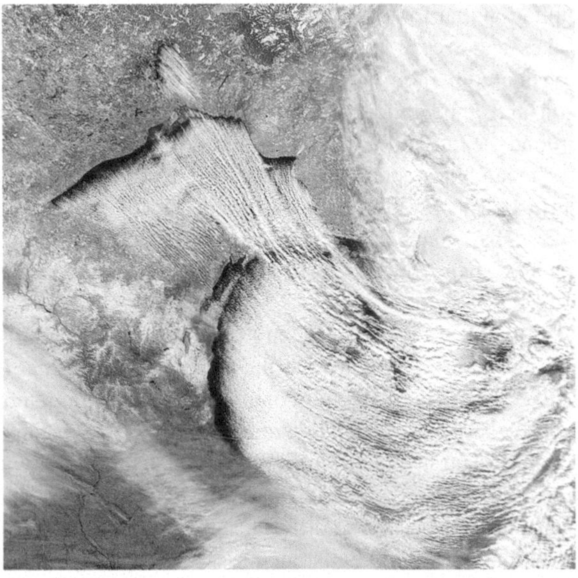

**Fig. 4.10** December 5, 2000. Multispectral image acquired by the SeaWiFS satellite sensor of lake-effect snow over the Great Lakes region. The wind is seen to be moving from the north-west. It picks up moisture from Lakes Nipigon, Superior, and Michigan and deposits this as snow further downwind. (Source: NASA)

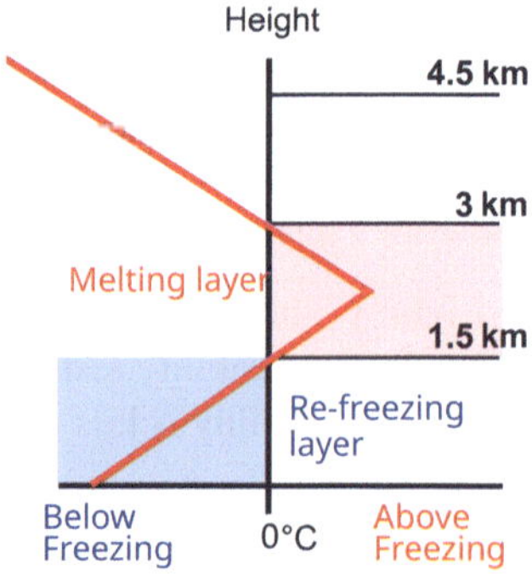


**Fig. 4.11** Schematic of vertical temperature profile for ice pellets formation

into the air near the surface increasing its relative humidity. If the latter increases enough, then ice crystals form. Very low temperatures between 0 and −39 °C are not always sufficient to form diamond dust since the water droplets may remain liquid as supercooled droplets.

Ice pellets (PL) or sleet are small, hard and translucent balls of ice made of frosty white opaque rime and develop from a mixture of rain and snow. They form when a layer of air above freezing is located between 1500 and 3000 m above the ground with subfreezing air above and below it. The snowflakes passing through the warm layer will melt and re-freeze when falling through the subfreezing air below forming the ice pellets (Fig. 4.11). In case the freezing layer close to the ground is not deep enough, the precipitation will not have sufficient time to re-freeze and will become freezing rain as we have seen above. A temperature profile with a warm layer above the ground is commonly observed ahead of a warm front and occasionally behind a cold front and with a stationary front.

Graupel (METAR code GS) are also called soft hail or snow pellets. This is precipitation that forms when supercooled droplets in the cloud are collected on snowflakes and freeze forming 2–5 mm balls of crisp, opaque rime. They are often considered by non-experts as hail or ice pellets, but they differ from them in formation and appearance. The droplets that collide with ice crystals have diameters around 10 μm and are supercooled down to −39 °C. The contact between a snow crystal and the supercooled droplets induces freezing of the droplets onto the crystal surface. The crystal becomes rimed, and the collection continues until the object takes a ball shape transforming in a graupel. During their fall graupels take often a conical shape due to a bell- shaped horizontal movement due to turbulence (Fig. 4.12).

Hail (METAR code GR) are balls or irregular lumps of ice falling mostly from summer storms, which take the name of hailstorms. Hailstones usually measure between 5 mm and 15 cm in diameter and can weigh more than 0.5 kg. Unlike graupels, hailstones are often layered (see Fig. 4.13) and can be irregular or even clumped together. The layers are normally about 1 mm thick with alternate transparent and translucent layers of ice, i.e. an onion-like structure. The transparent ice denser than the translucent one. The difference between the two types of layers is caused by the fact that the forming hailstone passes through layers of air where the concentration

**Fig. 4.12** A conical graupel particle collected on the palm of a bare hand prior to a spring snowstorm in Laramie, WY. (Public domain)

**Fig. 4.13** Large hailstone with concentric rings of translucent and opaque ice. (Public domain)

of humidity and supercooled water droplets varies. When the stone moves through a layer with a high concentration of water droplets, it captures them and acquires a translucent layer. Successively, it may pass through a layer where mostly water vapour is available, and the new layer will be opaquer. Former theories suggested an up and down movement of the hailstone through the cumulonimbus, but this has been recently shown to be not necessarily always the case. The vertical velocity of the updraft of the storm is sufficient to keep the hailstone aloft until its mass is not too large. This may take at least 30 min, and the stone will continue to grow when falling to the ground. Multiple trajectories of hailstones have been observed in multicell storms where the hailstone is ejected from the mother cell and captured in the updraft of a daughter cell. Hailstone growth stops gradually when the temperature falls below −30 °C since supercooled water becomes rarer.

The last type of precipitation is rare and not entirely understood. Megacryometeors (METAR code MC) are very large chunks of ice whose composition resembles closely that of hailstones. However, they do not originate from cumulonimbus clouds and their formation mechanism is still under investigation. More than 50 megacryometeors have been recorded since the year 2000. Their mass varies between 0.5 kilograms and several tens of kilograms. Because of their substantial mass they have been attributed to ice formed on aircraft, break off and then fall to the ground. However, they were observed also before the invention of aircraft. Recent studies indicate that the height fluctuations of the tropopause, associated with hydration of the lower stratosphere and stratospheric cooling, may be the real cause of such rare events. Sometimes they are confused with meteors because of the craters they leave on the ground, but they form in the atmosphere and not in outer space.

## 4.3 Precipitation Structure

### 4.3.1 Distribution of Precipitation

Precipitation amounts vary across the Earth's surface, fluctuating through different seasons and geographical locations. A global view of annual rainfall normal from the Global Precipitation Climatology Centre (GPCC; Schneider et al. 2017; Rustemeier et al. 2022) gauge analysis is reported in Fig. 4.14.

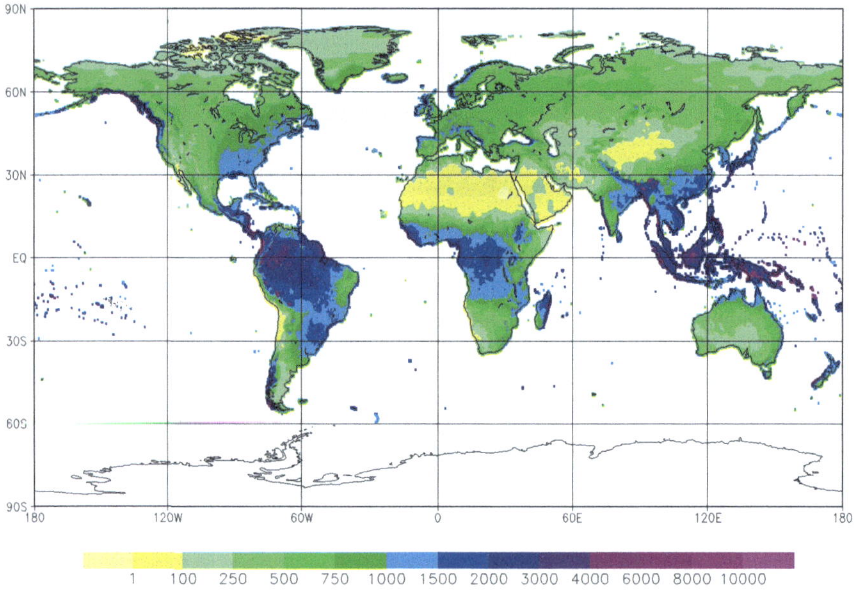


**Fig. 4.14** GPCC Precipitation normals in mm year$^{-1}$ (Version 2022) per 0.25-degree grid focused on the period 1951–2000. (Source: https://opendata.dwd.de/climate_environment/GPCC/html/gpcc_normals_v2022_doi_download.html, © The Authors under CC-BY-4.0 license)

Precipitation over the whole globe also include also oceanic precipitation and the only viable means of assessing precipitation amounts globally is through satellite sensors. The Global Precipitation Climatology Project (GPCP) maintains a database of precipitation measurements on a variety of space and time scales (Huffman et al. 2023). The annual average climatology for the period 1983–2020 of the GPCP at 0.5° resolution is shown in Fig. 4.15.

A few observations on the distribution of rainfall can be made. Generally, rainfall diminishes significantly when moving away from the equator toward the poles and this is due to the presence of the ITCZ. The sea, as a source of moisture, feeds more rainfall on windward coastal regions compared to the interior land areas (e.g., coastal deserts exist where offshore winds dominate); oceanic rainfall is higher with respect to landmasses, globally. The westerlies bring more rainfall on the western shores of the continents between 45 and 65°N and S with a decrease towards the east. Orography has a substantial influence on rainfall: the windward sides are rainier and the leeward experience less rain.

The world's major precipitation regimes are categorised based on annual precipitation levels:

- the equatorial belt, windward mountain slopes in the cold temperate zone, and coastal areas in monsoon regions receive over 200 cm $year^{-1}$;
- interior continental areas receive moderate rainfall, ranging from 100 to 200 cm $year^{-1}$;
- central tropical regions and the eastern and interior sections of temperate lands receive 50 to 100 cm $year^{-1}$;
- areas located in the rain shadow zones of interior continents (leeward areas of mountain chains) and high latitudes witness very low rainfall, less than 50 cm $year^{-1}$.

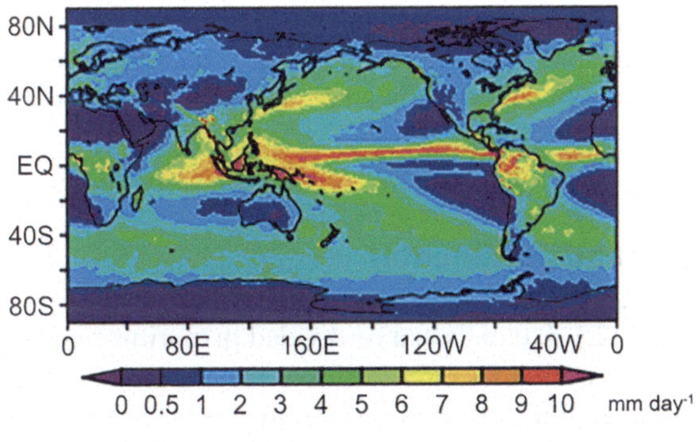


**Fig. 4.15** GPCP V3.2 climatology in mm $day^{-1}$ at 0.5° resolution for 1983–2020. (Source: NASA-GSFC, https://science.gsfc.nasa.gov/earth/mesoscale/researchareas/145)

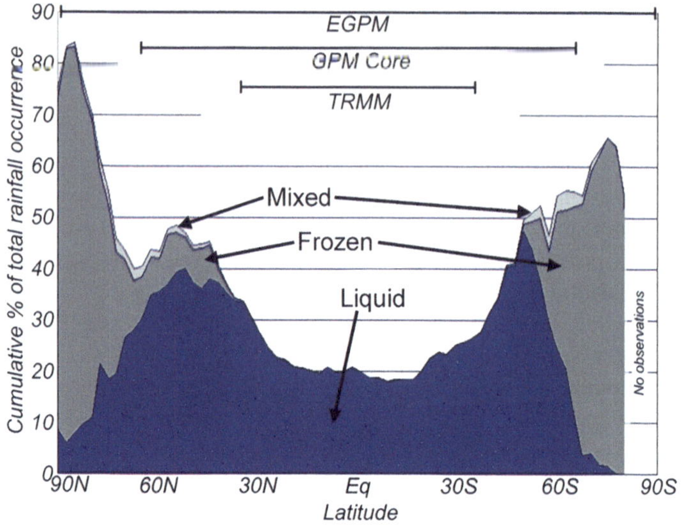


**Fig. 4.16** Latitudinal occurrence of the different light intensity (<1 mm h$^{-1}$) precipitation types as a percentage of the total precipitation occurrence derived from shipborne observations (1958–1991). Liquid precipitation is represented by blue shading, frozen precipitation by dark grey and mixed precipitation by light grey. (Source: Chris Kidd)

The seasonal distribution of rainfall is a crucial aspect and determines the annual cycle of the water cycle. Some regions, like the equatorial belt and the western parts of cool temperate regions, experience relatively even rainfall distribution throughout the year.

As we will see in the following chapter, precipitation is mostly measured in the liquid form and the available datasets have all this limitation. However, snowfall is the dominant form of annual light precipitation at high latitudes, especially in the northern hemisphere (e.g., Milani and Kidd 2023) as can be seen in Fig. 4.16. This creates biases when trying to quantify the effects on the global water cycle.

### 4.3.2 Maritime and Continental

As discussed in Sect. 2.1.3, clouds can be divided in maritime and continental. The two categories do not necessarily refer to clouds that develop over the ocean or over land. The two terms are linked to the origin of the air masses, especially that of the aerosol particles they contain. Satellite-based studies of Rosenfeld and Lensky (1998) have identified different microphysical processes at different heights in the convective clouds. From cloud base to top, the microphysical classification includes zones of diffusional droplet growth, coalescence droplet growth, rainout, mixed-phase precipitation, and glaciation. The authors have found that in maritime clouds, from base to top, there are zones of coalescence, rainout, a shallow mixed-phase region, and glaciation starting at −10 °C or even above this threshold. In contrast, continental clouds have a deep diffusional growth zone above their bases, followed by coalescence and mixed-phase zones, and glaciation at −15 to −20 °C. Highly continental clouds have a narrow or no coalescence zone, a deep mixed-phase zone, and glaciation occurring between −20 and −30 °C. Observations have shown that clouds undergo substantial transformations when moving from the sea inland, i.e. when they become more continental. Transformations occur also when the air mass moves across biomass burning or heavily polluted areas because the smaller aerosol particles suppress coalescence and thus precipitation.

In a context of climate change, however, the influence of moisture sources on precipitation over land will likely change so that the importance of oceanic water vapour will grow (e.g., Gimeno et al. 2020).

### 4.3.3 Convective and Stratiform

Another conventional way of dividing precipitation types is the convective-stratiform concept. Convective precipitation occurs when air rises vertically through the mechanism of convection. Stratiform precipitation occurs when large air masses rise as larger-scale winds and atmospheric dynamics force them to move over each other.

Convective precipitation is the result of cumulonimbus and cumulus congestus clouds and falls over limited areas and shows rapidly changing intensity. Convective clouds in general have a limited lifetimes and thus convective precipitation is of relatively short duration, in the form of showers. Most precipitation (by accumulation) in the Tropics occurs in the form of convective rainfall, even though this is not a completely general rule.

In frontal areas precipitation forms when warm air of tropical origin encounters cooler, subpolar air. The air mass rises over long distances and forms nimbostratus clouds that produce rainfall of low and medium intensity, or snowfall in wintry conditions.

The above is what it was thought before the advent of modern observation instruments, especially the radar on board satellites. Such radars showed large echoes composed of convective rain alongside stratiform precipitation, with the stratiform echoes covering great areas and accounting for a large portion of the tropical rainfall (Houze 1997). Young cumulonimbus clouds grow rapidly to produce stratiform rainfall, but as they age the convection is weaker and the hydrometeors drift downward growing by vapour diffusion. So, convective and stratiform precipitation categories are not always clearly separated.

Schumacher and Funk (2023) have produced an 8-year climatology of rain type observations from the Global Precipitation Measurement (GPM) radar. They identified two stratiform rain producing storm regimes differentiated by the relative importance of convection. The more convectively active regime occurs in the tropics and over warm season midlatitude land. The less convectively active regime dominates over midlatitude ocean and cold season midlatitude land. The ratio between cold-topped convective and stratiform rain area is highly distinct between the two regimes, with different precipitation amounts (and thus heating) for similar stratiform rain areas. A third tropics-only warm-topped convection regime exists, not associated with major stratiform rain production.

Identification of convective and stratiform precipitation from the ground is normally conducted using radar data. Several algorithms are available based on radar reflectivity values. An example is the Convective/Stratiform Precipitation Separation (CSPS) algorithm (Qi et al. 2013), which classifies single site radar reflectivity data. The purpose of the CSPS identification is to aid in adjusting reflectivity values in the melting layer, where mixed-phase hydrometeors can cause higher reflectivities and overestimate rain rates that assume liquid drop distributions. It is completed via three main steps:

- checking the hourly freezing level height;
- identifying convective cores via decision tree logic dependent on range from radar, 3D reflectivity structure, vertically integrated liquid, and environmental temperature data;
- determining the areal extent of convective precipitation by considering surrounding pixels.

More recent methods rely upon polarimetric radar data (Ryzhkov and Zrnić 2019) and vector machine algorithms (e.g., Wang et al. 2021) or X-band radars on small catchments (Guardiola-Albert et al. 2017). Several studies have concentrated on the subject (e.g., Deng et al. 2013; Wen et al. 2020; Romatschke and Dixon 2022; Wu et al. 2022; Kumar et al. 2024), but results are not conclusive. Additional information comes also from water isotopes studies (Aggarwal et al. 2016).

How important is it to partition convective and stratiform precipitation? A response come from the study of Yang and Smith (2008) that suggests that diurnal precipitation modes largely arise from distinct diurnal stratiform variations modulating convective variations. One more relevant application is in validating numerical weather prediction models (e.g., Langer and Reimer 2007).

### 4.3.4 Orographic Precipitation

Orographic precipitation is either rain or snow generally falls on the windward side of a mountain range (Fig. 4.17). The moist air is lifted on the slope of the mountains and reaches the condensation level where precipitation hydrometeors develop and grow. Rain (or snow, depending on the season and local temperature profile) falls on the mountains contributing to keep the slope green and vegetation grows. Rainfall is normally fed by warm and moist air coming from nearby oceans. On the other side of the mountain range the air sinks since it is colder than the environment and deprived of most of its moisture content. On the leeward side

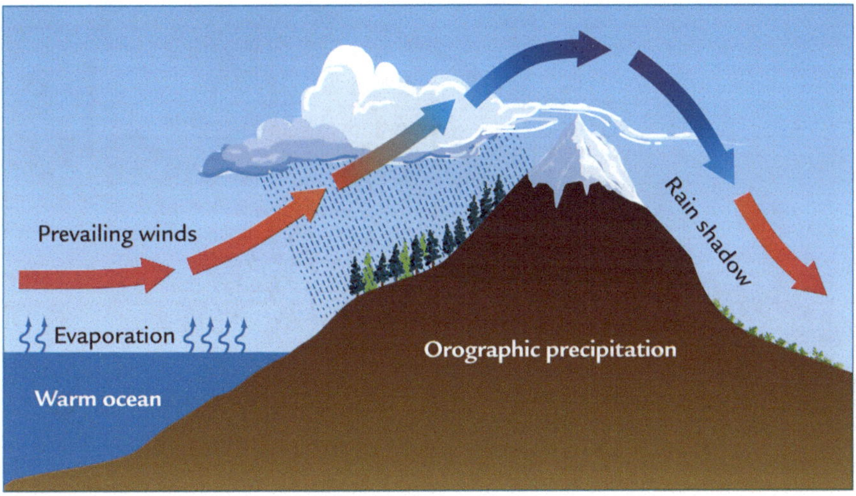


**Fig. 4.17** Orographic precipitation as the effect of rising moist air in the windward side of the mountains

a rain shadow effects takes place with dry soils and very little precipitation. Sometimes precipitation falls a short distance over the mountain top: this phenomenon is called spillover. Notable examples of this effect are the precipitation on the Sierra Nevada on the US Pacific coast and the Himalaya chain that makes rain fall on the plains of India and Bangladesh, while on northern side we find the Taklamakan desert.

A striking case of orographic effect is that of the interaction of an atmospheric river (see 4.4.4.3) with orography (Fig. 4.18). Atmospheric rivers are relatively long, narrow corridors in the atmosphere—like rivers in the sky—that transport most of the water vapour outside of the tropics. The term was first suggested by Zhu and Newell (1994) and the phenomenon has since then been investigated and observed all over the globe (Ralph et al. 2020). They are normally typically associated with a low-level jet stream ahead of the cold front of an extratropical cyclone and transport large moisture quantities, on average double of the flow of the Amazon River.

## 4.4 Weather and Climate

Precipitation together with temperature is the first meteorological variable that people look at in the weather forecasts. The reason is obvious in that water and ice heavily influence human activities every day. However, what weather forecasts not always tell is the full story on how precipitation may fall, depending on where we live and on the season of the year. This is even more true for climate simulations where the sub grid information content is normally very scarce. In the following an

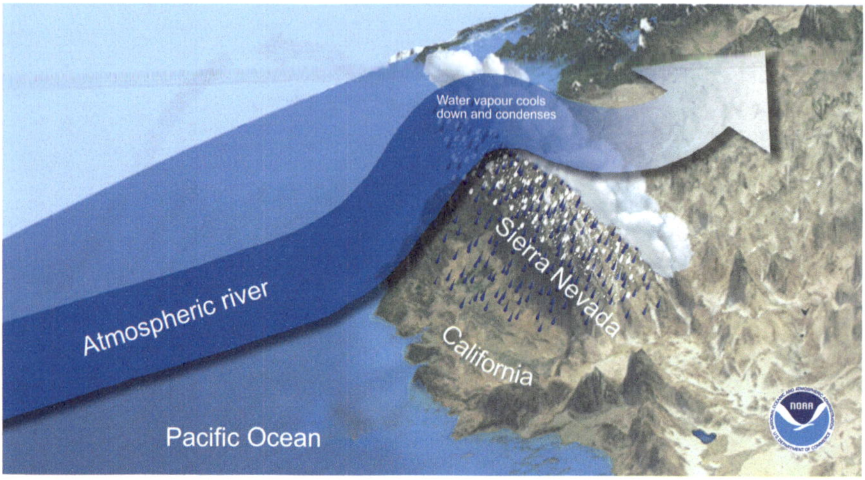


**Fig. 4.18** Interaction of an atmospheric river with the Sierra Nevada in California. (Adapted from NOAA, https://www.noaa.gov/stories/what-are-atmospheric-rivers, public domain)

overall description of precipitating systems at all latitudes is provided looking at the precipitation mechanisms.

### 4.4.1 Thunderstorms

The first precipitating system that comes to mind when describing convective precipitation is the thunderstorm. A thunderstorm is a storm characterised by the presence of lightning and thunder. When the storm is relatively weak it may be called thundershower. Thunderstorms are associated with cumulonimbus clouds and are normally accompanied by strong winds with heavy rains and sometimes snow, sleet, or hail.

The structure of thunderstorms is known since the middle of last century when, after the Second World War, in 1949 the Thunderstorm Project was conducted to investigate the science of stormy weather (Byers and Braham 1949). Perhaps the most important result of the project was the discovery that thunderstorms are structured in convective cells with a distinctive evolution in three main phases (Byers 1949; Braham 1996) (Fig. 4.19).

In the cumulus or developing stage moisture is lifted into the atmosphere and cools, condensing in liquid drops forming cumulus clouds. Latent heat is released, which warms the air, causing it to become less dense than the surrounding, drier air. In a typical thunderstorm, approximately $500 \times 10^6$ kg of water vapour are lifted into the Earth's atmosphere. In the mature stage the warmed air continues to rise until it reaches a level of warmer air where the rise stops (sometimes this boundary is the tropopause itself). The air spreads out forming an anvil shape, the cumulonimbus incus. The water droplets coalesce into larger and heavier droplets and freeze to

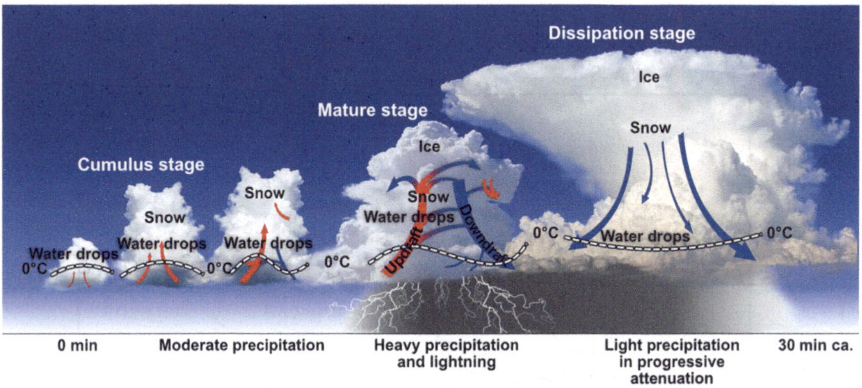


**Fig. 4.19** The three phases of the evolution of a convective storm cell: cumulus (or developing) stage, mature stage and dissipating stage

become ice particles. As these fall through the melting layer, they melt to become rain. If the updraft is strong enough, the droplets are held aloft long enough to participate to the formation of hail. With the updraft still present, rain drags down the surrounding air, creating downdrafts. The simultaneous presence of an updraft and a downdraft identifies the mature stage of the storm and produces cumulonimbus clouds. Finally, the dissipation stage is dominated by the downdraft. This stage occurs approximately 20–30 min into the life of the thunderstorm. The downdraft will push down out of the thunderstorm, hitting the ground and spreading out. The cool air carried to the ground by the downdraft cuts off the inflow of the thunderstorm, the updraft disappears, and the thunderstorm will dissipate.

Thunderstorms are not just single-cell storms. There are four main types of thunderstorms: single-cell, multi-cell, squall line and supercell. Which type forms depends on the instability and relative wind conditions at different layers of the atmosphere (wind shear) (Fig. 4.20). Organised thunderstorms and thunderstorm clusters/lines have longer life cycles as they form in environments of significant vertical wind shear, normally greater than 25 knots (13 m $s^{-1}$) in the lowest 6 km of the troposphere, which aids the development of stronger updrafts. The supercell is the strongest of the thunderstorms, mostly associated with large hail, strong winds, and tornado formation. Precipitable water values of greater than 31.8 mm favour the development of organised thunderstorm complexes. Those with heavy rainfall normally have precipitable water values greater than 36.9 mm. Upstream values of the Convective Available Potential Energy (CAPE) greater than 800 J $kg^{-1}$ are usually required for the development of organised convection. Note that CAPE is defined as follows:

$$CAPE = \int_{z_f}^{z_n} g\left(\frac{T_{v,parcel} - T_{v,env}}{T_{v,env}}\right) dz \qquad (4.1)$$

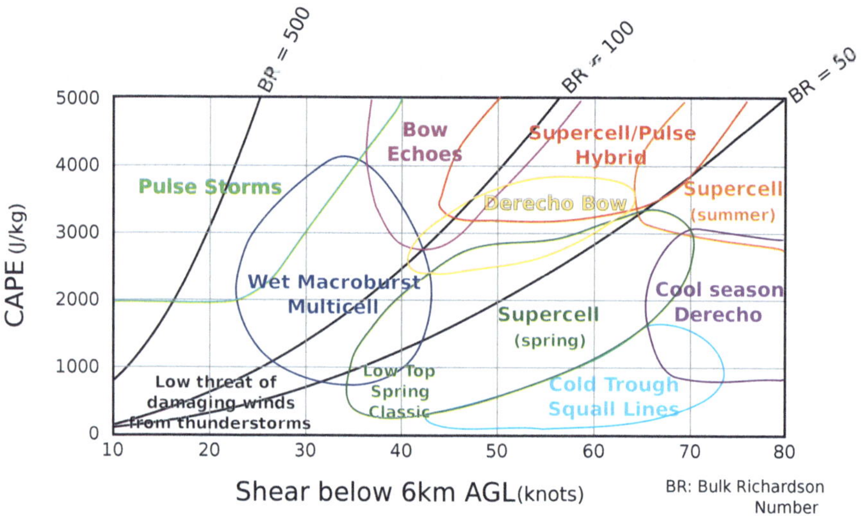


**Fig. 4.20** Conditions favourable for the development of thunderstorm types. See text for variable definitions. (Adapted from image in the public domain)

where $z_f$ is the height of the level of free convection, $z_n$ the height of the equilibrium level (neutral buoyancy), $T_{v,parcel}$ the virtual temperature of the rising parcel and $T_{v,env}$ the virtual temperature of the environment (both in Kelvin). It is a measure of the capacity of the atmosphere to support rising air eventually leading to cloud formation and storms. The Bulk Richardson (BR) number is defined as

$$BR = \frac{(g / T_v)\Delta\theta_v \Delta z}{(\Delta U)^2 + (\Delta V)^2} \tag{4.2}$$

where $g$ is the gravitational acceleration, $T_v$ the absolute virtual temperature, $\Delta\theta_v$ the virtual potential temperature difference across a layer, $\Delta z$ the vertical depth, and $\Delta U$ and $\Delta V$ the changes in horizontal wind components across that layer. High BR values indicate unstable and/or weakly-sheared environments; low values indicate weak instability and/or strong vertical shear. Values in the range of around 10 to 50 suggest environmental conditions favourable to supercell development.

Single cell storms occur occasionally and are summer storms with a single updraft. A single cell is named also as a component of a more complex cluster of storms. These storms are rarely severe and are a result of local atmospheric instability; hence the term "air mass thunderstorm"; they normally last 20–30 min.

Multicell clusters are the most common type of thunderstorm. Mature thunderstorms are located near the centre of the cluster, while dissipating thunderstorms are found on the downwind side. Multicell storms form as clusters of storms but may then evolve into one or more squall lines. While each cell of the cluster may only last 20 min, the cluster itself may persist for hours at a time. They often arise from convective updrafts in or near mountain ranges and linear weather boundaries, such as strong cold fronts or troughs of low pressure. Obviously, their duration implies that the precipitation production is much higher than that of single cells.

Squall lines are elongated lines of severe thunderstorms that can form along or ahead of a cold front. They are associated with heavy precipitation, hail, frequent lightning, strong straight-line winds, and possibly tornadoes and waterspouts.

The most severe form of thunderstorms are supercell storms (Fig. 4.21), i.e. large, usually severe, quasi-steady-state storms that form in an environment where wind speed or wind direction varies with height (wind shear), and they have separate downdrafts and updrafts (i.e., where its associated precipitation is not falling through the updraft) with a strong, rotating updraft (the "mesocyclone"). This separation of updraft and downdraft locations favours a relatively long duration of the storm since the downdraft cannot cut the updraft at its base as rapidly as it happens in other kinds of thunderstorms. These storms normally have such powerful updrafts that the top of the supercell storm cloud (or anvil) can break through the troposphere and reach into the lower levels of the stratosphere. Supercell storms can be 24 km wide (or more). Supercells can produce destructive tornadoes, extremely large hailstones (10 cm diameter), straight-line wind speeds >130 km $h^{-1}$, and flash floods.

The attribution of unambiguous percentages of occurrence of thunderstorm precipitation over the globe is still a matter of investigation. The already mentioned observational study of Schumacher and Funk (2023) over the Tropics argues that the more convectively active regimes occur in the Tropics and over warm-season midlatitude land, where warm- and cold-topped convection accounts for 55% (40%) of the rain (rain area). Li et al. (2024) have shown that convective precipitation prevails in Southern China, one of the wettest regions of the world: their results show that the precipitation frequency and accumulation in South China are dominated by stratiform precipitation and deep convective precipitation, respectively. Shallow

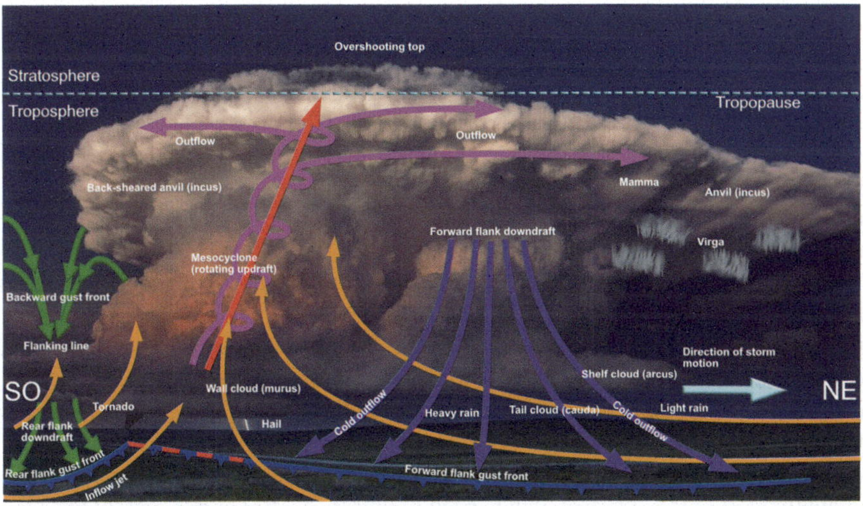


**Fig. 4.21** Schematic representation of a supercell thunderstorm

convective precipitation has comparable occurrence ratio to deep convective precipitation, but the mean intensity is much lower. Such studies indicate that the vertical evolution and the drop size distribution characteristics of different precipitation types need to be considered for improving the accuracy of near surface quantitative precipitation estimate.

When we consider torrential rain, what is the rainfall budget difference between convective and stratiform rain? Huang et al. (2019) have shown that in such cases the magnitude of precipitation budget processes in the convective region was one order larger of magnitude higher than that in the stratiform region. In convective/stratiform updraft regions, precipitation was mainly produced from the contribution of moisture-related processes, with a small negative contribution from cloud-related processes. In convective/stratiform downdraft regions, cloud-related processes played positive roles in precipitation, while moisture-related processes made a negative contribution.

Global change is expected to play a significant role in modifying the climatology of rainfall regimes. Global climate models suggest that the frequency of thunderstorms and their intensity is likely to increase in the future. This seems to be confirmed by observations (e.g., Chernokulsky et al. 2019). However, how the global frequency of thunderstorms and their accompanying hazards has changed over the past four decades remains unclear (Taszarek et al. 2021). The reason is that we still do not have access to observational datasets that are long enough. CAPE values, which are normally derived from rawinsonde data, are missing in the Tropics and reanalysis data are to be used with caution, given the incomplete representation of convection in numerical models. Satellite data (e.g., Hong et al. 1999) and radar techniques (e.g., Penide et al. 2013) are considerably advancing our capabilities of separating thunderstorm precipitation contributions from those from other sources.

### 4.4.2 Tropical Cyclones

A tropical cyclone is a rapidly rotating storm system with a marked low-pressure area, a closed low-level atmospheric circulation, high-speed winds, and a spiral arrangement of thunderstorms that produce heavy rain and squalls (Fig. 4.22). Depending on the regions and its strength, a tropical cyclone is called a hurricane (Atlantic Ocean), typhoon (Pacific Ocean), tropical storm, cyclonic storm, tropical depression, or simply cyclone (Indian Ocean).

Several factors are required for these storms to fully develop, including sea surface temperatures of around 27 °C and low vertical wind shear surrounding the system, atmospheric instability, high humidity in the lower to middle levels of the troposphere, sufficient Coriolis force to develop a low-pressure centre, and a pre-existing low-level focus or disturbance. There is a limit on tropical cyclone intensity which is strongly related to the water temperatures along its path and upper-level divergence. An average of 86 tropical cyclones of tropical storm intensity form annually worldwide: 47 reach wind speed higher than 119 km $h^{-1}$, and 20 become

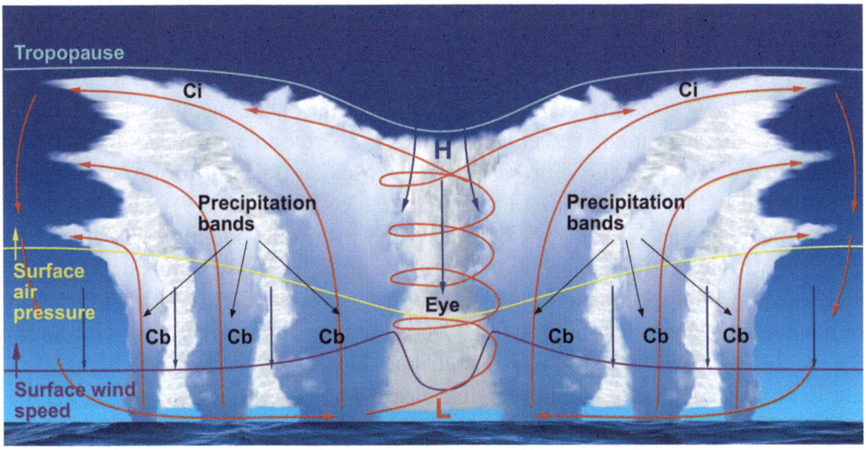


**Fig. 4.22** Schematic structure of a tropical cyclone. Note the location of the low (L) pressure centre at the surface and of the high (H) pressure centre aloft. Indicated is also the central eye and the precipitation storm bands. Red arrows depict the air flow

intense tropical cyclones, of at least Category 3 intensity on the Saffir–Simpson scale (https://www.nhc.noaa.gov/aboutsshws.php).

The precipitation structure of tropical cyclones has long since been observed, but it is only with the advent of the TRMM and GPM radars in space that observations became more widespread, and their quality increased significantly. In particular, tropical cyclone precipitation was always deemed to be convective with lines of spiralling thunderstorms around the system eyewall. This is mostly true, but the structure is much more complex as shown by Hence and Houze (2012) who used 10 years of TRMM radar observation and found that rainbands have a two-layered structure, with distinct modes separated by the melting layer. The ice layer is a combination of particles imported from the eyewall and ice left aloft as convective cells collapse. This layering is most pronounced in the inner region of the storm, and the layering is enhanced by storm strength. The inner-region rainbands are vertically confined by outflow from the eyewall but nevertheless are a combination of strong embedded convective cells and robust stratiform precipitation, both of which become more pronounced in stronger cyclones. Further insights into the structure of the stratiform precipitation of tropical cyclones is given by the airborne radar observations of Didlake and Houze (2013) who show that the stratiform rainband sector is a mesoscale feature consisting of nearly uniform precipitation and weak vertical velocities from collapsing convective cells.

More recently, Brauer et al. (2024) have observed the precipitation microphysics in 112 tropical cyclones between 2014 and 2020 using the GPM DPR on a global scale. They found that differences exist in microphysical processes in tropical cyclones by distance from the storm centre, ocean basin, and shear-relative quadrant.

Satellite radar data have also provided information on the contribution of tropical cyclones to global precipitation. Jiang and Zipser (2010) focused on the period

1998–2006 and six basins: Atlantic, east-central Pacific, northwest Pacific, north Indian Ocean, south Indian Ocean, and South Pacific. Their results show that: (1) tropical cyclones contribute, respectively, 8%–9%, 7%, 11%, 5%, 7%–8%, and 3%–4% of the seasonal rainfall to the entire domain of the mentioned basins; (2) regionally, the maximum percentage of tropical cyclone rainfall contribution is located in the east-central Pacific basin near the Mexico Baja California coast (about 55%), south Indian Ocean close to the Australian coast (about 55%), and northwest Pacific near Taiwan (about 35%–40%); (3) the maximum monthly percentage of tropical cyclone rainfall contribution is in September for the Atlantic basin, August and September for east-central Pacific, August for northwest Pacific, May for north Indian Ocean, March for south Indian Ocean, and January and February for South Pacific; (4) the percentage of rainfall contributed by tropical cyclones is higher during El Niño years than La Niña years for east-central Pacific and northwest Pacific basins. The study of Khouakhi et al. (2017) took daily precipitation data from 18,607 global rain gauges with at least 25 complete years of data between 1970 and 2014. The authors found that the highest tropical cyclone-induced rainfall totals occur in East Asia (>400 mm year$^{-1}$) and northeastern Australia (>200 mm year$^{-1}$), followed by the southeastern United States and along the coast of the Gulf of Mexico (100–150 mm year$^{-1}$). Tropical cyclones account for 35%–50% of the mean annual rainfall in northwestern Australia, southeastern China, the northern Philippines, and Baja California, Mexico. Seasonally, between 40% and 50% of tropical cyclone-induced rain is recorded along the western coast of Australia and in islands of the south Indian Ocean in the austral summer and in East Asia and Mexico in boreal summer and fall.

In general terms, tropical cyclone rainfall makes up a larger percentage of total rainfall during years when global rainfall is low, and more tropical cyclone rainfall is recorded in the Northern Hemisphere (10–17% of global rain from 15–25°N compared with 5–10% of global rain from 15–25°S). Concerning the factors that influence rainfall from tropical cyclones, here are some important ones: slow forward motion can produce more rain; the larger the storm, the greater the area typically receiving rain; the storm track determines the location of the rain; heaviest rainfall is generally located near the storm centre overnight, and on the outer bands during the day; topography enhances rainfall in upslope areas, but decreases rainfall past the mountain ridges; entrainment of dry air can redistribute and/or reduce the amount of precipitation while increased moisture can increase rainfall; interaction with other meteorological features (troughs, fronts, jets) and extratropical transition can greatly modify rainfall distribution.

However, all these analyses bring no conclusive results since the various datasets do not agree with each other due to the different observational or modelling means (e.g., Mazza and Chen 2023). This obviously reflects on the rainfall climatology and consequently on the derived flood potential.

Forecasting tropical cyclone rainfall is based on numerical models, but systems also exist that combine model output with satellite-based microwave rainfall data, such as the Ensemble Tropical Rainfall Potential (eTRaP, Ebert et al. 2011) method. Other forecast and modelling approaches are, among others: parametric models based on rainfall climatology and persistence models (Lonfat et al. 2007), machine learning methods (Lockwood et al. 2024), and deep learning and satellite image sequencing (Yang et al. 2024).

### 4.4.3 Mesoscale Convective Systems and Complexes

Mesoscale Convective Systems (MCS) are complexes of thunderstorms organised on a scale larger than the individual thunderstorms but smaller than extratropical cyclones and generally persists for several hours or more (Houze 2004; see Fig. 4.23). They are round or linear in shape, on the order of 100 km or more across in one direction, and include tropical cyclones, squall lines, and mesoscale convective complexes (MCCs). More specifically, MCCs are a unique kind of MCS defined by characteristics observed in infrared satellite imagery: their area of cold cloud tops exceeds 100,000 km$^2$ with temperature $\leq -32$ °C and an area of cloud top of 50,000 km$^2$ with temperature $\leq -52$ °C. Such size definitions must be met for 6 h or more. Their system dynamics is discussed in detail in Houze (2004) and Markowski and Richardson (2010).

Early studies on the precipitation characteristics of MCSs (Houze 1989) and MCCs (Kane et al. 1987) provide the necessary theoretical framework for identifying their structure, including the split between convective and stratiform precipitation regimes.

The GPM DPR has proved to be very instrumental for MCS detection at the global scale, particularly over remote regions without a dense observation network (Wang et al. 2019). A recent study based on geostationary satellite infrared brightness temperature and precipitation feature characteristics from GPM precipitation data sets have produced high resolution MCS databases (Feng et al. 2021). The

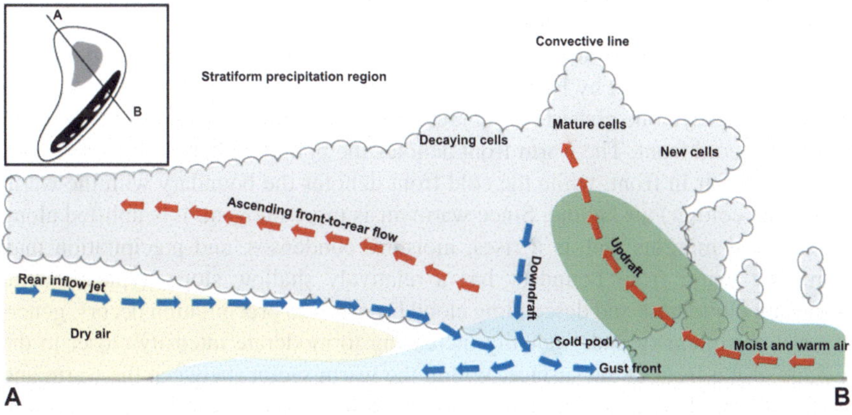


**Fig. 4.23** Scheme of the vertical cross section of a mesoscale convective system. The A-B section is indicated in the radar image on the top left. (Adapted from A. J. Punkka and R. A. Houze Jr., public domain)

authors show that MCSs account for over 50% of annual total rainfall across most of the tropical belt and in selected regions of the midlatitudes, with a strong seasonality over many regions of the globe. The longest-lived MCSs preferentially occur over the subtropical oceans. The land MCSs have higher cloud-tops associated with more intense convection, and oceanic MCSs have much higher rainfall production. Other studies focus on specific areas of the world such as Europe (Da Silva and Haerter 2023), the United States (Hu et al. 2021), Asia (Guo et al. 2023) and the Third Pole (Hindu-Kush-Karakoram-Himalayan) regions (Kukulies et al. 2023).

The availability of data from GPM has prompted for unprecedented studies on the characteristics and precipitation features of MCSs in tropical cyclones of different categories (Wang and Lyu 2024). The authors found that the development and evolution of active MCSs in tropical cyclones are influenced by the degree of organization of the system. Such studies increase our understanding of the structure and precipitation features of tropical cyclones.

### 4.4.4 Extra-Tropics

Precipitation in the extra-tropics is dominated by large scale weather systems that proliferate along the boundary between the tropical and polar airmasses. The differences in the characteristics of these air masses provide the necessary moisture and uplift needed for the formation of clouds and precipitation. Over the oceans and windward coastal regions much of the precipitation originates from cyclones, or low-pressure systems, which have major socio-economic impacts due to extreme (intensity and/or longevity) precipitation (Ulbrich et al. 2003). Over land areas convection and orographic enhancement provides much of the rainfall: the distribution of precipitation particularly over land is the result of a complex mix of wind direction, topography, diurnal variations and seasonality.

#### 4.4.4.1 Extra-Tropical Cyclones

The basic form of an extratropical cyclone (or low-pressure system) was described in the conceptual model by Bjerknes and Solberg (1922), as shown in Fig. 4.24. In this model the cyclone consists of an area of low pressure with two fronts, a warm front and a cold front. The warm front denotes the boundary between the warm air and the cold air in front, while the cold front denotes the boundary with the warm air and the cold(er) air behind. Since warm air is more buoyant, it is uplifted along these frontal margins and as it rises, moisture condenses, and precipitation may form. The warm front boundary has a relatively shallow slope, typically with increasing cloudiness and descending cloud base before precipitation occurs, generally of light intensity and gradually increasing to moderate intensity closer to the passage of the front at the surface. Within the warm sector (between the warm and the cold fronts) light precipitation may occur in the form of showers. The cold front is characterised by the cold air undercutting the warm air, forcing the warm air to rise with the associated clouds and precipitation. Typically, the precipitation along cold front will be more intense that that along the warm front. While the basic model

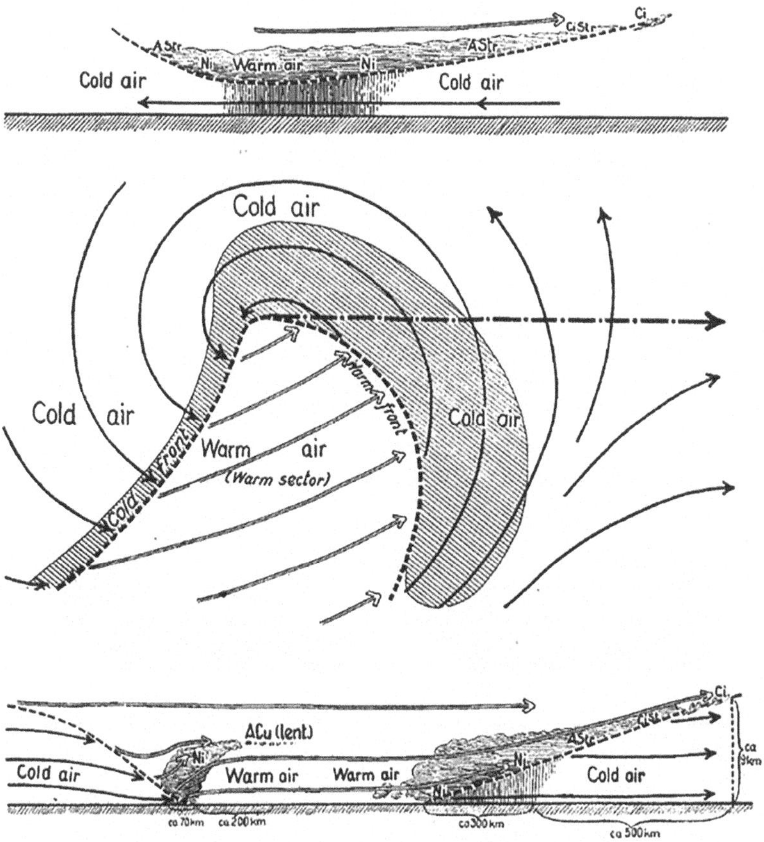


**Fig. 4.24** The conceptual model of a cyclone as described by Bjerknes and Solberg (1922). (Public domain)

is useful for understanding the basic form and evolution of such weather systems, such systems are more complex, differing in scale, intensity and life cycle duration.

Along the boundary between the moist tropical and the dry polar air mass boundaries cyclones have the potential to form. In this region the tropical winds generally flow eastwards and poleward, while the polar winds generally flow westwards and equatorward. The warmer tropical air will rise above the colder polar air and, in doing so, condensation will occur and possibility precipitation when the conditions are conducive. The uplifted air will cause a surface low pressure to develop resulting in circulation, anticlockwise in the northern hemisphere, and clockwise in the southern hemisphere. Marked frontal boundaries may develop (see Fig. 4.25a). Further development of the cyclone depends on several factors, including upper-level winds and surface interactions. The position of the jet stream is important in determining the strengthening of surface low pressure systems: if the surface low is close to the diverging part of the jet stream, the system will develop further, often

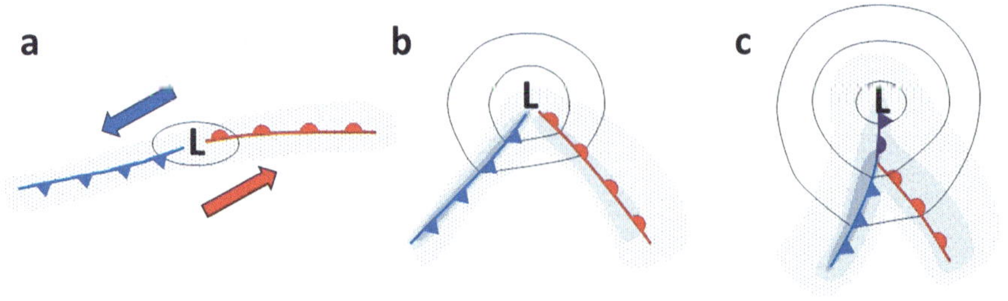


**Fig. 4.25** Illustration of the development of an extra-tropical cyclone, or low-pressure system (northern hemisphere) from (**a**) initiation, (**b**) development, and (**c**) maturity. The blue cones/line indicate the cold front while the red semi-circles/line indicate the warm front. The occluded front is denoted by the purple alternating cone/semi-circle and purple line. The shading represents the indicative areas of precipitation and relative intensities

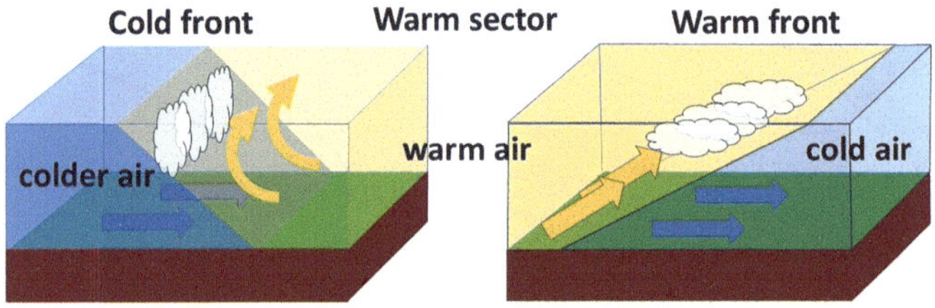


**Fig. 4.26** Illustration of the cold and warm fronts an extra-tropical cyclone. At the warm front the warm air rises above the cold air, while at the cold front the colder air undercuts the warm cut. In both situations the warm air is forced to rise, resulting in clouds and precipitation

rapidly, to produce well-defined warm and cold fronts (Fig. 4.25b). Since the cold front usually travels faster than the warm front, the warm air is lifted off the surface and an occlusion occurs (Fig. 4.25c). The low-pressure system finally loses its identity when all the warm air is lifted off the surface or the impetus for the uplift has decayed (e.g., the relative positions of the surface low and upper-level divergence).

Some regions are prone to the development of low-pressure systems. Over land these tend to be in the lee of mountain ranges, such as the Rockies in the U.S., where the terrain impacts both the low- and the high-level airflow. Low pressures may also develop along land/sea boundaries due to the differences in the friction in airflow close to the surface. The availability of moisture is clearly crucial for the formation of clouds and precipitation. Over land where little moisture is available, systems may be relatively dry, although many draw in moisture: for example, low-pressure systems forming in the lee of the Rockies often draw moisture from the Gulf of Mexico to fuel further development. Over the oceans, Atlantic, Pacific and Southern Oceans, moisture is readily available leading to the development of large cyclones. In certain situations, strings of cyclones may form where the fronts of one system may merge into those of a trailing system.

The fronts associated with these cyclones (Fig. 4.26) provide most of the precipitation at the surface, although some fronts may be "dry fronts" due to lack of available moisture, or evaporation occurs before the precipitation reaches the surface. At the warm front (Fig. 4.26 right), the warmer air overrides the cooler air, causing condensation and precipitation. The approach of a warm front is often marked by thin high-level cloud, then mid- to low-level cloud with precipitation gradually increasing towards the passage of the warm front at the surface. Behind the warm front, depending on the characteristics of the air mass of the warm sector, light precipitation or showers may occur. At the cold front the colder air undercuts the warm air forcing it to rise and creating instability. Here the onset of precipitation may be quite rapid with typically more intense precipitation than experienced at the passage of the warm front. In certain situations, a narrow cold-frontal rainband (NCFR; Hobbs and Biswas 1979) may develop associated with a low-level jet that lies ahead and parallel to the surface cold front, generating a (often discontinuous) band (Moore 1985) of intense but relatively shallow convection (Koch and Kocin 1991). At an occluded front, the precipitation is typically of light to moderate intensity.

Hawcroft et al. (2012) found that extra tropical cyclones contribute more than 60% of the precipitation north of 30°N. Using a feature tracking algorithm (based on vorticity, longevity and radius) combined with precipitation values derived from the GPCP and EMCWF ERA-Interim reanalysis, they showed that extra-tropical cyclones contribute as much as 90% of the accumulated precipitation in DJF and more than 85% in JJA, with 20% of the precipitation falling from the 10% most intense storms. The distribution of this contribution for the northern latitudes (Fig. 3, Hawcroft et al. 2012) shows the main storm tracks across the Pacific and Atlantic Oceans, together with the dominance of such systems during the northern hemisphere winter. Also of note is the high contribution of cyclonic precipitation events in the eastern Mediterranean Sea.

#### 4.4.4.2 Tropical-Like Cyclones

There is a continuum of cyclone characteristics between the tropical cyclones and extra tropical cyclones, with no clear separation between them (Hart and Evans 2001). While not sharing the full set of tropical storm characteristics, tropical-like cyclones are similar to tropical cyclones since they typically have warm cores together with significant wind and precipitation. Indeed, the characteristics of such storms may often fit one particular type of cyclone, but not necessarily have all the required characteristics. Gaertner et al. (2007) considered the tropical-like cyclones of the Mediterranean Sea as part of a group encompassing subtropical cyclones which also include those over the Atlantic Ocean, Pacific Ocean and off the east coast of Australia. Over the Mediterranean such storms have been termed "Medicanes", an example of which is shown in Fig. 4.27. Their formation is possible with sea surface temperatures <26 °C and often show a resemblance to tropical cyclones with calm winds at the centre, warm core, weak vertical shear, strong rotation around the region of minimum pressure, formation of an eyewall (albeit at relatively low speeds) with convective cells with spiralling rain bands and storm surges (Miglietta and Rotunno 2019). However, their tropical characteristics are short-lived, often lasting only a few days due to the limited expanse of the Mediterranean

**Fig. 4.27** NOAA-20 VIIRs false colour composite imagery of Cyclone Ianos on 17 September 2020. Aktion reported 62 and Kalamata 147 mm of rain on 18 September after landfall. (Source: NASA)

**Fig. 4.28** TERRA/MODIS image of a tropical like cyclone over the western Black Sea on 27 September 2005. (Source: NASA)

Sea and rarely exceeding a level 1 category of a hurricane: most of their lifetime they have extra-tropical characteristics. Cavicchia et al. (2014) identified 100 such storms between 1947 and 2011, mostly in the western or central Mediterranean, occurring mostly from September through January. Similar systems have also been noted over the Black Sea (see Fig. 4.28).

Medicanes are of a primary concern/issue in the Mediterranean regions due to high winds and torrential rain falling on areas that often have little annual precipitation and results in flash floods. Notable precipitation events can have substantial impact particularly over regions where such events contribute a significant proportion (or in excess) of the annual precipitation. Cyclone Daniel, 4–12 September 2024, generated significant precipitation both in Greece and Libya. Zagora, Greece, received 754 mm of rainfall on 5 September and a storm total of 1096 mm. Daniel then moved southwards and strengthened, making landfall in north Africa at Benghazi, Libya. Heavy rainfall, exacerbated by the collapse of two dams used for water management, led to significant flooding in the town/city of Derna, killing 5900 people.

#### 4.4.4.3 Atmospheric Rivers

Atmospheric rivers are elongated corridors of moisture that flow from the Tropics to the mid-latitudes. Atmospheric rivers have beneficial and detrimental societal impacts: they contribute precipitation (both rain and snow) to water resources that are crucial in many regions, although extreme precipitation can lead to floods together with wind hazards associated with such weather systems. They are also of great importance to the Earth System due to mass of water (as water vapour, and therefore energy) moved over short periods of time: Ralph et al. (2018) put the flow contributed by atmospheric rivers as twice that of the Amazon River. Rhoades et al. (2021) studied the potential change due to global warming and showed an increasing trend in atmospheric rivers per year, resulting in larger and longer-lived systems, with a doubling of resulting damage from a 2 °C rise in temperatures, and more than three times increase for a 3 °C rise in temperature. They also note that current ecosystems that have adapted to the current environmental conditions may not cope well under changed states.

The impact of atmospheric rivers depends on the water vapour amount, wind direction and speed, temperature and antecedent conditions. While the amount of water vapour is the primary control on precipitation amounts, the other variables influence the distribution and impacts of the precipitation. Albano et al. (2020) studied the vertically integrated water vapour transport with precipitation and found that these are strongly correlated along the western coastal areas of the US, although other factors such as uplift affect the precipitation efficiency, while wind direction relative to the terrain is also important. The seasonal occurrence of atmospheric rivers is also important since this affects the annual regeneration of the snowpacks, particularly in the west of the US, which then provides mass long term storage of fresh water. Over the west coast of the US, particularly California, atmospheric rivers are the primary drivers of hydrological extremes and flood loss. These systems deliver up to half of California's water annually, which is a crucial part of the hydroclimatology in this region. Bowers et al. (2024) noted that the most impactful atmospheric rivers events are associated with sequences of storms which multiply the effects of precipitation and subsequent flooding and damage. During the January/February 2017 an atmospheric river event approximately 1000 mm of rain fell over California, or 174 $km^3$ of water (Wen et al. 2018).

Over Europe, Eiras-Barca et al. (2021), using an "AR scale" based upon intensity and impacts (Ralph et al. 2018), showed that Ireland is most affected by lowest AR category (AR1), with stronger atmospheric rivers shifting southwards so that the strongest (AR5) affects mostly Brittany (France). In terms of accumulated precipitation, AR1s had greatest impact on the Norwegian coast, AR2 and AR3 along the Dalmatian coast, while AR4 have the greatest impact on western Scotland and the west of the Iberian Peninsula. Although the AR1 are the lowest category, these provided the greatest contribution to the precipitation across western Europe, amounting to 16% of the precipitation over Ireland and western France. Extreme precipitation and flooding events can be traced to atmospheric rivers. For example, over Cumbria (UK) from 18–20 November 2009 a broad flow of tropical maritime air was drawn over the UK in a "warm conveyor" leading to prolonged, steady, orographically enhanced precipitation. Seathwaite (Cumbria) received 378 mm in 34 h. The intensity and longevity of the precipitation lead to significant flooding and damage across the region.

Although atmospheric rivers reaching Polar Regions are less common, they have been shown to be particularly impactful. While atmospheric rivers reaching the Antarctica are relatively rare a few intense events provide a large percentage of the precipitation, with 10–20% of the snowfall in eastern Antarctica being from atmospheric rivers (Wille et al. 2021). In addition, atmospheric rivers affect the temperature and thus melting of sea ice which, coupled with strong winds and swell, help disintegration of the ice edge: 60% of the calving events (2000–2020) resulted from atmospheric rivers (Wille et al. 2022). In the Arctic, atmospheric rivers mostly have negative effects of the sea ice, reducing the ability of the ice to grow during the winter months (Zhang et al. 2023). Although the amount of precipitation modelled under climate change scenarios is variable, the number of identified atmospheric rivers is likely to increase.

#### 4.4.4.4 Convection

Convective systems also have the potential to contribute significant precipitation amounts across the mid-latitude regions. While most of these are not on the same scale as those in the more tropical regions, some can be severe and persistent, and specific regions are prone to severe weather that generated substantial precipitation, lightning, hail and damaging winds, resulting in flash floods and other damage.

Convective and mesoscale convective systems, as described above, affect many parts of the extra tropics, although are usually seasonal in nature since they typically result when hot humid air is advected polewards and mixes with cooler air. Over the US, particularly in spring, severe weather can develop when unstable warm moist air from the Gulf of Mexico is drawn northwards and interacts with cooler dry air over the US mainland. The instability results in significant convection, either as a line of storms associated with a cold front, or through the development of mesoscale systems. MCSs are also not uncommon over Europe: e.g., southern England on 11 June 2023, producing frequent lightning and heavy rainfall (Pike and Smart 2024) and over Norfolk (England) in 2020 when up to 240 mm fell over a limited area. In both these cases a high precipitation efficiency (i.e., much of the available moisture is converted to precipitation) arose from high specific and relative humidity and warm rain collision-coalescence processes.

Heat lows resulting from the rapid heating of the Earth's surface can result in thermal lows, particularly over regions with little or no vegetation (e.g., the Iberian Peninsula and the southwest of the US) may lead to convection, particularly when these systems are advected over the surrounding region. A good example is the

"Spanish Plume" where a plume of warm (hot) air moves northwards from northern Africa or Iberia over northwest Europe. Additional moisture entrained from the Mediterranean and eastern Atlantic allows the development of significant weather, such as thunderstorms, bringing high winds and intense precipitation (e.g., Martinstown in Dorset, England, with 279 mm of rainfall on 18 July 1955).

Other extreme precipitation events have convective elements, although the combination of convection with local conditions, such as topography and convergence are crucial. For example, in 2004 severe floods at Boscastle and along the north Cornwall/Devon coastline (England) were the result of small but intense thundery showers that propagated just inshore along a stalled sea freeze front. The air flow and topography resulted in a storm total of 200 mm at Otterham, with maximum intensities of 85.7 mm falling in 1 h and 32.6 mm falling in just 15 min (Burt 2005). Furthermore, such events are not limited to warm seasons: Ottery St Mary (Devon, England) had a significant hailfall on 30 October 2008. In this case an anomalously layer of cold air and upper-level trough led to the development of a prolonged convective event that generated over 200 mm of rainfall and a covering of hail 25 cm deep which resulted in ice floes causing significant disruption (Grahame et al. 2009). A similar event in Padstow (north Cornwall, England) on 16 October 2006 generated hail 20–22 mm in diameter (Webb et al. 2009).

Thundersnow is also not uncommon in the extra tropics if the potential differences between the surface and cold airmasses result in unstable temperature profiles that are conducive for generating convective storms, and when the near surface air is sufficiently cold, snow will result. While not as productive as warm season storms (generating 100–1000 cloud-ground strikes per hour), thundersnow storms can generate up to 50 strikes per hour.

### 4.4.5 Polar

#### 4.4.5.1 Polar Precipitation

Precipitation in the Polar Regions is typically dominated by the sub-zero (freezing) temperatures near the surface often with strong inversions. The cold temperatures limit the ability of the air to hold water vapour and therefore reduce the development of hydrometeor-sized particles. In addition, the strong temperature inversion limits the development of any in-situ weather system. Consequently, large regions across the Arctic and Antarctic regions experience limited precipitation: much of the Antarctic is classified as a desert region due to the lack of precipitation, although precipitation occurs around the coastal margins. Over the Arctic there tends to be more precipitation due to the proximity to and seasonal fluctuations of the ocean/sea ice margins. Weather systems over the open water may transport moisture over the nearby land and frozen surfaces (including sea ice) to generate precipitation as rain, sleet or snow. Importantly, with warming temperatures and diminishing sea ice extent and coverage, more open water allows the availability of more moisture for precipitation to form.

Under certain conditions precipitation can form directly from water vapour in the air. Ice crystals, or diamond dust, form at temperatures below 0 °C when supercooled droplets freeze, often in an environment that is devoid of any condensation nuclei. Although commonly observed in the Arctic and Antarctic, these may form in other regions, particularly where pollution allows the formation of the ice crystals at temperatures ≤ −10 °C. While the contributions of diamond dust to annual precipitation is generally small, in some regions it can be a major contributor to the annual total.

In certain regions (not necessarily confined to Polar Regions) significant snowfall may result from "lake effect snow" which occurs when cold dry air passes over open water, gaining moisture and often leads to deep convection capable of producing lightning and heavy precipitation in the form of snow. The name, lake effect, derives from the Great Lakes between the US and Canada where in early winter cold air picks up moisture from the unfrozen lake surfaces, creating bands of heavy snowfall. However, similar situations can arise around the margins of any large water body, such as the North Sea in Europe, and the Sea of Japan. Figure 4.29 shows a situation (28 February 2018) where strong easterly winds extended from the Urals in Russia to across the UK, drawing in cold winds that picked up moisture of the Baltic and North Sea, resulting in significant snowfall across Denmark, the UK and Ireland.

#### 4.4.5.2 Polar Lows

A more vigorous weather system found in Polar Regions is the Polar Low. These are mesoscale low-pressure systems found at the higher latitudes which can generate significant snowfall events. Historical records of these lows are difficult using conventional observations due to the remoteness of their locations, although satellite imagery has more recently greatly helped detect and monitor these systems. Characteristic signatures of Polar lows include cloud bands wrapped around a central low, with some producing an eye-like feature surrounded by convection, but

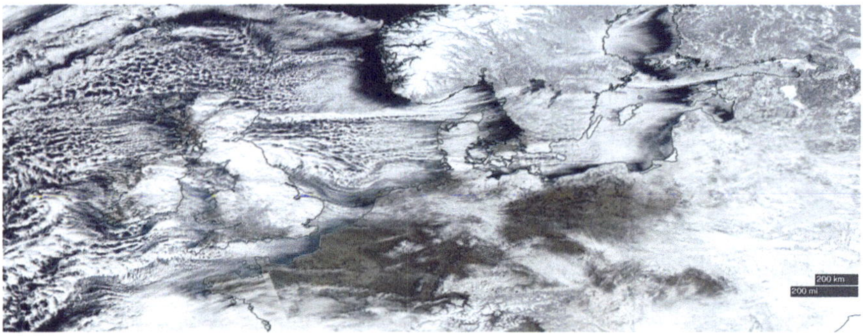

**Fig. 4.29** Example of the formation of Lake Effect snow across Europe on 28 February 2018, showing the long fetch of cold air from the Urals in Russia westwards across the Baltic, southern Sweden, Denmark and the North Sea. The increase in moisture generating clouds over the Baltic and North Sea is evident, as is the organisation of the clouds. (Source: NASA)

also comma-shaped systems. If mid- to upper-level temperatures are very cold, over open water deep convection is also possible. Classification of such systems varies, with some described as cold-low-type polar lows when cold air is advected over warmer seas (Businger and Reed 1989). They are typically generated in shallow baroclinic waves that form in strongly baroclinic northerly (in northern hemisphere) flows, typical of the eastern Atlantic in winter. Characteristics include limited vertical and horizontal (300–400 km) extent with no discernible frontal features, a 12–18-h lifetime, a path less than 750 km long and maximum wind speeds typically 19–21 m $s^{-1}$, but up to 29 m $s^{-1}$.

Rojo et al. (2015) investigated the distribution of Polar Lows over the Nordic Seas (north and west of Scandinavia) over a 14-winter period using AVHRR imagery to identify and track the storms. They noted a strong inter-annual and inter-seasonal variation in the formation of Polar Lows, but that warmer seas enhanced the potential for their formation. The distribution of Polar Lows in the North Atlantic between 2015 and 2017 was mapped by Golubkin et al. (2021), noting that virtually all occurred between November and April, over the Labrador Sea (between Greenland and North America) and in the northern part of the Norwegian Sea and Barents Sea. Similar storms are also found over the Sea of Japan (Yanase et al. 2016) and the Sea of Okhotsk. Farjami and Kazemi (2024) noted that Polar Lows also occur around Antarctica but are weaker due to the smaller land-sea differences in temperature and stronger meridional flow.

The precipitation from Polar Lows usually falls as snow since the air mass is cold, despite the sea surface temperatures and the near-surface air temperatures being above freezing. As Polar Lows move south (northern hemisphere) the surface precipitation is more likely to be rain or sleet, particular at sea level. While many of these storms remain over the ocean, the ones that do make landfall often bring significant precipitation, usually in the form of snow, which can cause significant disruption. (e.g., Harrold and Browning 1969), or even early snowfall (28–29 October 2008) as occurred over the UK resulting from a Polar Low over the North Sea and cold northerly airstream that generated heavy showers with thunder, hail and snow (Pritchard 2009).

## 4.5 What We Know and What We Need to Understand

Efforts are needed to understand the stochastic character of precipitation associated with cloud organisation (e.g., Shamekh et al. 2023). Moreover, there is growing interest in understanding the spontaneous aggregation of convective storms, called self-aggregation (Muller et al. 2022). This process creates moist cloudy regions with intense deep convective storms surrounded by extremely dry subsiding air devoid of deep clouds. Model simulations suggest that self-aggregation may have a substantial effect convective extremes (Moseley et al. 2016) and on extreme precipitation intensities (Da Silva et al. 2021). Angulo-Umana and Kim (2023) have recently found that in the Tropics strongly clustered oceanic convection precipitates more intensely than weakly clustered convection; this calls for a more correct representation of mesoscale organised convective systems in weather and climate models for improving predictions of extreme precipitation events.

One more very important point for a thorough physical understanding of the organisation of precipitation systems concerns the statistics of tropical rainfall. Studies have shown that the meteorological states producing the organisation of precipitation in the Tropics are associated with the critical point of a continuous phase transition (Arakawa 2006; Peters and Neelin 2006). A tendency for the atmospheric state to organize around this critical region leads to self-organised criticality (SOC), which is proposed as an explanation for scale-free behaviour in a variety of physical systems and will possibly lead to new avenues in climate modelling. This is an important finding since observations seem to confirm that most of the regional increase in tropical precipitation is associated with changes in the frequency of organised deep convection (Tan et al. 2015).

The warming climate is expected to contribute to the expansion of the global footprint of tropical cyclone precipitation, which calls for more studies (Qin et al. 2024). Moreover, insights into the structure of post tropical cyclones are very important for flood forecasting since these cyclones tend to contribute significant amounts of rainfall. More research is needed since only few studies are available (e.g., Ali et al. 2025).

One more important problem is the understanding of the environmental controls on tropical mesoscale convective systems to develop metrics that drive metrics model diagnosis and guiding physical intuition regarding expected changes to precipitation extremes with anthropogenic warming (Schiro et al. 2020). High resolution modelling efforts are also required such as the one of Tao and Chern (2017) to enhance the degree to which MCSs are resolved in multiscale modelling frameworks, which can impact their global surface rainfall amount. These studies will help in the ongoing investigations on the hypothesis that the proportion of MCS precipitation to the total precipitation will increase (e.g., Haberlie et al. 2023).

Atmospheric rivers are now investigated by several groups. Their structure and associated water vapour transport mechanism are reasonably known (Ralph et al. 2020). However, what we seem to gather from the observations is that they may be behind most of the intense/extreme precipitation events. This is not fully demonstrated, but there is evidence that it may be the case. Several studies focus on the influence of atmospheric rivers on severe events (e.g., Lavers and Villarini 2015; Blamey et al. 2018; Viale et al. 2018; Davolio et al. 2020, 2023; Massoud et al. 2020; Lauer et al. 2023; Kim et al. 2024; van der Breggen and Hudson 2024; Pradhan et al. 2025) and on future climate change-driven scenarios (e.g., Payne et al. 2020; Shields et al. 2023; Wang et al. 2023; Zhang et al. 2024; Henny and Kim 2025). Studies also suggest that precipitation from atmospheric rivers is enhanced by climate change (Michaelis et al. 2022) and is expected to significantly enhance river flood damages (Corringham et al. 2022). Finally, research should be focused on the response of the upper ocean to atmospheric rivers under climate change conditions (e.g., Shields et al. 2024).

One more field of investigation where we do not yet have reached sufficient firm knowledge is tropical-like cyclones. We have seen above that medicanes have been investigated by several authors, but their genesis and evolution mechanisms are far from being completely understood. In particular, the role of moist and dry air advection seems very important (e.g., Miglietta et al. 2017, 2021; Coll-Hidalgo et al. 2022). Attempts to ensure a better understanding of the characteristics of these cyclones are being made using reanalysis data (e.g., Gutiérrez-Fernández et al. 2024). Also forecasting studies are done using ensemble date (e.g., Saraceni et al. 2023). Air-sea interactions are one more field where research should be carried out for example as regards to sea surface temperature (e.g., Avolio et al. 2024) and wave production (Davison et al. 2024).

In a more general context, we do not know enough on the modification of global precipitating systems under the action of global warming. The hydrological cycle is expected to intensify with global warming, which likely increases the intensity of extreme precipitation events and the risk of flooding. However, these changes may differ from those predicted by the theory, especially as regards to the water-holding capacity of the atmosphere in warmer conditions (e.g., Papalexiou and Montanari 2019; Tabari 2020). Models seem to be generally in agreement when predicting more extreme events and a decrease in the number of light precipitation events (e.g., Giorgi et al. 2019; Sun et al. 2024). The question on the frequency of the extreme events in such changed conditions is not quite answered and subject to investigation (e.g., Myhre et al. 2019) as are the constraints on the pattern and magnitude of projected extreme precipitation (Kotz et al. 2023). Interesting research topics concern the annual range of precipitation (i.e., the difference between the maximum and the minimum precipitation within a year) (Chou and Lan 2012) and the morphology of global precipitation systems (Zhang and Wang 2024). In all these directions new observational datasets will significantly help as will do new reanalysis tools (e.g., Soci et al. 2024).

## References

Aggarwal, P.K., Romatschke, U., Araguas-Araguas, L., Belachew, D., Longstaffe, F.J., Berg, P., Schumacher, C., Funk, A.: Proportions of convective and stratiform precipitation revealed in water isotope ratios. Nat. Geosci. **9**, 624–629 (2016). https://doi.org/10.1038/ngeo2739

Albano, C.M., Dettinger, M.D., Harpold, A.A.: Patterns and drivers of atmospheric river precipitation and hydrologic impacts across the western United States. J. Hydrometeorol. **21**(1), 143–159 (2020). https://doi.org/10.1175/JHM-D-19-0119.1

Ali, H., Wong, L.C.Y., Prein, A.F., Fowler, H.J.: Characteristics of precipitation associated with post-tropical cyclones in the North Atlantic. Weather Clim. Extrem. **47**, 100742 (2025). https://doi.org/10.1016/j.wace.2024.100742

AMS: Precipitation. Glossary of Meteorology. https://glossary.ametsoc.org/wiki/Precipitation (2024a). Last accessed 10 Mar 2025

AMS: Rain. Glossary of Meteorology. https://glossary.ametsoc.org/wiki/Rain (2024b). Last accessed 10 Mar 2025

AMS: Snow. Glossary of Meteorology. https://glossary.ametsoc.org/wiki/Snow (2024c). Last accessed 10 Mar 2025

Angulo-Umana, P., Kim, D.: Mesoscale convective clustering enhances tropical precipitation. Sci. Adv. **9**(2) (2023). https://doi.org/10.1126/sciadv.abo5317

Arakawa, A.: Scaling tropical rain. Nat. Phys. **2**, 373–374 (2006). https://doi.org/10.1038/nphys323

Avolio, E., Fanelli, C., Pisano, A., Miglietta, M.M.: Unveiling the relationship between Mediterranean tropical-like cyclones and rising sea surface temperature. Geophys. Res. Lett. **51**(20), e2024GL109921 (2024). https://doi.org/10.1029/2024GL109921

Baumgardner, D., McFarquhar, G.M., Heymsfield, A.J. (eds.): Ice Formation and Evolution in Clouds and Precipitation - Measurement and Modelling Challenges. American Meteorological Society (2018)

Beck, H.E., Zimmermann, N.E., McVicar, T.R., Vergopolan, N., Berg, A., Wood, E.F.: Present and future Köppen-Geiger climate classification maps at 1-km resolution. Sci. Data. **5**, 180214 (2018). https://doi.org/10.1038/sdata.2018.214

Bjerknes, J., Solberg, H.: Life cycle of cyclones and the polar front theory of atmospheric circulation. Kristiania, Geophysisks Publikationer. **3**, 1–18 (1922). https://doi.org/10.1002/qj.49704920609

Blamey, R.C., Ramos, A.M., Trigo, R.M., Tomé, R., Reason, C.J.C.: The influence of atmospheric rivers over the South Atlantic on winter rainfall in South Africa. J. Hydrometeorol. **19**(1), 127–142 (2018). https://doi.org/10.1175/JHM-D-17-0111.1

Bowers, C., Serafin, K.A., Baker, J.W.: Temporal compounding increases economic impacts of atmospheric rivers in California. Sci. Adv. **10**(3), eadi7905 (2024). https://doi.org/10.1126/sciadv.adi7905

Braham Jr., R.R.: The thunderstorm project. 18th conference on severe local storms luncheon speech. Bull. Am. Meteorol. Soc. **77**(8), 1835–1846 (1996). https://doi.org/10.1175/1520-0477-77.8.1835

Brauer, N.S., Kirstetter, P.E., Basara, J.B., Hristova-Veleva, S., Tanelli, S., Turk, F.J.: Precipitation microphysics in tropical cyclones: a global perspective using the NASA global precipitation measurement mission dual-frequency precipitation radar. J. Geophys. Res. Atmos. **129**, e2023JD038709 (2024). https://doi.org/10.1029/2023JD038709

Burt, S.: Cloudburst upon Hendraburnick Down: the Boscastle storm of 16 August 2004. Weather. **60**(8), 219–227 (2005). https://doi.org/10.1256/wea.26.05

Businger, S., Reed, R.J.: Cyclogenesis in cold air masses. Weather Forecast. **4**(2), 133–156 (1989). https://doi.org/10.1175/1520-0434(1989)004<0133:CICAM>2.0.CO;2

Byers, H.R.: Structure and dynamics of the thunderstorm. Science. **110**, 291–294 (1949). https://doi.org/10.1126/science.110.2856.291

Byers, H.R., Braham Jr., R.R.: The Thunderstorm: Final Report of the Thunderstorm Project. US Department of Commerce (1949)

Cavicchia, L., von Storch, H., Gualdi, S.: Mediterranean tropical-like cyclones in present and future climate. J. Clim. **27**(19), 7493–7501 (2014). https://doi.org/10.1175/JCLI-D-14-00339.1

Chernokulsky, A., Kozlov, F., Zolina, O., Bulygina, O., Mokhov, I.I., Semenov, V.A.: Observed changes in convective and stratiform precipitation in Northern Eurasia over the last five decades. Environ. Res. Lett. **14**, 045001 (2019). https://doi.org/10.1088/1748-9326/aafb82

Chou, C., Lan, C.-W.: Changes in the annual range of precipitation under global warming. J. Clim. **25**(1), 222–235 (2012). https://doi.org/10.1175/JCLI-D-11-00097.1

Coll-Hidalgo, P., Pérez-Alarcón, A., Nieto, R.: Moisture sources for the precipitation of tropical-like cyclones in the Mediterranean Sea: a case of study. Atmos. **13**(8), 1327 (2022). https://doi.org/10.3390/atmos13081327

Comino, J.R.: Precipitation, Earth Surface Responses and Processes. Elsevier (2021)

Corringham, T.W., McCarthy, J., Shulgina, T., Gershunov, A., Cayan, D.R., Ralph, F.M.: Climate change contributions to future atmospheric river flood damages in the western United States. Sci. Rep. **12**, 13747 (2022). https://doi.org/10.1038/s41598-022-15474-2

Da Silva, N.A., Haerter, J.O.: The precipitation characteristics of mesoscale convective systems over Europe. J. Geophys. Res. Atmos. **128**(23), e2023JD039045 (2023). https://doi.org/10.1029/2023JD039045

Da Silva, N.A., Muller, C., Shamekh, S., Fildier, B.: Significant amplification of instantaneous extreme precipitation with convective self-aggregation. J. Adv. Model. Earth Syst. **13**(11), e2021MS002607 (2021). https://doi.org/10.1029/2021MS002607

Davison, S., Benetazzo, A., Barbariol, F., Ricchi, A., Ferretti, R.: Characterization of extreme wave fields during Mediterranean tropical-like cyclones. Front. Mar. Sci. **10**, 1268830 (2024). https://doi.org/10.3389/fmars.2023.1268830

Davolio, S., Della Fera, S., Laviola, S., Miglietta, M.M., Levizzani, V.: Atmospheric river in the Mediterranean basin during the 29 October 2018 "Vaia" storm over Italy. Mon. Weather Rev. **148**, 3571–3588 (2020). https://doi.org/10.1175/MWR-D-20-0021.1

Davolio, S., Vercellino, M., Miglietta, M.M., Drago Pitura, L., Laviola, S., Levizzani, V.: Heavy precipitation over the Alps and the influence of an atmospheric river. Weather Clim. Extrem. **39**(7), 100542 (2023). https://doi.org/10.1016/j.wace.2022.100542

Deng, M., Kollias, P., Feng, Z., Zhang, C., Long, C.N., Kalesse, H., Chandra, A., Kumar, V.V., Protat, A.: Stratiform and convective precipitation observed by multiple radars during the DYNAMO/AMIE experiment. J. Appl. Meteorol. Climatol. **53**(11), 2503–2523 (2013). https://doi.org/10.1175/JAMC-D-13-0311.1

Didlake Jr., A.C., Houze Jr., R.A.: Dynamics of the stratiform sector of a tropical cyclone rainband. J. Atmos. Sci. **70**(7), 1891–1911 (2013). https://doi.org/10.1175/JAS-D-12-0245.1

Dohring, H., Dixon, J. (eds.): Precipitation: Prediction, Formation and Environmental Impact. Nova Science Pub. Inc. (2012)

Ebert, E.E., Turk, M., Kusselson, S.J., Yang, J., Seybold, M., Keehn, P.R., Kuligowski, R.J.: Ensemble Tropical Rainfall Potential (eTRaP) forecasts. Weather Forecast. **26**, 213–224 (2011). https://doi.org/10.1175/2010WAF2222443.1

Eiras-Barca, J., Ramos, A.M., Algarra, I., Vázquez, M., Dominguez, F., Miguez-Macho, G., Nieto, R., Gimeno, L., Taboada, J., Ralph, F.M.: European West Coast atmospheric rivers: a scale to characterize strength and impacts. Weather Clim. Extrem. **31**, 100305 (2021). https://doi.org/10.1016/j.wace.2021.100305

Farjami, H., Kazemi, A.F.: Spatiotemporal patterns of polar low activity over the Southern Ocean. J. Geophys. Res. Atmos. **129**(4), e2023JD039832 (2024). https://doi.org/10.1029/2023JD039832

Feng, Z., Leung, L.R., Liu, N., Wang, J., Houze Jr., R.A., Li, J., Hardin, J.C., Chen, D., Guo, J.: A global high-resolution mesoscale convective system database using satellite-derived cloud tops, surface precipitation, and tracking. J. Geophys. Res. Atmos. **126**(8), e2020JD034202 (2021). https://doi.org/10.1029/2020JD034202

Gaertner, M.A., Jacob, D., Gil, V., Domínguez, M., Padorno, E., Sánchez, E., Castro, M.: Tropical cyclones over the Mediterranean Sea in climate change simulations. Geophys. Res. Lett. **34**(14) (2007). https://doi.org/10.1029/2007GL029977

Gay, D.A., Davis, R.E.: Freezing rain and sleet climatology of the southern USA. Clim. Res. **3**(3), 209–220 (1993). https://doi.org/10.3354/cr003209

Gimeno, L., Nieto, R., Sorí, R.: The growing importance of oceanic moisture sources for continental precipitation. NPJ Clim. Atmos. Sci. **3**, 27 (2020). https://doi.org/10.1038/s41612-020-00133-y

Giorgi, F., Raffaele, F., Coppola, E.: The response of precipitation characteristics to global warming from climate projections. Earth Syst. Dynam. **10**(1), 73–89 (2019). https://doi.org/10.5194/esd-10-73-2019

Golubkin, P., Smirnova, J., Bobylev, L.: Satellite-derived spatio-temporal distribution and parameters of North Atlantic polar lows for 2015–2017. Atmos. **12**(2), 224 (2021). https://doi.org/10.3390/atmos12020224

Grahame, N., Riddaway, B., Eadie, A., Hall, B., McCallum, E.: Exceptional hailstorm hits Ottery St Mary on 30 October 2008. Weather. **64**(10), 255–263 (2009). https://doi.org/10.1002/wea.458

Guardiola-Albert, C., Rivero-Honegger, C., Monjo, R., Díez-Herrero, A., Yagüe, C., Bodoque, J.M., Tapiador, F.J.: Automated convective and stratiform precipitation estimation in a small mountainous catchment using X-band radar data in Central Spain. J. Hydroinf. **19**(2), 315–330 (2017). https://doi.org/10.2166/hydro.2016.225

Guo, Y., Fu, Q., Leung, L.R., Na, Y., Lu, R.: Trends in warm season mesoscale convective systems over Asia in 2001–2020. J. Geophys. Res. Atmos. **128**(17), e2023JD038969 (2023). https://doi.org/10.1029/2023JD038969

Gutiérrez-Fernández, J., Miglietta, M.M., González-Alemán, J.J., Gaertner, M.A.: A new refinement of Mediterranean tropical-like cyclones characteristics. Geophys. Res. Lett. **51**(8), e2023GL106429 (2024). https://doi.org/10.1029/2023GL106429

Haberlie, A.M., Ashley, W.S., Gensini, V.A., Michaelis, A.C.: The ratio of mesoscale convective system precipitation to total precipitation increases in future climate change scenarios. NPJ Clim. Atmos. Sci. **6**, 150 (2023). https://doi.org/10.1038/s41612-023-00481-5

Harrold, T.W., Browning, K.A.: The polar low as a baroclinic disturbance. Q. J. R. Meteorol. Soc. **95**(406), 710–723 (1969). https://doi.org/10.1002/qj.49709540605

Hart, R.E., Evans, J.L.: A climatology of the extratropical transition of Atlantic tropical cyclones. J. Clim. **14**(4), 546–564 (2001). https://doi.org/10.1175/1520-0442(2001)014<0546:ACOTET>2.0.CO;2

Hartmann, D.L.: Global Physical Climatology. Elsevier (2016)

Hawcroft, M.K., Shaffrey, L.C., Hodges, K.I., Dacre, H.F.: How much Northern Hemisphere precipitation is associated with extratropical cyclones? Geophys. Res. Lett. **39**(24) (2012). https://doi.org/10.1029/2012GL053866

Hence, D.A., Houze Jr., R.A.: Vertical structure of tropical cyclone rainbands as seen by the TRMM Precipitation Radar. J. Atmos. Sci. **69**, 2644–2661 (2012). https://doi.org/10.1175/JAS-D-11-0323.1

Henny, L., Kim, K.-M.: The changing nature of atmospheric rivers. J. Clim. **38**(6), 1435–1456 (2025). https://doi.org/10.1175/JCLI-D-24-0234.1

Hobbs, P.V., Biswas, K.R.: The cellular structure of narrow cold-frontal rainbands. Q. J. R. Meteorol. Soc. **105**(445), 723–727 (1979). https://doi.org/10.1002/qj.49710544516

Hong, Y., Kummerow, C.D., Olson, W.S.: Separation of convective and stratiform precipitation using microwave brightness temperature. J. Appl. Meteorol. Climatol. **38**(8), 1195–1213 (1999). https://doi.org/10.1175/1520-0450(1999)038<1195:SOCASP>2.0.CO;2

Houze Jr., R.A.: Observed structure of mesoscale convective systems and implications for large-scale heating. Q. J. R. Meteorol. Soc. **115**, 425–461 (1989). https://doi.org/10.1002/qj.49711548702

Houze Jr., R.A.: Stratiform precipitation in regions of convection: a meteorological paradox? Bull. Am. Meteorol. Soc. **78**(10), 2179–2196 (1997). https://doi.org/10.1175/1520-0477(1997)078<2179:SPIROC>2.0.CO;2

Houze Jr., R.A.: Mesoscale convective systems. Rev. Geophys. **42**(4), RG4003 (2004). https://doi.org/10.1029/2004RG000150

Hu, H., Leung, L.R., Feng, Z.: Early warm-season mesoscale convective systems dominate soil moisture–precipitation feedback for summer rainfall in Central United States. PNAS. **118**(43), e2105260118 (2021). https://doi.org/10.1073/pnas.2105260118

Huang, Y., Wang, Y., Cui, X.: Differences between convective and stratiform precipitation budgets in a torrential rainfall event. Adv. Atmos. Sci. **36**, 495–509 (2019). https://doi.org/10.1007/s00376-019-8159-1

Huffman, G.J., Adler, R.F., Behrangi, A., Bolvin, D.T., Nelkin, E.J., Gu, G., Ehsani, M.R.: The new version 3.2 Global Precipitation Climatology Project (GPCP) monthly and daily precipitation products. J. Clim. **36**(21), 7635–7655 (2023). https://doi.org/10.1175/JCLI-D-23-0123.1

Jiang, H., Zipser, E.J.: Contribution of tropical cyclones to the global precipitation from eight seasons of TRMM data: regional, seasonal, and interannual variations. J. Clim. **23**(6), 1526–1543 (2010). https://doi.org/10.1175/2009JCLI3303.1

Kane Jr., R.J., Chelius, C.R., Fritsch, J.M.: Precipitation characteristics of mesoscale convective weather systems. J. Appl. Meteorol. Climatol. **26**(10), 1345–1357 (1987). https://doi.org/10.1175/1520-0450(1987)026<1345:PCOMCW>2.0.CO;2

Khouakhi, A., Villarini, G., Vecchi, G.A.: Contribution of tropical cyclones to rainfall at the global scale. J. Clim. **30**(1), 359–372 (2017). https://doi.org/10.1175/JCLI-D-16-0298.1

Kim, T.-J., Kim, J., Kim, J.-U., Chung, C.-Y., Byun, Y.-H.: Impacts of climate change on atmospheric rivers over East Asia. Asia-Pac. J. Atmos. Sci. **60**, 639–654 (2024). https://doi.org/10.1007/s13143-024-00372-6

Koch, S.E., Kocin, P.J.: Frontal contraction processes leading to the formation of an intense narrow rainband. Meteorog. Atmos. Phys. **46**, 123–154 (1991). https://doi.org/10.1007/BF01027339

Köppen, W.: Das geographische System der Klimate. In: Köppen, W., Geiger, R. (eds.) Handbuch der Klimatologie, vol. 1. Borntraeger (1936)

Kotz, M., Lange, S., Wenz, L., Levermann, A.: Constraining the pattern and magnitude of projected extreme precipitation change in a multimodel ensemble. J. Clim. **37**(1), 97–111 (2023). https://doi.org/10.1175/JCLI-D-23-0492.1

Kukulies, J., Lai, H.-W., Curio, J., Feng, Z., Lin, C., Li, P., Ou, T., Sugimoto, S., Chen, D.: Mesoscale convective systems in the third pole region: characteristics, mechanisms and impact on precipitation. Front. Earth Sci. **11**, 1143380 (2023). https://doi.org/10.3389/feart.2023.1143380

Kumar, S., Jash, D., Resmi, E.A., Sumesh, R.K., Gopan, G., Kumar, M., Andrews, A., Sukumar, N.: Hydrometeor categorization in stratiform and convective precipitation systems over a tropical coastal region. Theor. Appl. Climatol. **155**, 9305–9325 (2024). https://doi.org/10.1007/s00704-024-05185-1

Langer, I., Reimer, E.: Separation of convective and stratiform precipitation for a precipitation analysis of the local model of the German Weather Service. Adv. Geosci. **10**, 159–165 (2007). https://doi.org/10.5194/adgeo-10-159-2007

Lauer, M., Rinke, A., Gorodetskaya, I., Sprenger, M., Mech, M., Crewell, S.: Influence of atmospheric rivers and associated weather systems on precipitation in the Arctic. Atmos. Chem. Phys. **23**(15), 8705–9726 (2023). https://doi.org/10.5194/acp-23-8705-2023

Lavers, D.A., Villarini, G.: The contribution of atmospheric rivers to precipitation in Europe and the United States. J. Hydrol. **522**, 382–390 (2015). https://doi.org/10.1016/j.jhydrol.2014.12.010

Li, D., Qi, Y., Li, H.: Statistical characteristics of convective and stratiform precipitation during the rainy season over South China based on GPM-DPR observations. Atmos. Res. **301**, 107267 (2024). https://doi.org/10.1016/j.atmosres.2024.107267

Lockwood, J.W., Loridan, T., Lin, N., Oppenheimer, M., Hannah, N.: A machine learning approach to model over-ocean tropical cyclone precipitation. J. Hydrometeorol. **25**(1), 207–221 (2024). https://doi.org/10.1175/JHM-D-23-0065.1

Lonfat, M., Rogers, R., Marchok, T., Marks Jr., F.D.: A parametric model for predicting hurricane rainfall. Mon. Weather Rev. **135**(9), 3086–3097 (2007). https://doi.org/10.1175/MWR3433.1

Lull, H.W.: Soil compaction on forest and range lands. U.S. Dept. of Agriculture, Forestry Service, Misc. Publication No. 768. Available at https://archive.org/details/soilcompactionon768lull (1959)

Markowski, P., Richardson, Y.: Mesoscale Meteorology in Midlatitudes. Wiley (2010)

Martín, O.E., Roberts, T.M. (eds.): Rainfall: Behavior, Forecasting and Distribution. Nova Science Pub. Inc. (2013)

Massoud, E., Massoud, T., Guan, B., Sengupta, A., Espinoza, V., De Luna, M., Raymond, C., Waliser, D.: Atmospheric rivers and precipitation in the Middle East and North Africa (MENA). Water. **12**(10), 2863 (2020). https://doi.org/10.3390/w12102863

Mazza, E., Chen, S.S.: Tropical cyclone rainfall climatology, extremes, and flooding potential from remote sensing and reanalysis datasets over the Continental United States. J. Hydrometeorol. **24**(9), 1549–1562 (2023). https://doi.org/10.1175/JHM-D-22-0199.1

Michaelides, S. (ed.): Precipitation: Advances in Measurement, Estimation and Prediction. Springer Nature (2008)

Michaelis, A.C., Gershunov, A., Weyant, A., Fish, M.A., Shulgina, T., Ralph, F.M.: Atmospheric river precipitation enhanced by climate change: a case study of the storm that contributed to California's Oroville dam crisis. Earth's Future. **10**(3), e2021EF002537 (2022). https://doi.org/10.1029/2021EF002537

Miglietta, M.M., Rotunno, R.: Development mechanisms for Mediterranean tropical-like cyclones (medicanes). Q. J. R. Meteorol. Soc. **145**(721), 1444–1460 (2019). https://doi.org/10.1002/qj.3503

Miglietta, M.M., Cerrai, D., Laviola, S., Cattani, E., Levizzani, V.: Potential vorticity patterns in Mediterranean "hurricanes". Geophys. Res. Lett. **44**, 2537–2545 (2017). https://doi.org/10.1002/2017GL072670

Miglietta, M.M., Carnevale, D., Levizzani, V., Rotunno, R.: Role of moist and dry air advection in the development of Mediterranean tropical-like cyclones (Medicanes). Q. J. R. Meteorol. Soc. **147**, 876–899 (2021). https://doi.org/10.1002/qj.3951

Milani, L., Kidd, C.: The state of precipitation measurements at mid-to-high latitudes. Atmos. **14**(11), 1677 (2023). https://doi.org/10.3390/atmos14111677

Moore, G.W.K.: The organization of convection in narrow cold-frontal rainbands. J. Atmos. Sci. **42**(17), 1777–1791 (1985). https://doi.org/10.1175/1520-0469(1985)042<1777:TOOCIN>2.0.CO;2

Morbidelli, R. (ed.): Rainfall: Physical Process, Measurement, Data Analysis and Usage in Hydrological Investigations. Elsevier (2022)

Moseley, C., Hohenegger, C., Berg, P., Haerter, J.O.: Intensification of convective extremes driven by cloud–cloud interaction. Nat. Geosci. **9**, 748–752 (2016). https://doi.org/10.1038/ngeo2789

Muller, C., Yang, D., Craig, G., Cronin, T., Fildier, B., Haerter, J.O., Hohenegger, C., Mapes, B., Randall, D., Shamekh, S., Sherwood, S.C.: Spontaneous aggregation of convective storms. Annu. Rev. Fluid Mech. **54**, 133–157 (2022). https://doi.org/10.1146/annurev-fluid-022421-011319

Myhre, G., Alterskjær, K., Stjern, C.W., Hodnebrog, Ø., Marelle, L., Samset, B.H., Sillmann, J., Schaller, N., Fischer, E., Schulz, M., Stohl, A.: Frequency of extreme precipitation increases extensively with event rareness under global warming. Sci. Rep. **9**, 16063 (2019). https://doi.org/10.1038/s41598-019-52277-4

Papalexiou, S.M., Montanari, A.: Global and regional increase of precipitation extremes under global warming. Water Resour. Res. **55**(6), 4901–4914 (2019). https://doi.org/10.1029/2018WR024067

Payne, A.E., Demory, M.-E., Leung, L.R., Ramos, A.M., Shields, C.A., Rutz, J.J., Siler, N., Villarini, G., Hall, A., Ralph, F.M.: Responses and impacts of atmospheric rivers to climate change. Nat. Rev. Earth Environ. **1**, 143–157 (2020). https://doi.org/10.1038/s43017-020-0030-5

Peel, M.C., Finlayson, B.L., McMahon, T.A.: Updated world map of the Köppen-Geiger climate classification. Hydrol. Earth Syst. Sci. **11**(5), 1633–1644 (2007). https://doi.org/10.5194/hess-11-1633-2007

Penide, G., Protat, A., Kumar, V.V., May, P.T.: Comparison of two convective/stratiform precipitation classification techniques: radar reflectivity texture versus drop size distribution–based approach. J. Atmos. Ocean. Technol. **30**(12), 2788–2797 (2013). https://doi.org/10.1175/JTECH-D-13-00019.1

Peters, O., Neelin, J.D.: Critical phenomena in atmospheric precipitation. Nat. Phys. **2**, 393–396 (2006). https://doi.org/10.1038/nphys314

Pike, W.S., Smart, D.J.: Thunderstorms coming up against the wind on 11 June 2023–the development of a mesoscale convective system over southern England. Weather. **79**(8), 262–268 (2024). https://doi.org/10.1002/wea.4552

Pradhan, S., Wasko, C., Peel, M.C.: Atmospheric rivers and Australian precipitation: impact of detection algorithm choice. J. Hydrol. **651**, 132586 (2025). https://doi.org/10.1016/j.jhydrol.2024.132586

Pritchard, B.: The story of a cold pool. Weather. **64**(4), 98 (2009). https://doi.org/10.1002/wea.404

Qi, Y., Zhang, J., Zhang, P.: A real-time automated convective and stratiform precipitation segregation algorithm in native radar coordinates. Q. J. R. Meteorol. Soc. **139**, 2233–2240 (2013). https://doi.org/10.1002/qj.2095

Qin, L., Zhu, L., Liu, B., Li, Z., Tian, Y., Mitchell, G., Shen, S., Xu, W., Chen, J.: Global expansion of tropical cyclone precipitation footprint. Nat. Commun. **15**, 4824 (2024). https://doi.org/10.1038/s41467-024-49115-1

Ralph, F.M., Dettinger, M.D., Cairns, M.M., Galarneau, T.J., Eylander, J.: Defining "atmospheric river": how the glossary of meteorology helped resolve a debate. Bull. Am. Meteorol. Soc. **99**(4), 837–839 (2018). https://doi.org/10.1175/BAMS-D-17-0157.1

Ralph, F.M., Dettinger, M.D., Rutz, J.J., Waliser, D.E. (eds.): Atmospheric Rivers. Springer Nature (2020)

Randall, D.: Atmosphere, Clouds and Climate. Princeton University Press (2012)

Rhoades, A.M., Risser, M.D., Stone, D.A., Wehner, M.F., Jones, A.D.: Implications of warming on western United States landfalling atmospheric rivers and their flood damages. Weather Clim. Extrem. **32**, 100326 (2021). https://doi.org/10.1016/j.wace.2021.100326

Rojo, M., Claud, C., Mallet, P.E., Noer, G., Carleton, A.M., Vicomte, M.: Polar low tracks over the Nordic Seas: a 14-winter climatic analysis. Tellus A: Dyn. Meteorol. Oceanogr. **67**(1), 24660 (2015). https://doi.org/10.3402/tellusa.v67.24660

Romatschke, U., Dixon, M.J.: Vertically resolved convective–stratiform echo-type identification and convectivity retrieval for vertically pointing radars. J. Atmos. Ocean. Technol. **39**, 1705–1716 (2022). https://doi.org/10.1175/JTECH-D-22-0019.1

Ropelewski, C.F., Arkin, P.A.: Climate Analysis. Cambridge University Press (2019)

Rosenfeld, D., Lensky, I.M.: Satellite-based insights into precipitation formation processes in continental and maritime convective clouds. Bull. Am. Meteorol. Soc. **79**(11), 2457–2476 (1998). https://doi.org/10.1175/1520-0477(1998)079<2457:SBIIPF>2.0.CO;2

Rustemeier, E., Hänsel, S., Finger, P., Schneider, U., Ziese, M.: GPCC Climatology version 2022 at 0.25°: monthly land-surface precipitation climatology for every month and the total year from rain-gauges built on GTS-based and historical data. (2022). https://doi.org/10.5676/DWD_GPCC/CLIM_M_V2022_025

Ryzhkov, A.V., Zrnić, D.: Radar Polarimetry for Weather Observations. Springer Nature (2019)

Saraceni, M., Silvestri, L., Bechtold, P., Bongioannini Cerlini, P.: Mediterranean tropical-like cyclone forecasts and analysis using the ECMWF ensemble forecasting system with physical parameterization perturbations. Atmos. Chem. Phys. **23**(21), 13883–13909 (2023). https://doi.org/10.5194/acp-23-13883-2023

Schiro, K.A., Sullivan, S.C., Kuo, Y.-H., Su, H., Gentine, P., Elsaesser, G.S., Jiang, J.H., Neelin, J.D.: Environmental controls on tropical mesoscale convective system precipitation intensity. J. Atmos. Sci. **77**(12), 4233–4249 (2020). https://doi.org/10.1175/JAS-D-20-0111.1

Schneider, U., Finger, P., Meyer-Christoffer, A., Rustemeier, E., Ziese, M., Becker, A.: Evaluating the hydrological cycle over land using the newly-corrected precipitation climatology from the Global Precipitation Climatology Centre (GPCC). Atmosphere. **8**(3), 52 (2017). https://doi.org/10.3390/atmos8030052

Schumacher, C., Funk, A.: Assessing convective-stratiform precipitation regimes in the tropics and extratropics with the GPM satellite radar. Geophys. Res. Lett. **50**(14), e2023GL102786 (2023). https://doi.org/10.1029/2023GL102786

Shamekh, S., Lamb, K.D., Huang, Y., Gentine, P.: Implicit learning of convective organization explains precipitation stochasticity. PNAS. **120**(20), e2216158120 (2023). https://doi.org/10.1073/pnas.2216158120

Shields, C.A., Payne, A.E., Shearer, E.J., Wehner, M.F., O'Brien, T.A., Rutz, J.J., Leung, L.R., Ralph, F.M., Marquardt Collow, A.B., Ullrich, P.A., Dong, Q., Gershunov, A., Griffith, H., Guan, B., Lora, J.M., Lu, M., McClenny, E., Nardi, K.M., Pan, M., Qian, Y., Ramos, A.M., Shulgina, T., Viale, M., Sarangi, C., Tomé, R., Zarzycki, C.: Future atmospheric rivers and impacts on precipitation: overview of the ARTMIP Tier 2 high-resolution global warming experiment. Geophys. Res. Lett. **50**(6), e2022GL102091 (2023). https://doi.org/10.1029/2022GL102091

Shields, C.A., Li, H., Castruccio, F.S., Fu, D., Nardi, K., Liu, X., Zarzycki, C.: Response of the upper ocean to Northeast Pacific atmospheric rivers under climate change. Commun. Earth Environ. **5**, 603 (2024). https://doi.org/10.1038/s43247-024-01774-0

Soci, C., Hersbach, H., Simmons, A., Poli, P., Bell, B., Berrisford, P., Horányi, A., Muñoz-Sabater, J., Nicolas, J., Radu, R., Schepers, D., Villaume, S., Haimberger, L., Woollen, J., Buontempo, C., Thépaut, J.-N.: The ERA5 global reanalysis from 1940 to 2022. Q. J. R. Meteorol. Soc. **150**, 4014–4048 (2024). https://doi.org/10.1002/qj.4803

Strangeways, I.: Precipitation. Theory, Measurement and Distribution. Cambridge University Press (2007)

Sumner, G.N.: Precipitation. Process and Analysis. Wiley (1988)

Sun, W., Li, J., Yu, R., Li, N., Zhang, Y.: Exploring changes of precipitation extremes under climate change through global variable-resolution modeling. Sci. Bull. **69**(2), 237–247 (2024). https://doi.org/10.1016/j.scib.2023.11.013

Tabari, H.: Climate change impact on flood and extreme precipitation increases with water availability. Sci. Rep. **10**, 13768 (2020). https://doi.org/10.1038/s41598-020-70816-2

Tan, J., Jakob, C., Rossow, W.B., Tselioudis, G.: Increases in tropical rainfall driven by changes in frequency of organized deep convection. Nature. **519**, 451–454 (2015). https://doi.org/10.1038/nature14339

Tao, W.-K., Chern, J.-D.: The impact of simulated mesoscale convective systems on global precipitation: a multiscale modeling study. J. Adv. Model. Earth Syst. **9**(2), 790–809 (2017). https://doi.org/10.1002/2016MS000836

Taszarek, M., Allen, J.T., Marchio, M., Brooks, H.E.: Global climatology and trends in convective environments from ERA5 and rawinsonde data. NPJ Clim. Atmos. Sci. **4**, 35 (2021). https://doi.org/10.1038/s41612-021-00190-x

Testik, F.Y., Gebremichael, M. (eds.): Rainfall: State of the Science. Wiley (2013)

Ulbrich, U., Brücher, T., Fink, A.H., Leckebusch, G.C., Krüger, A., Pinto, J.G.: The central European floods of August 2002: part 1—rainfall periods and flood development. Weather. **58**(10), 371–377 (2003). https://doi.org/10.1256/wea.61.03A

van der Breggen, N.N., Hudson, P.F.: Influence of atmospheric rivers on extreme rainfall and high streamflow events in northwestern Europe: Rur (Roer) River basin. J. Hydrol.: Reg. Stud. **51**, 101644 (2024). https://doi.org/10.1016/j.ejrh.2023.101644

Viale, M., Valenzuela, R., Garreaud, R.D., Ralph, F.M.: Impacts of atmospheric rivers on precipitation in southern South America. J. Hydrometeorol. **19**(10), 1671–1687 (2018). https://doi.org/10.1175/JHM-D-18-0006.1

Wang, K., Lyu, X.: Comparison of the characteristics and precipitation features of mesoscale convective systems in tropical cyclones of different intensity categories over the western North Pacific. Mon. Weather Rev. **152**(8), 1963–1976 (2024). https://doi.org/10.1175/MWR-D-23-0214.1

Wang, J., Houze Jr., R.A., Fan, J., Brodzik, S.R., Feng, Z., Hardin, J.C.: The detection of mesoscale convective systems by the GPM Ku-band spaceborne radar. J. Meteorol. Soc. Jpn. **97**(6), 1059–1073 (2019). https://doi.org/10.2151/jmsj.2019-058

Wang, Y., Tang, L., Chang, P.-L., Tang, Y.-S.: Separation of convective and stratiform precipitation using polarimetric radar data with a support vector machine method. Atmos. Meas. Tech. **14**(1), 185–197 (2021). https://doi.org/10.5194/amt-14-185-2021

Wang, S., Ma, X., Zhou, S., Wu, L., Wang, H., Tang, Z., Xu, G., Jing, Z., Chen, Z., Gan, B.: Extreme atmospheric rivers in a warming climate. Nat. Commun. **14**, 3219 (2023). https://doi.org/10.1038/s41467-023-38980-x

Webb, J.D.C., Elsom, D.M., Meaden, G.T.: Severe hailstorms in Britain and Ireland, a climatological survey and hazard assessment. Atmos. Res. **93**(1–3), 587–606 (2009). https://doi.org/10.1016/j.atmosres.2008.10.034

Wen, Y., Behrangi, A., Chen, H., Lambrigtsen, B.: How well were the early 2017 California Atmospheric River precipitation events captured by satellite products and ground-based radars? Q. J. R. Meteorol. Soc. **144**, 344–359 (2018). https://doi.org/10.1002/qj.3253

Wen, L., Zhao, K., Yang, Z., Chen, H., Huang, H., Chen, G., Yang, Z.: Microphysics of stratiform and convective precipitation during Meiyu season in Eastern China. J. Geophys. Res. Atmos. **125**, e2020JD032677 (2020). https://doi.org/10.1029/2020JD032677

Wille, J.D., Favier, V., Gorodetskaya, I.V., Agosta, C., Kittel, C., Beeman, J.C., Jourdain, N.C., Lenaerts, J.T.M., Codron, F.: Antarctic atmospheric river climatology and precipitation impacts. J. Geophys. Res. Atmos. **126**(8), e2020JD033788 (2021). https://doi.org/10.1029/2020JD033788

Wille, J.D., Favier, V., Jourdain, N.C., Kittel, C., Turton, J.V., Agosta, C., Gorodetskaya, I.V., Picard, G., Codron, F., Leroy-Dos Santos, C., Amory, C., Fettweis, X., Blanchet, J., Jomelli, V., Berchet, A.: Intense atmospheric rivers can weaken ice shelf stability at the Antarctic peninsula. Commun. Earth Environ. **3**(1), 90 (2022). https://doi.org/10.1038/s43247-022-00422-9

Wu, Z., Zhang, Y., Zhang, L., Zheng, H., Huang, X.: A comparison of convective and stratiform precipitation microphysics of the record-breaking typhoon In-Fa (2021). Remote Sens. **14**(2), 344 (2022). https://doi.org/10.3390/rs14020344

Yanase, W., Niino, H., Watanabe, S.I.I., Hodges, K., Zahn, M., Spengler, T., Gurvich, I.A.: Climatology of polar lows over the Sea of Japan using the JRA-55 reanalysis. J. Clim. **29**(2), 419–437 (2016). https://doi.org/10.1175/JCLI-D-15-0291.1

Yang, S., Smith, E.A.: Convective–stratiform precipitation variability at seasonal scale from 8 yr of TRMM observations: implications for multiple modes of diurnal variability. J. Clim. **21**(16), 4087–4114 (2008). https://doi.org/10.1175/2008JCLI2096.1

Yang, N., Wang, C., Li, X.: Improving tropical cyclone precipitation forecasting with deep learning and satellite image sequencing. J. Geophys. Res.: Mach. Learn. Comput. **1**(2), e2024JH000175 (2024). https://doi.org/10.1029/2024JH000175

Zhang, Y., Wang, K.: The changing morphology of global precipitation systems during the last two decades. Bull. Am. Meteorol. Soc. **105**(10), E1861–E1880 (2024). https://doi.org/10.1175/BAMS-D-23-0106.1

Zhang, P., Chen, G., Ting, M., Ruby Leung, L., Guan, B., Li, L.: More frequent atmospheric rivers slow the seasonal recovery of Arctic Sea ice. Nat. Clim. Chang. **13**(3), 266–273 (2023). https://doi.org/10.1038/s41558-023-01599-3

Zhang, L., Zhao, Y., Cheng, T.F., Lu, M.: Future changes in global atmospheric rivers projected by CMIP6 models. J. Geophys. Res. Atmos. **129**(3), e2023JD039359 (2024). https://doi.org/10.1029/2023JD039359

Zhu, Y., Newell, R.E.: Atmospheric rivers and bombs. Geophys. Res. Lett. **21**(18), 1999–2002 (1994). https://doi.org/10.1029/94GL01710

# Precipitation Observations

# 5

Water is a fundamental component of the Earth's energy and water cycle, helping to redistribute energy across the Earth's surface and atmosphere through the processes of evaporation, condensation and deposition (precipitation). Precipitation ultimately provides fresh, potable water that is essential to life on Earth—the only planet we know of that has abundant, usable water. Our social and economic well-being is moulded by where, when and how much precipitation occurs. On a practical level managing water within our environment is essential since too much or too little can be disastrous requiring the mitigation of droughts or floods through civil defence. Furthermore water, especially precipitation is increasingly important politically since water resources transcends national boundaries and has significant economic value for agriculture through agrobusinesses such as crop forecasting. Precipitation also impacts our health through the removal of particulate matter from the atmosphere (Thornes et al. 2010). The monitoring and measurement of precipitation is therefore essential, not only to our understanding and knowledge of the Earth System, but also economically and socially.

In terms of measuring precipitation, the first question posed is: "what is precipitation?" Precipitation is defined as deposition of water from the atmosphere in solid or liquid form. While this may appear to be straightforward, gauge measurements usually include surface-generated liquid, such as dew, frost and condensation, whereas weather radars and satellites only provide a bulk measure of precipitation within the atmosphere which is assumed to fall to the surface.

Precipitation measurements are usually defined as the liquid (equivalent) falling on a unit area over a period of time from the atmosphere. Measuring precipitation, both rainfall and snowfall, its accuracies and errors is however, extremely problematic. High-quality measurements that cover a wide range of temporal and spatial scales are needed by the user community. Climate studies require long-term, consistent data records not only to monitor changes in precipitation amounts and distribution, but also to assess and quantify the changing character of precipitation at fine scales. This is particularly important for the validation of climate models due to



V. Levizzani, C. Kidd, *Precipitation*, Geophysics and Environmental Physics,
https://doi.org/10.1007/978-3-031-97096-2_5

their poor representation of precipitation. Precipitation measurements are also important to a range of different scientific disciplines beyond the atmospheric community, including the hydrological, oceanic, cryospheric, environmental, ecological, and biological communities.

Measurements made (primarily) by observations/reports or by rain/snow gauges are considered "conventional" measures and often relate to the direct physical observation or measures of precipitation at or over certain periods of time. Newer technologies, such as weather radars and information gathered from satellite observations tend to be more comprehensive in terms of spatial and temporal sampling but are also less direct. Fundamental to the measurement of any phenomena is the directness and accuracy of the observation: providing accurate measurements is dependent upon the characteristics of the precipitation (intensity, phase, etc), the observational capability of the instrument, how the measurements/observations are interpreted and, the perceived versus real requirements of the measurements.

However, precipitation itself is relatively rare, often occurring probably less than 1% of the time (<5% within 68°N–68°S) and is highly variable in both time and space. Spatially, precipitation is governed by a range of factors related to individual weather systems (e.g. stratiform vs convection) and interactions with the surface (e.g., land/ocean, orography etc). When precipitation does occur, intensities may range from very light to very heavy with the range of intensities for instantaneous precipitation being highly skewed toward lighter intensities. The spatial and temporal variability makes it difficult to measure satisfactorily as an environmental parameter, particularly on a global basis. Dense observational networks are necessary to adequately capture this variability at fine temporal and spatial scales, although averaging over time and space generally results in the measure of accumulated precipitation being more normally distributed and more representative (Bell et al. 1990). However, even for climatological studies, less dense measurement networks may not necessarily faithfully capture small-scale, extreme events or the variability, particularly over regions with complex terrain where large gradients in the distribution of precipitation may occur (e.g., near/over islands or close to coasts).

A report by the Group on Earth Observations (GEO 2014) that polled users of precipitation data noted a significant range of user requirements (see Fig. 5.1), with spatial resolutions ranging from 300 m to 50 km, temporal scales ranging from 18 min to 15 days and acquisition/latency (observation to user) ranging from 6 min to 24 h. For flash flood studies precipitation measurements are required at local, fine scales with rapid access to the data (low latency) while for drought, longer-term measurements will suffice, with less stringent spatial, temporal, and latency requirements. For climate studies the accuracy of the measurements and the homogeneity of the data record are perhaps paramount over other criteria to enable the assessment of the subtleties in the distribution and amounts due to climate change. Crucially, the measurement requirements differ between the different user communities, therefore (ideally) providing observations at the best possible resolutions should encompass all users.

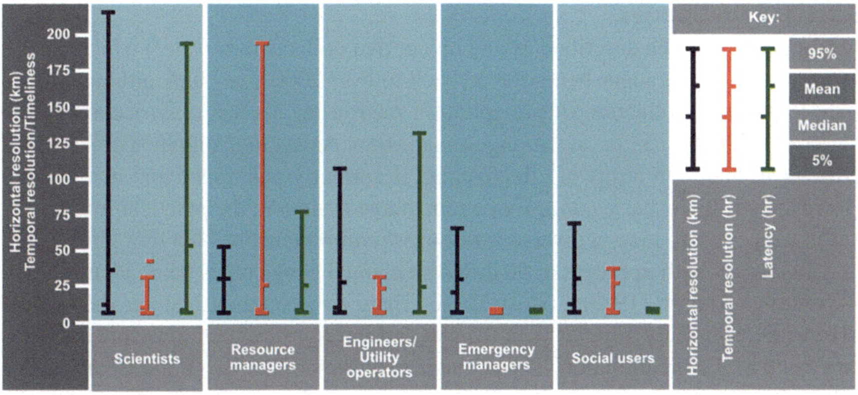


**Fig. 5.1** User requirements for precipitation data for spatial resolution, temporal resolution and data latency as reported in a user survey by the Group on Earth Observations. (Source: GEO 2014, their Figure 2, public domain)

## 5.1 Surface Measurements

The measurement of precipitation at the surface is vital to many applications across the hydrological, meteorological and water resources community. These at-surface measurements provide direct information on the intensity and amount of precipitation, with networks providing information on the distribution of precipitation. Weather radars have now become an integral part of the precipitation observation network in many parts of the world, offering good spatial and temporal sampling and low data latency. In addition, at-surface instrumentation is often deployed for research purposes, such as studying the micro-physical structure of precipitation, spatial and temporal characteristics of precipitation as well as providing more detailed data for use in, for example, hydrological models.

### 5.1.1 Gauge Measurements

Gauge measurements have provided the mainstay of precipitation observations, not least due the longevity of the data records and their ability to provide climate-scale data sets. They also provide the de facto data record due to their directness of measuring precipitation at the surface. Although most gauges are designed for measuring liquid precipitation (i.e., rain), a rain gauge will also capture snowfall, even if not optimised to do so. In appropriate climates gauges may be fitted with thermostatically controlled heaters to melt snow and thus measure the liquid water equivalent. To measure snowfall specifically, standard rain gauges may be adapted through the provision of "fences" or through snowfall-specific gauges.

#### 5.1.1.1 Rain Gauges

A rain gauge may be described as any object that collects rain(water) which can be measured. Simple gauges being traced back to over 2000 years ago, although long-term systematic collection of precipitation measurements has only occurred over the last hundred years or so (Strangeways 2010). As such, a simple "can" may be difficult to match in terms of effectiveness if correctly positioned and maintained. However, while the basic concept of a rain gauge is simple, the practical implementation necessary to meet a diverse set of user requirements has led to a great range of gauge types with in excess of 50 different manual gauge types being identified by Sevruk and Klemm (1989a). Most gauges share one common feature: the orifice. The orifice is usually circular with the rim and interior designed to ensure an accurate catch of the precipitation. The actual size of the orifice does not appear to critically affect the gauge accuracy (Strangeways 2003) with most official gauges having orifices between about 127 and 400 $cm^2$. A larger orifice does however result in more precipitation entering the gauge and thus allows finer measurements to be made.

Gauges can generally be categorised by their physical design of the gauge, the mechanisms used to collect and quantize the rainfall, and the technology used to report the rainfall, namely non-recording gauges and recording gauges.

Non-recording, or accumulation, gauges provide measurements that make up the "standard" climatological records. Such gauges require manual intervention to measure and record the amount of rain captured by the gauge, usually once a day at a specific time. Readings may be made more frequent (e.g., 3- or 6-hourly) where the gauge is part of a main meteorological or climatological station or can be longer particularly in more remote regions where weekly or even monthly observations of the collected precipitation are made. The time these observations are made are usually early morning local time, which on a global scale makes the synchronisation of global daily precipitation awkward. More frequent observations are often made to coincide with the synoptic reporting times at 0000, 0600, 1200, and 1800 UTC. Where possible the siting and operation of these gauges conform to the WMO standard practices to ensure regularity and consistency of the observations and measurements, and where possible the recording and transmission of the measurements to central meteorological services to allow timely access and further analysis.

Despite the simplicity of the accumulation gauge many applications require information on precipitation intensity (i.e., rate/flux) or accumulation over short periods of time (e.g., minutes). Mechanisms have therefore been devised to quantise precipitation over a period. Mechanically recording gauges, such as the tipping siphon gauge and weighing gauges record the gauge catch over time on paper using pen/paper trace mechanically linked to a float. However, these gauges have generally been superseded by electrically recording devices such as tipping-bucket gauges, electronic weighing gauges, capacitance gauges, and drop counting gauges (Strangeways 2010). These (quasi-automatic) electrically recording gauges can measure, record, and report the rainfall in near–real time through electronic dataloggers and communication systems (satellite or phone networks). The availability of gauge measurements in near–real time greatly enhances the usefulness of such measurements for meteorological and hydrological applications.

Tipping Bucket Raingauges (TBR) are designed to provide quantised measures of precipitation. A TBR uses a metal or plastic twin bucket to measure the water collected falling into the gauge. The buckets are designed to collect a certain weight (volume) of water: once that weight is exceeded the bucket mechanism tips, emptying the full bucket and allowing the other bucket to fill with water. Since each bucket is of a fixed quantity of precipitation, if the time of the tip or the number of tips per unit time, is recorded, the amount and intensity of the precipitation may be calculated. The provision of electrical connections allows these tips to be recorded electronically allowing transmission of the information remotely. The number of tips per 15-min is usually recorded, although a time-event logger can be used to accurately record the time of each tip. The latter allows improved quality control/assurance of the data since any asynchronous timings of the tips may be identified. In addition, the water may also be collected which allows the total accumulated precipitation to be checked against the total tip amounts. The precision of TBRs depends upon the design of the gauges: some gauges will provide 0.1 mm/tip measures (i.e., Young's gauge, see Fig. 5.2), others 0.2 mm/tip (or 0.1" or 0.25 mm/tip). The finer-resolution tip measures are however liable to greater mechanical errors, although these can provide better rain intensity measurements if paired with time-event data loggers.

Weighing precipitation gauges measure the amount of precipitation over time by determining its weight. A high-capacity collection bucket captures liquid, solid, or mixed precipitation types, and measures the weight of the collected water as a function of time. A load cell sensor converts the force of the water in the collection bucket into electrical signals that can then be recorded. Measurements of rain, snow,

**Fig. 5.2** An external view of a Young's tipping bucket raingauge of metal and plastic construction with a leaf filter within the orifice. On the right is an internal view showing the twin-bucket, screws for adjusting the tip amount and levelling of the platform, and electrical connections for the reed-relay and for a thermostatic heater if required. (Photos by Chris Kidd)

hail, and sleet amounts may be made to an accuracy of ±0.05 mm and ±0.1 mm $min^{-1}$ intensity, operating up to a 1-min sampling interval. The precipitation intensity may be derived by measuring the difference between the weights of each collection period, with the accumulated precipitation being the sum of all the weights. Algorithms are used to reduce the effects of wind, temperature, and evaporation on the measurement accuracy.

Despite the directness of gauge measurements, there are numerous errors sources that need to be considered and reduced. These errors can be attributed to the instrument (e.g., design, mechanical), micro-scale issues (e.g., local effects of the site) and macro-scale issues (e.g., representativeness).

Much work has gone into the reducing errors associated with the rain gauge. Errors such as mechanical issues, wetting losses (particularly at the onset of precipitation), evaporation, especially in hot/dry climates and perhaps most importantly, wind effects. The latter accounts for substantial under-catch in conditions with high winds and/or light precipitation: the interaction between the wind and the gauge orifice causes some of the precipitation to miss the gauge and therefore go unrecorded. A number of gauge designs have been proposed to reduce the wind/gauge interaction and, together with shielding (particularly in regions with snow) and proper siting of the gauge, these effects may be reduced. To further improve at-surface measurements the "pit gauge" (Fig. 5.3) may be considered the reference standard. These gauges have their office level with the surface to minimise any wind effects. Such gauges with suitable corrections for wetting and evaporation can provide reliable and consistent measurements, typically to within ±2%. However, the practicalities of installing pit gauges restricts their use to specific validation sites. Most gauges are placed to meet user requirements while minimising errors as much as possible resulting in the gauge orifices typically being about 40–50 cm above ground surface. Such gauges are subject to systematic errors primarily due to wind loss that can amount to 4–6% depending upon the gauge type, types of weather systems and exposure.

In most circumstances a rain gauge is considered to represent the rainfall falling in its immediate vicinity, over distances of a few metres and preferably a few kilometres. However, gauge measurements have their limitations given the spatial and temporal variability of precipitation and the fact that gauges are (small) point measurements. Standards set by the WMO (2008) are designed to ensure consistency between gauge measurements to reduce some of the inherent errors, such as those caused by siting or exposure. For example, a gauge should not be overshadowed by structures or trees and, conversely, should not be too exposed as to increase wind losses. Nevertheless, even under ideal situations the representativeness or autocorrelation length of precipitation is surprisingly small, often from less than 100 metres to a few kilometres for instantaneous to hourly measurements respectively. This correlation length is dependent upon the meteorology of the precipitation event and the local topography. Fortunately, accumulating precipitation over time increases the correlation length; over longer periods, the gauges become more representative of the regional precipitation regime.

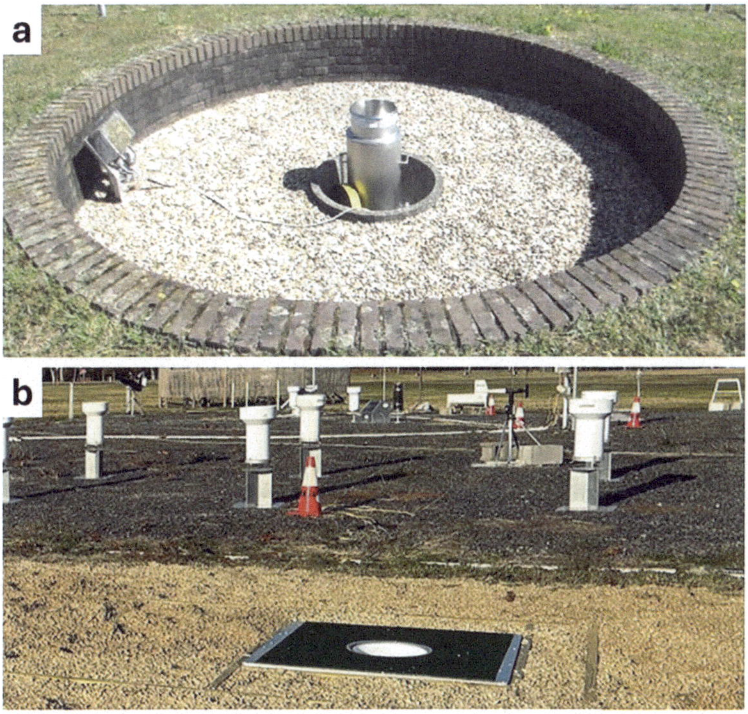


**Fig. 5.3** An open pit gauge (top) which has the orifice of the gauge level with the raised lip of the pit, compared with a pit gauge level with the surrounding area (bottom) which has splash-reducing surface surrounding to gauge. (Source: top-public domain, bottom-photo by D. Wolff, NASA/WFF)

At the broader scale, to properly represent the spatial distribution of precipitation it is necessary to provide sufficient gauges to faithfully capture the variability of rainfall over the area or region. However, the number of gauges, even in well-represented regions tends to be small (Fig. 5.4). The total number of gauges globally varies over time as do their measurement capabilities and operability. Sevruk and Klemm (1989b) and New et al. (2001) put the number of gauges worldwide at about 150,000, Groisman and Legates (1995) noted as many as 250,000, while Strangeways (2003) identified at least 123,014 gauges that can be used for monthly accumulations. Variations in the number of gauges depends upon the criteria used to count the gauges, such as whether all "stations" that have existed and have provided some form of precipitation measurements at some time in their observational record, while others will only report locations that currently report precipitation measurements. While the number of sub-daily rainfall gauge observations available in near–real time is small, more observations tend to become available over time if the user is willing to wait longer for the data. Daily measurements represent perhaps the greatest number of official data entries in line with the WMO recommendations and are therefore most easily implemented by the national meteorological agencies. It should be noted that not all gauge observations are available to the public or even to

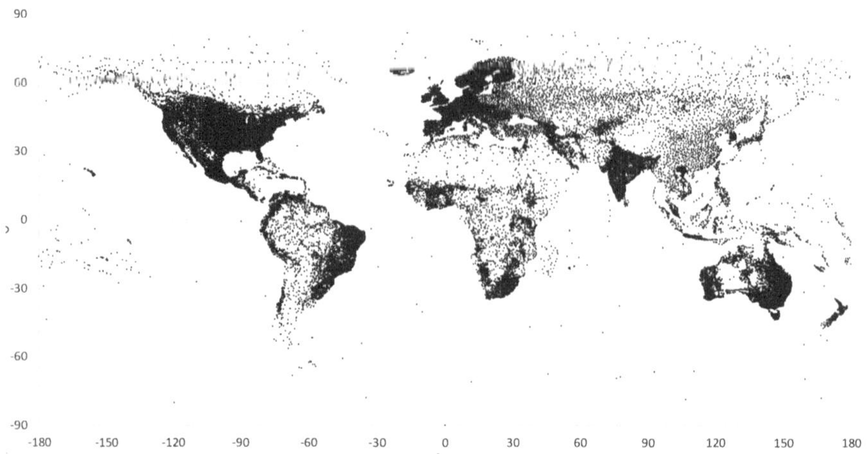


**Fig. 5.4** Worldwide distribution of the GPCC stations illustrating the large variations in station density over the land areas, and the paucity of station data over the oceans

researchers. A potentially large number of additional gauge observations are collected through commercial networks (e.g., water companies) but typically have restricted access to their data.

Finally, obtaining gauge measurements of the type required (e.g., temporal sampling, latency, etc) may be difficult, not least since the provision of such data varies between different countries/regions depending upon the organization within each country, region, or locality. In many cases the accessible gauge data may not necessarily be available for all temporal samples (i.e., 3 hourly, daily, etc.) or with adequate data latency for applications, such as flood monitoring and forecasting which requires the timely delivery of data to be truly useful. Importantly, while it is certain that many gauges exist, not all gauges have operated continuously or simultaneously. Finally, although many schemes exist for the interpolation of precipitation, great care is needed since different interpolation schemes generate different distributions and when applied to instantaneous or monthly precipitation data can produce undesired results. The interpolation of instantaneous gauge data should be avoided where possible due to the inherent heterogeneity of precipitation at fine temporal and spatial scales.

#### 5.1.1.2 Snowfall Gauges

The measurement of snowfall (i.e., falling snow rather than snow on the ground) is difficult. When measuring rainfall, different types of gauges will often measure different amounts even when installed at the same location: this is particularly true for snowfall measurements since falling snow is affected more by the wind. The interception of precipitation by a rain gauge is largely affected by the wind effect around the orifice, an effect that is exacerbated with increased exposure (Duchon and Essenberg 2001; Goodison et al. 1998). This greatly impacts the measurement of snow owing to the nature of falling/blowing snow, the variety of snow gauges used, and the inefficiencies of the catch by the gauges. Assessing the impact of this has been the focus of the WMO Solid Precipitation Intercomparison Experiment (SPICE) project (Nitu and Wong 2010b; Rasmussen et al. 2012).

Importantly, many precipitation measurements in cold regions are relatively poor since they are made by automated systems such as weighing or tipping-bucket gauges (Nitu and Wong 2010a; Goodison et al. 1998). Although most automated snow gauges are heated to permit the liquid equivalent to be measured, only about 28% of precipitation gauges are equipped with shields to mitigate the effect of airflow over the orifice (Nitu and Wong 2010a). While rainfall can typically be measured to within 10%–20% (Vuerich et al. 2009), wind effects may result in less than 25% of the snowfall being collected (Goodison et al. 1998). SPICE found that a shielded precipitation gauge is deemed the most appropriate type for cold-season precipitation measurements, with the Russian double fence deemed to be the most suitable. This consists of a Tretyakov gauge (see Fig. 5.5) encircled by two octagonal lath-fences, the first with a diameter with 4 m and the second with a diameter of 12 m, at a height of 3 and 3.5 m, respectively. Clearly the construction and operation of such a gauge structure is expensive, consequently such gauges are often restricted to particular locations for use as a reference or for specific application requirements.

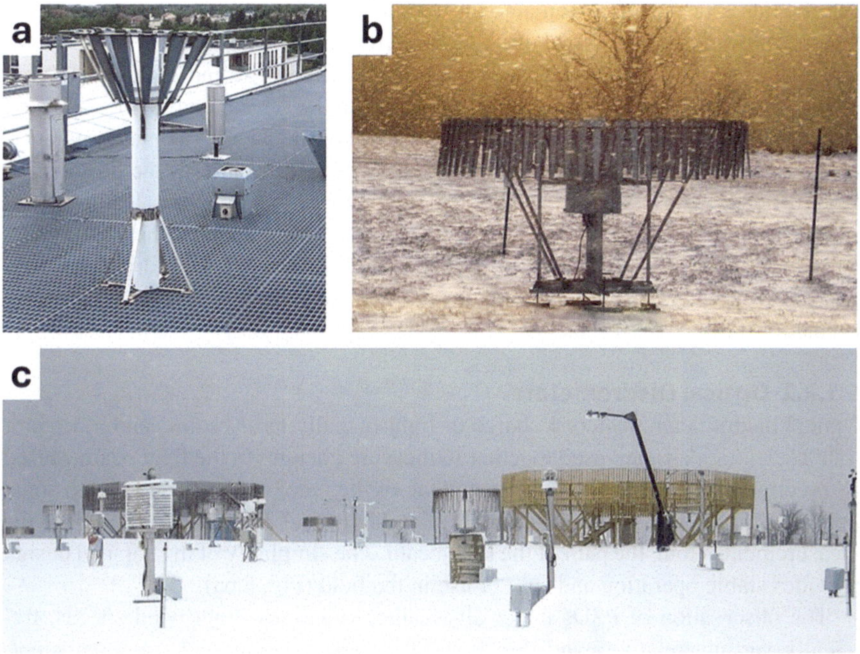


**Fig. 5.5** (**a**) Tretyakov gauge with wind shield on the roof of the Finnish Meteorological Institute, Kumpula, Finland, (**b**) Tretyakov gauge with outer wind shields installed, CARE site, Canada, and (**c**) array of meteorological instruments, including two Russian (-type) double fence enclosures, CARE site Canada. (Photos by Chris Kidd)

## 5.1.2 Disdrometers

Disdrometers are instruments that can measure the particle size distribution (PSD) and velocity of falling hydrometeors, including rain, snow, graupel, and hail. For rainfall the PSD is the Drop Size Distribution (DSD). Establishing PSDs are important for understanding the microphysical processes governing the formation of precipitation and its processes, such as the coalescence and ice-phase nucleation, breakup of large droplets into numerous small ones, evaporation of falling droplets, size sorting of particles in convective drafts, or wind shear effects. They can be used to establish radar-reflectivity and precipitation rate (Z–R) relationships to improve the retrieval from precipitation radars, particularly under different climatological regions with different PSDs (Tokay et al. 2001).

The measurements are made using light beams (usually laser) or active microwave (radar) and as such the sensors does not come into contact the precipitation particles. From the measurements of DSD and fall velocity the precipitation intensity may be calculated. Three basic types of disdrometers are described below.

### 5.1.2.1 Impact Disdrometers

As their name suggest, these rely upon the recording of the impact of the precipitation particle(s) with a surface. Early work with precipitation falling onto fine powder or filter paper over a given amount to time, together with necessary correction factors, allowed the size distribution of the particles to be analysed. The first automatic sensor was the Joss and Waldvogel disdrometer (Joss and Waldvogel 1969, 1970), which measured the impact of raindrops on the sensor, converting this to an electrical signal from which the drop sizes could be calculated. The main limitation of this sensor was the sensitivity to small droplets, particularly in the presence of large droplets whose signal masked that of the smaller drops, thus leading to an underestimation of these drops and of the rainfall. Mitigation of the background noise is continuously adjusted to avoid false reporting. Despite these drawbacks, the Joss and Waldvogel disdrometer remains popular due to its simplicity and extensive evaluation.

### 5.1.2.2 Optical Disdrometers

Optical disdrometers, where a source of light (usually laser beams) and a detector (photo-electric cell), are used together to measure changes in the light beam caused by precipitation passing between them. One of the more common one-dimensional optical disdrometers is the Parsivel (Löffler-Mang and Joss 2000) which provide measurements along the path of the light beam. The simplicity of the Parsivel design provides stable operation and ease of use in the field (Fig. 5.6a).

The observation of PSDs under all-weather situations (high wind speeds, frequently varying relative wind directions, flow distortion around the ships superstructure, ship movement in high sea states, and frozen precipitation) over remote ocean areas in different climatic regimes has seen the development of specialist optical disdrometers (Klepp 2015). An automated disdrometer (ODM470) was specifically designed for shipboard use, capable of measuring precipitation occurrence,

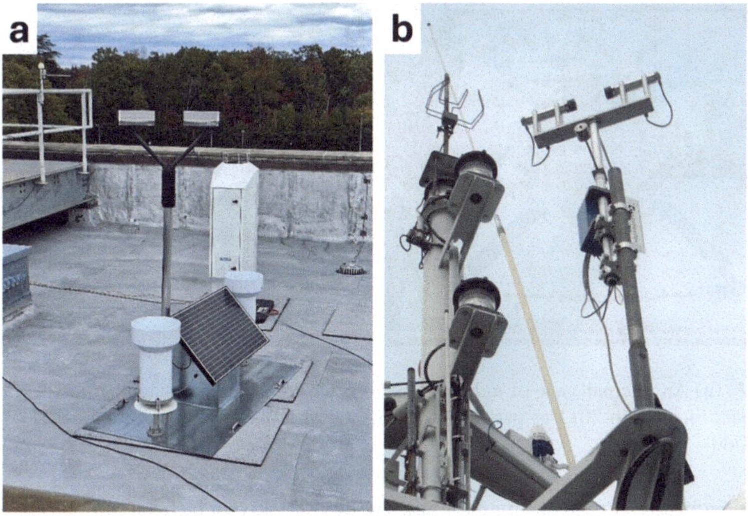


**Fig. 5.6** (**a**) Parsivel 1D distrometer with two reference gauges (NASA/GSFC) and (**b**) an ODM470-300 laser distrometer with wind vane as installed on the RV Aranda in 2011 as part of the NASA GPM Light Precipitation Validation Experiment (LPVEx) ground validation exercise. (Photos by Chris Kidd)

intensity and accumulation and discriminates for rain, snow and mixed phase precipitation through minute-based particle size distributions. The instrument pivots around a vertical axis through the aid of a wind vane thus keeping the sensitive volume perpendicular to the local wind direction (Fig. 5.6b). A total of 128 size bins allows for a wide dynamic range of the PSD which, together with modifications improved the measure of snowfall rate by means of logarithmic instead of linear size binning of the measured particles (Großklaus et al. 1998; Bumke et al. 2004).

The principles behind the optical disdrometer have led to the development of the two-dimensional video disdrometer that provides measurements as particles passes through a 2-D viewing area as observed by two video cameras (Fig. 5.7). By continuously capturing imagery of the falling particle the fall velocity and 3-D shape of the hydrometeor can be determined. The 2DVD also has the benefit of being able to observe multiple particles at once (Kruger and Krajewski 2002). The 2DVD consists of two internal lamps that create two light planes that cut across an approximate 10 cm × 10 cm measurement window and then projected onto two high speed line-scan cameras. Hydrometeors that fall through these light planes cast a shadow which is recorded by the two cameras. The two light planes are separated by a set distance which allows the velocity of the falling particle(s) to be measured. The high-speed scanning of the cameras samples each plane every 18 microseconds such that the falling hydrometeor is captured by several scan lines of each camera and from two sides. This allows precise measurements to be made of both the shape and fall speed of the particle, together with derived measures of the drop-size-distribution, rain rate and precipitation type.

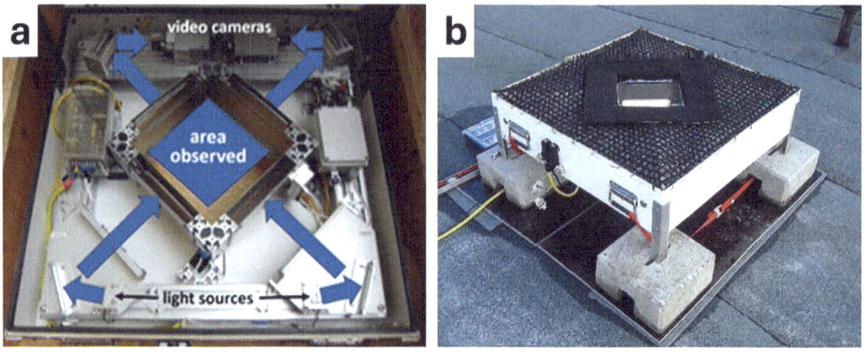


**Fig. 5.7** (**a**) An internal view of a 2DVD disdrometer with schematics and (**b**) as deployed at Järvenpää, Finland, in 2011 as part of the NASA GPM ground validation exercise. (Photos by Chris Kidd)

The accuracy of measurements made by the 2DVD can be affected by other non-hydrometeor particles, such as in-splashing of water droplets and multiple droplets falling between the cameras at once which are accounted for through quality control measures. The instrument also has measuring limitations, such as droplets that exceed (below/above) certain sizes or certain velocities.

#### 5.1.2.3 Radar-Based Disdrometers

Radar-based disdrometers measure the strength and Doppler frequencies of the backscattered return signal allowing the fall velocities and size distribution of the droplets to be determined. The advantages of such devices are the large sampling volume of the instrument (many $m^3$) that reduce in the sampling uncertainties, high temporal sampling (usually ca. 10 s) and the ability to provide measurements along a vertical beam up to several thousand metres. However, the accuracy of the calculated DSD is dependent upon an assumption of the vertical air velocities, with both updraft and downdrafts affecting the relative fall velocity as measured at the surface. The frequency of the instrument depends upon the type of particles being studied. Since the intensity of Rayleigh (back)scattering increases with the inverse-fourth power of the wavelength, precipitation-size hydrometeors are best sensed using lower frequencies (e.g., 24 GHz), while cloud-size particles are best sensed using higher frequencies (e.g., 35 and 94 GHz).

Micro rain radars (MRRs) are relatively inexpensive ground-based vertically pointing radars (Peters et al. 2002). The radars operate at 24.1 GHz, with Doppler capability. The MRR-2 provides vertical measurements averaged over 1 min with 31 range bins, which can be set from 35 m through to 200 m providing a total possible range of between 1085 and 6200 m, while the MRR-Pro provides 10 s measurements over 64, 128 or 256 range bins, with the range bins set to between 10 and 200 m. Figure 5.8 shows a MRR-Pro set up to observe and measure light precipitation in a coastal marine environment located at the Plymouth Marine Laboratory in Plymouth, UK, together with an example of these measurements, showing the fall

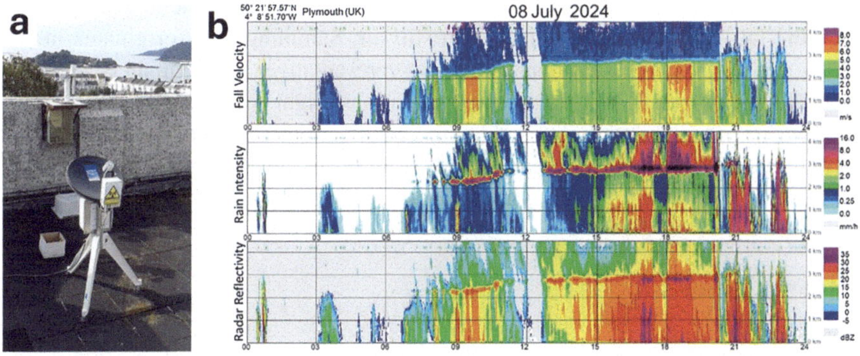


**Fig. 5.8** (**a**) An MRR-Pro set up to observe and measure light precipitation in a coastal marine environment located at the Plymouth Marine Laboratory in Plymouth, UK. The MRR-Pro at this location is set to provide measurements every 10 s from the surface in 35 m range bins to an altitude of 4480 m. (**b**) An example of these measurements is shown for 8 July 2024. The top plot shows the fall speed; the centre plot shows the rain (precipitation) intensity; the bottom plot shows the radar reflectivity. The brightband/melting layer can be clearly discerned at about 3 km, as evidenced by an increase in the reflectively and the change in the fall speed of the hydrometeors

speeds, the precipitation intensity and the radar reflectivity. Importantly, since the radar MRR is vertically pointing there is little surface clutter, although the very lowest range bins are usually best avoided. Crucially, MRR observations are very sensitive to light precipitation and are available at fine resolution time scales allowing the vertical distribution of precipitation to be measured together with the fall velocity of the hydrometeors. The main drawback is that path-integrated attention becomes significant in moderate to heavy precipitation and is sufficient to block returns at higher altitudes.

Clouds are best observed at the higher frequencies of around 35 and 94 GHz. The Chilbolton Observatory (UK) operates a 34.96 GHz (9 mm wavelength) Doppler cloud radar together with a 35.1 GHz (8.55 mm wavelength) MIRA mobile scanning pulsed Doppler cloud radar. While these radars will sense precipitation, they are primarily designed for observing and measuring the microphysics of clouds and fog. These Doppler radars with high sensitivity allow cloud observations, even very thin clouds. Dual polarisation provides the ability to differentiate between rain types, hail, snow and sleet by determining the size and shape of precipitation particles and can identify the melting layer. A high-frequency (94 GHz) radar provides complementary observations to the 35 GHz radars, and through a comparison of measurements at the 2 frequencies (the dual-wavelength ratio, DWR) they can provide the size of the scattering particles, together with liquid water content and ice water content.

#### 5.1.2.4 Disdrometer Errors and Uncertainties

Errors associated with disdrometers can be categorised into instrument errors, sampling errors and observation errors (Habib et al. 2010). The instrument errors are linked to the inherent design of the disdrometer and measurement assumptions, such the fall velocity of the particles in still air. Sampling error originates in the area/volume of the measurement and how representative: this ultimately depends upon the final application where some applications might need detailed information on (almost) individual particles, while other applications need bulk statistics on the particles. Observational errors may result from differential sampling of the DSDs over time, not least since small particles will take longer to fall than large ones.

### 5.1.3 Radars

The development of weather radars allows frequent spatial observations of precipitation, and when multiple radars are used in a network, over very large areas (Whiton et al. 1998a, b). Unlike rain gauge measurements, radars can provide frequent, 3-dimensional observations of precipitation to about 250 km away from the radar location (Zhang et al. 2011, 2016). This is, of course, of great importance to many users such as hydrologists and civil defence.

Ground-based scanning weather radars transmit pulses of electromagnetic radiation in a beam at set elevations. In between the transmitted pulses the radar measures any returned, backscattered radiation from particles (or objects) present within the outgoing beam. By measuring the time between the outgoing pulse and the return signal the distance to the particles can be calculated, along with the strength of the return. As the radar rotates and changes elevation a 3-dimensional map of the particles can be generated. The length of the outgoing pulse defines the range resolution, while the beam width defines the radial and vertical resolution: for convenience the radial measurements of the radar are usually plotted onto a x,y,z Cartesian coordinate system.

While the basic principle of a weather radar is simple, a few assumptions are made, such as that the returned power is the sum of the scattered power from particles within a uniformly filled volume, that these targets exhibit Rayleigh scattering and that the energy of the radar beam is Gaussian centred on beam's axis.

The measured reflectivity Z ($mm^6\ m^{-3}$) is proportional to the 6th moment of the diameter of the raindrop; therefore, the reflectivity increases greatly with an increase in the drop size. N(D) ($mm^{-1}\ m^{-3}$) is used to denote the raindrop size distribution (DSD) in a unit volume ($m^3$) with D denoting the diameter of the raindrop (mm). A power law formula (Atlas and Ulbrich 1977) approximates the terminal velocity $v_t(D)$ ($m\ s^{-1}$) of a raindrop of diameter D, where $v_t(D) = 3.80\ D^{0.67}$: the rain rate is proportional to the raindrop size. The DSD is key in radar rainfall estimation as it determines the Z-R relationships, where $Z = AR^b$. For the most accurate Z-R relationships the DSD within the volume of atmosphere being observed by the radar needs to be known. However, the DSD encountered where the radar samples the atmosphere is not necessarily the same as that at the surface: precipitation may be modified by various processes while falling through the air, such as growth, breakup, evaporation, coalescence, collision, etc. as well as changes in phase, typically (but not necessarily just) from ice to liquid. Even for field campaigns this is an almost impossible task, consequently approximations need to be made with the values of A and b being varied depending upon the perceived meteorological regimes, such as those shown in Table 5.1.

The frequency used by radars determines their sensitivity to the size of particles being observed. A radar operating at wavelengths between 1 and 10 cm sends out a beam of energy and due to Rayleigh scattering, precipitation-size particles (1–10 mm) will scatter some of this radiation back towards the radar, which can then be measured. This backscattered energy is dependent upon the wavelength of the radar pulse and the size of the precipitation particle. As shown in Fig. 5.9 (see also Wallace and Hobbs 2006), the radar bands sit at the part of the spectrum where hydrometeors are at the interface between the Mie and Rayleigh scattering. Most weather radars use longer wavelengths, such as 10 cm (S-band) or 5 cm (C-band) since these can operate over longer distances with less (clear air) atmospheric attenuation yet still be responsive to precipitation-size particles. Shorter wavelengths, e.g. 3 cm or X-band, are more sensitive to smaller particles, but suffer from increased atmospheric attenuation, thus limiting their effective range. Still shorter wavelengths (e.g., 1 cm/Ka-band, or 3 mm/W-band) are used, but tend to be more for research or retrieving cloud properties rather than for operational purposes. A list of most common radars used in meteorology is provided in Table 5.2.

Development of weather radars have provided additional capabilities including Doppler, polarization and dual polarization measurements. By measuring and comparing the frequency of the backscattered radiation compared with the original emitted radiation, the Doppler "shift" (difference) can be calculated. This is a measure the speed of the atmospheric particles relative to the radar location. While the main Doppler component is the movement of the weather towards/away from the radar, small scale features can be identified and analysed such as local circulation such as tornadic activity or downbursts. Polarization and dual polarization relate to the orientation of the emitted signal and the measured return. Small raindrops are spherical so both components of the polarization will be the same, however, large drops become oblate and therefore will return a greater difference in the components: polarization can therefore be used to measure the size of the rain drops, as well as reduce errors in the return signal and from non-precipitation particles. When considering precipitation measurements from weather radars a holistic view of the atmosphere is needed. The hydrometeors within the volume of atmosphere being viewed may be a combination of liquid, solid particles. These may fall to the surface

**Table 5.1** Examples of Z-R relationships used for the US WSR-88D radar network

| | Relationship | Meteorological situation |
|---|---|---|
| Marshall-Palmer | $Z = 200\ R^{1.6}$ | General stratiform precipitation |
| East: cool stratiform | $Z = 130\ R^{2.0}$ | Winter stratiform precipitation: east of continental divide |
| West: cool stratiform | $Z = 75\ R^{2.0}$ | Winter stratiform precipitation: west of continental divide |
| WSR-88D convective | $Z = 300\ R^{1.4}$ | Summer deep convection |

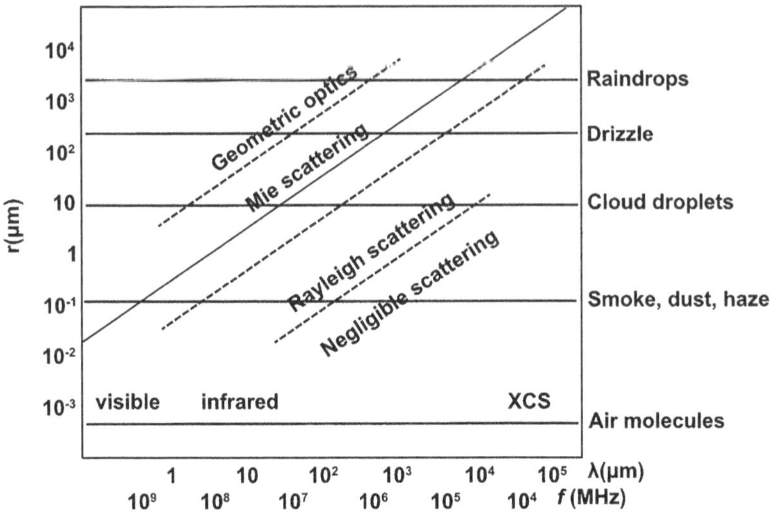


**Fig. 5.9** Radiation scattering regimes as a function of particle size and wavelength/frequency. (Public domain)

**Table 5.2** Radar bands, wavelengths and frequencies use for weather radars

| Band name | Wavelength | Frequency band |
|---|---|---|
| S-band | 10 cm | 2700–2900 MHz |
| C-band | 5 cm | 5250–5725 MHz (mainly 5600–5650 MHz) |
| X-band | 3 cm | 9300–9500 MHz |
| Ka-band | 1 cm | 35 GHz |
| W-band | 3 mm | 94 GHz |

Source: ITU-WMO met radars

as rainfall, snowfall, or a mix of these two depending on several other meteorological factors. It should be noted that a radar with single polarization cannot easily differentiate between the backscatter from liquid and frozen particles. Furthermore, providing accurate measurements of frozen particles is difficult due to the complex nature of the interactions of the radar signal and the snowflakes and ice crystals.

Above the freezing level in stratiform clouds the radar observes ice particles. These produce a much smaller effective backscatter than raindrops of comparable size. As these ice particles fall through freezing level, they begin to melt from the tips of the crystals inwards, generating water coated ice particles. This produces significantly larger reflectivity resulting in the bright band, as can be clearly seen in Fig. 5.8. For a vertically pointing radar, the bright band can usually be determined quite easily, although when this is close to the surface, making accurate measurements of “real” precipitation becomes difficult. This is more complex for a scanning radar where the elevated beam intercepts the bright band over a broad area, particularly where the beam is broader than the thickness of the melting layer.

#### 5.1.3.1 Errors and Uncertainties

While radars provide spatial measurements of precipitation, these measurements are less direct than those provided by gauges since they rely upon assumptions to convert the backscatter of the radiation from the hydrometeors (liquid or frozen) to a rain equivalent. They also suffer from several artifacts, such as beam blockage, false-returns, range effects and imperfect backscatter to rainfall relationships. As the radar range increases so the radar beam gets broadens and thus the volume sampled will increase. The practical implication is that at far range the measured return signal will represent a greater variety of particles than at shorter ranges. Features or objects at or close to the surface can block the propagation of the radar beam or generate a false return. Consequently, the radar beam is usually elevated (e.g. 0.5 degrees for the WSR-88D network) which, together with the curvature of the Earth, results in the height of the beam above the Earth's surface increasing with range, meaning that the precipitation measured is no longer at or near the surface (Mimikou and Baltas 1996). This leads to several possible errors, such as not identifying low level precipitation, incorrect measures of precipitation due to low-level enhancement or evaporation, or incorrect detection of the phase (liquid/ice) of the precipitation which is crucial when the brightband is in the lowest beam elevation.

Many of the reflectivity and the sampling geometries of radars assume that the radar beam travels through the standard atmosphere. However, the actual propagation of the radar beam depends on the density of the atmosphere which is affected by the temperature and moisture. Anomalous propagation, including super-refraction and ducting, results from strong low-level temperature and/or moisture inversion, and may result in the radar returns being from the surface rather than within the atmosphere. These irregular patterns of anomalous reflectivity, or ground clutter, may also be generated by the radar side lobes resulting in more subtle false returns. Typically, all operational weather radars used for quantitative precipitation measurements require calibration with gauge data. These gauges are often dedicated for this use and provide data in real time to constantly update the calibration of the Z-R relationships.

### 5.1.4 Other Measures

Several other surface-based precipitation measurements are available, although these are not used routinely. The attenuation from microwave communication links have been studied and shown to have a good correlation with surface radar, rain gauges and satellite data sets (Overeem et al. 2016). The exploitation of such networks is seen as augmenting precipitation measurements in regions where current surface data is poor. Other measurements include acoustic gauges: Ma et al. (2005) deployed such gauges over the ocean with good success helping to address the paucity of precipitation measurements over these regions.

Other studies have looked at the used of third-party measurements as surrogate measures of precipitation. For example, the (positive) changes in soil moisture can be an indicator of where and how much precipitation has fallen, although the characteristics of the soil and its structure needs to be carefully incorporated into any retrieval technique. Likewise, the use of stream or river flow has been studied, however, careful consideration is needed since many water courses are managed to a certain degree. On a larger scale, changes in the water storage as seen by missions like Surface Water and Ocean Topography (SWOT) or Soil Moisture Active Passive (SMAP) can be related to the precipitation over large regions and catchments.

## 5.2 Satellites

The measurement of precipitation by conventional means (i.e., gauge, radar, etc) is primarily constrained by the limited extent of the observations. Most gauge and radar measurements are limited to the more populated regions of the Earth, with few or no measurements over inhospitable regions or over the oceans. Satellite-based instruments can provide observations on a truly global scale and, in particular, over data-sparse regions such as the world's oceans, allowing observations of the whole Earth System and interconnections between its different elements.

The range of sensors used for the observation and retrieval of precipitation cover the region of the spectrum from the visible and infrared wavelengths through to the microwave region. Radiation in this part of the electromagnetic spectrum is governed by the scattering and/or absorption by components of the atmosphere, including clouds and precipitation (see Fig. 5.10). At the visible end of this part of the

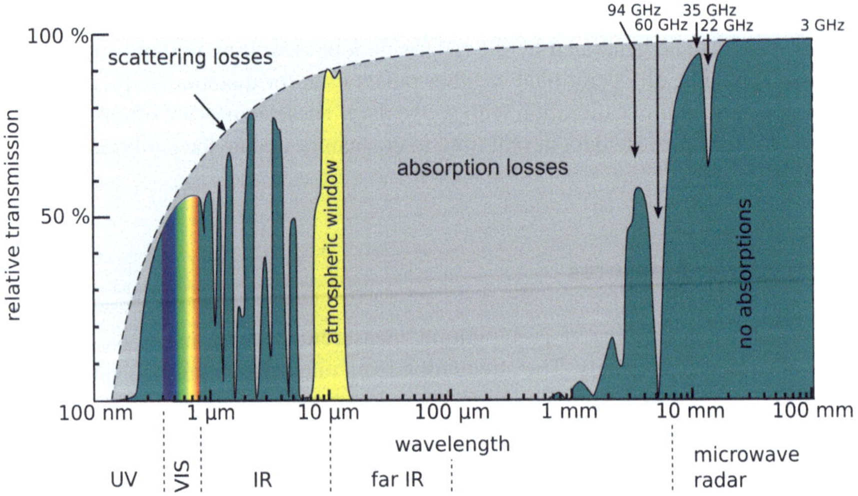


**Fig. 5.10** Relative transmission of radiation through the atmosphere across the electromagnetic spectrum from the ultraviolet through to the microwave regions. (Public domain)

spectrum (about 0.4 μm) scattering of radiation is a major factor since the wavelengths are similar in size to the cloud particles: the presence/absence of clouds is most obvious in this region. Within the near to mid-infrared region there are several absorption lines primarily associated with water vapour that may obscure the background surface. The thermal infrared (at about 10 μm) lies at an atmospheric window channel allowing the relatively free passage of radiation, although water vapour and clouds may limit this. Between the thermal infrared and the microwave lies a broad region where atmospheric absorption is caused primarily by water vapour. At millimetre wavelengths the atmosphere becomes more transparent, particularly so at the longer wavelengths (100 mm, or about 3 GHz).

Sensors have been designed to exploit these wavelengths/frequencies to provide the bulk of the observations that can be used for precipitation estimation. These sensors rely upon the reflected or emitted radiation: sensors using the visible part of the spectrum are largely reliant upon the solar radiation to illuminate the Earth and its' atmosphere. From the mid-infrared through to the microwave, emitted radiation is measured, although in much smaller amounts than that of the visible radiation. The implications are that the resolutions from the visible sensors tend to finer than those of infrared, and much finer than those from (passive) microwave sensors. Nevertheless, since the interactions between the radiation and the atmosphere differs across these wavelengths, each wavelength has its place in precipitation estimation. Table 5.3 provides an overview of the characteristics of these bands.

### 5.2.1 Satellite Orbits

Satellites provide the platform on which a sensor, or number of sensors are placed. These sensors may provide complementary observations for particular geophysical features, or different observations to maximise the utility of the satellite. Satellites are placed into orbits to provide the optimum conditions for the sensors to provide their observations. For precipitation-capable sensors, the satellites are usually placed into a circular orbit, such that their perigee (nearest point to the Earth) and the apogee (furthest point form the Earth) are the same. However, most orbits are not truly circular due to changes in the Earth's surface topography, magnetic fields and atmospheric drag. Each satellite sits in an orbital plane, with the Earth rotating beneath: it should be noted that once a satellite is in an orbital plane it is extremely difficult to change it, although the satellite may increase/decrease its altitude and speed within that plane.

The altitude of the satellite's orbits above the surface is determined by the user requirements and capability of the instrumentation, but generally meteorological satellite orbits range from about 350–800 km for low Earth orbits to 38,000 km for Geostationary orbits. The lowest usable altitude is governed by the atmosphere and the drag it exerts, which limits the lifetime of the satellite. The atmosphere expands and contracts due to variations in temperature because of diurnal heating/cooling, or solar activity, affecting the drag, and hence the lifetime of missions. Satellites placed in lower Earth orbits (<450 km) may experience increased atmospheric drag, thus

**Table 5.3** Characteristics of present-day LEO and GEO satellites and their observational capabilities

| | Orbit | LEO[a] | GEO[b] |
|---|---|---|---|
| Band | Characteristics | 400–800 km altitude<br>c. 14–16 orbits/day<br>Polar orbiting Sun-synchronous (up to 2 overpasses/day), or low inclination precessing orbits | c. 36,000 km altitude orbits<br>Geosynchronous orbit—satellite remains stationary relative to sub-satellite point |
| VIS/IR | Cloud top properties<br>Reflection, emission (T)<br>Texture & particle sizes | Multi-spectral<br>Orbital swaths<br><1 m to 1 km | Multi-spectral<br>Frequent/regular<br>Samples over sectors or full disk coverage<br><1–4 km footprints at sub-satellite point |
| PMW | Hydrometeor column<br>Liquid, ice and water vapour | Multi-channel<br>Orbital swath<br>c. 5–70 km footprints | *Not possible at present*<br>*Several feasibility studies* |
| AMW (Radar) | Vertical profiles of Hydrometeors<br>Backscatter<br>Liquid, ice | Single/dual frequency (13.6/35/94 GHz)<br>Single beam (CloudSat with 1.4/5.4 km footprints) or narrow swath (DPR 245 km swath and 5.4 km × 5.4 km footprints) | *Not possible at present*<br>*Several feasibility studies* |

[a] LEO: Low Earth Orbiting VIS/IR observations provided by a large group of sensors ranging from land surface monitoring missions (e.g., Landsat) through to meteorological missions (e.g., AVHRR) with resolutions typically matched to user requirements. PMW channels used for the retrieval of geophysical and meteorological parameters, including precipitation

[b] GEO: Geostationary constellation of satellites provides global coverage with frequent/regular observations, disseminated at a nominal 3 hourly interval, but with sensors typically providing operational 10–15 min. Data collection, and rapid scans <1 min

reducing their lifetime to a few years, but satellites placed in higher orbits, for example, around 800 km, may remain in orbit for centuries. The altitude of the orbit also governs the orbital period, and how often the satellite retraces its ground track (the repeat cycle).

Low Earth Orbiting (LEO) satellites typically orbit the Earth at an altitude of between 400 and 800 km, completing about 14–16 orbits per day (Fig. 5.11). Most of these satellites operate in a Sun-synchronous polar orbit (one that passes close to the Poles) such that the local time of the overpass for each orbit is the same, allowing similar solar illumination conditions. A small number of satellites operate in low-inclination, non-Sun-synchronous orbits to provide observations at different times of day as the orbits precesses over periods of weeks or months. The sensors onboard LEO satellites generally provide a broad swath of data providing global coverage twice each day. Data from these missions support a broad range of environmental monitoring applications related to climate and weather (e.g., analysis, forecasting and prediction), sea surface characteristics (temperature and ocean

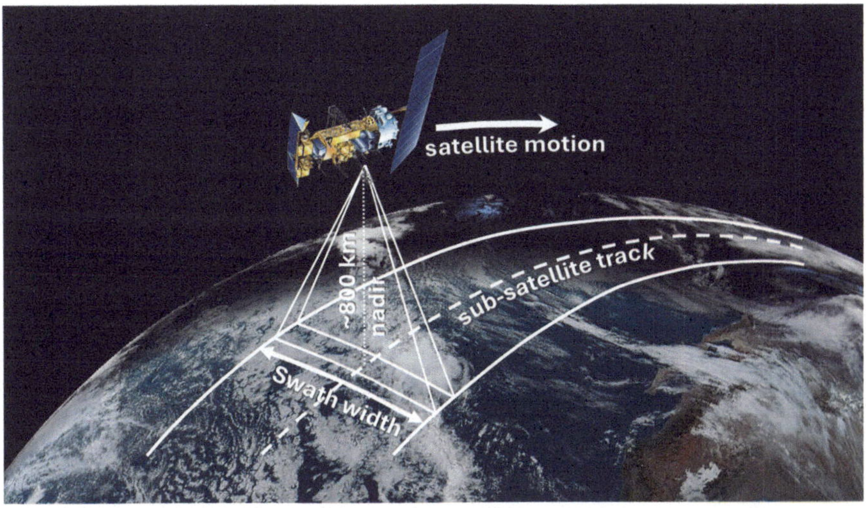


**Fig. 5.11** Schematic of a low Earth orbiting satellite scan

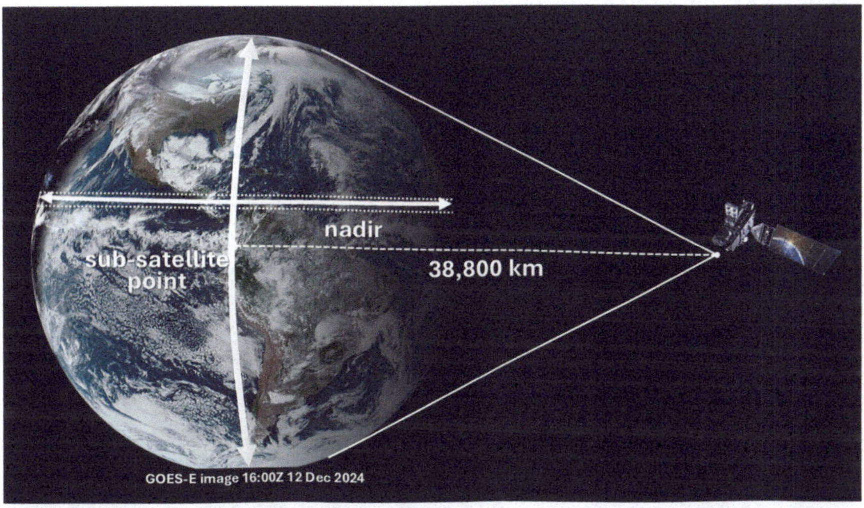


**Fig. 5.12** Schematic of geostationary satellites scan

dynamics) as well as land surface applications (e.g., monitoring volcanic eruptions, forest fire detection and global vegetation analysis).

Geostationary or Geo-Synchronous (GEO) satellites occupy orbits which are synchronised with the Earth's rotation so that they appear stationary over a fixed location on the Equator. This, combined with their much higher orbit of about 35,800 km, allows their sensors to gather frequent and regular observations over a full Earth disc, or about 1/3 of the Earth's surface (Fig. 5.12). They also experience the same day/night cycle as the Earth so provide observations across the full diurnal cycle. Since no single GEO satellite can observe the whole globe a minimum of 5 satellites are needed to contribute to a network of satellites that provide (near) global coverage.

The main meteorological missions and their orbits are illustrated in Fig. 5.13. Typically, no one system can adequately deliver all the necessary observations, consequently these missions provide complementary observations to aid the monitoring and forecasting of meteorological conditions.

While LEO and GEO missions form the bulk of meteorological satellites, two additional types of similar orbits are worth mentioning. Both these orbits are highly elliptical, with a perigee of about 1000 km and apogee at about 40,000 km: satellites in orbit must obey the law of the conservation of energy, potential energy (due to the altitude) is converted to kinetic energy (due to velocity) during the satellite orbit. Thus, at the perigee the satellite is moving quickly, while at the apogee the satellite is moving slowly. This allows the satellite to assume an almost geostationary position at the apogee, providing an ideal platform for gathering observations, albeit for a limited time period during its orbit.

### 5.2.1.1 Low Earth Orbit Missions

LEO satellites provide the mainstay of observations for precipitation studies, carrying visible (VIS), Infrared (IR), passive microwave (PMW) and active microwave (AMW) sensors (Kidd and Levizzani 2011). The operational meteorological series have included the National Oceanographic and Atmospheric Administration (NOAA) Television Infrared Observation Satellite (TIROS) series of satellites, the Advanced TIROS-N (ATN) and the current Joint Polar Satellite System (JPSS).

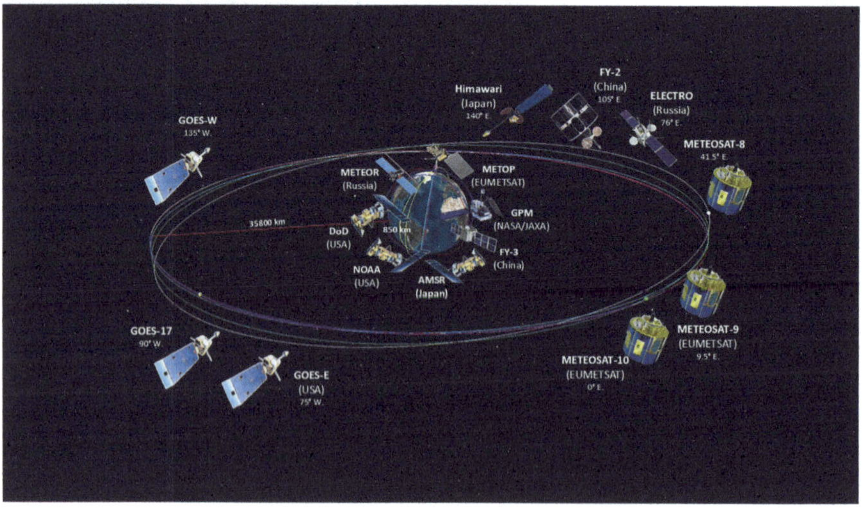


**Fig. 5.13** Satellites forming the meteorological observing system showing the operational low Earth orbiting missions and the geostationary missions

Complementary observations are provided by the European Meteorological Satellite Organisation (EUMETSAT) Meteorological Operational (MetOp) satellites and the China Meteorological Agency (CMA) Feng-Yun program satellites. Additional satellites that provide precipitation-capable observations include the Defense Meteorological Satellite Program (DMSP) satellites as well as more specialist mission such at the Tropical Rainfall Measuring Mission (TRMM), the Global Precipitation Measurement (GPM) mission, the Megha-Tropiques satellite and a secession of Japanese missions.

The operational meteorological satellites carry a suite of sensors capable of providing detailed information on the Earth's atmosphere (as shown in Fig. 5.14). The NOAA and EUMETSAT satellites carry the Advanced Very High Resolution Radiometer (AVHRR) which provides visible and infrared imagery across 5 bands at resolutions from 1 km for visible imagery to 4 km for infrared imagery. Complementing these are observations from passive microwave sounding

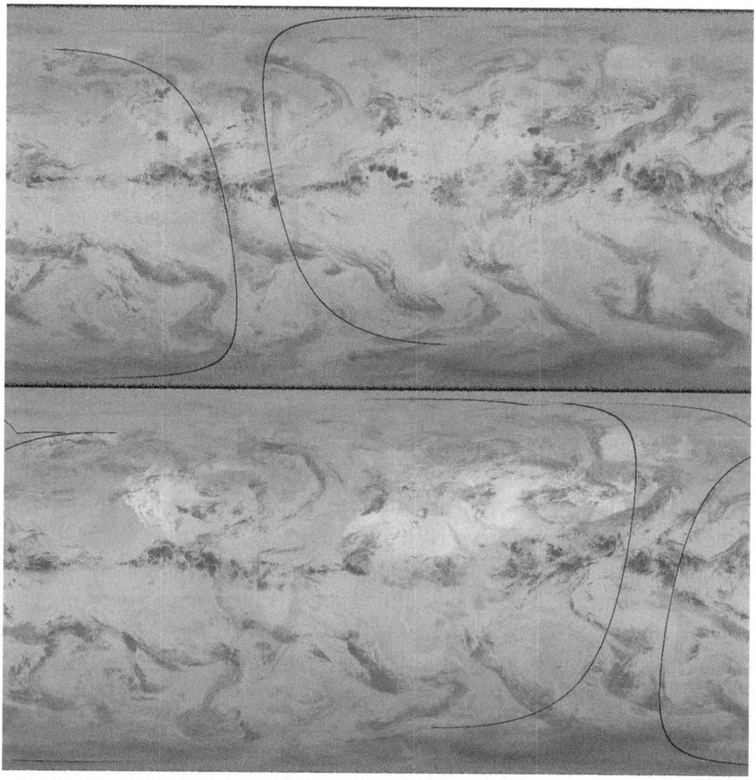

**Fig. 5.14** Composite of infrared imagery from all ascending (top) and descending (bottom) orbits for 1 day, 5 June 2018. from the EUMETSAT MetOp-B AVHRR sensor. Note the grey scale represents the "true" radiative scale where dark grey indicates low radiation (temperature) and light grey indicates higher radiation (temperature). The black lines indicate the extent of data from a single orbit. (Data from NOAA CLASS)

instruments designed to provide information on the vertical distribution of atmospheric parameters, in particular water vapour and temperature, albeit with poorer resolutions of about 16 km at nadir. Passive microwave imaging sensors (or "imagers") such as those carried by the DMSP, and GPM satellites provide observations primarily within the atmospheric window frequencies with little or no atmospheric attenuation, with resolutions up to 3 km × 5 km.

Since these satellites provide observations required by operational agencies, data acquisition and distribution are a key feature of the mission architecture. For example, NOAA has Command and Data Acquisition (CDA) stations located near Fairbanks, Alaska and Wallops Island, Virginia, allowing data collected by their satellites to be collected and distributed such that data for any region of the Earth are no more than 6 h old. The NASA GPM mission uses a satellite data relay system that allows the core GPM satellite to send its data to the processing system on the ground within minutes of observation.

#### 5.2.1.2 Geostationary Missions

Operational geostationary satellites include the US Geostationary Operational Environmental Satellites (GOES) series, the European Meteosat series, and the Japanese Geostationary Meteorological Satellite (GMS) and Multifunctional Transport Satellite (MT-SAT) series, together with the Chinese Feng-Yun-2 (FY-2) series and the Indian National Satellite (INSAT) series.

The sensors on board these GEO satellites vary with the satellite operators but share a number of common attributes, although the altitude and requirements for sufficiently fine-resolution measurements restrict these sensors to VIS/IR observations. Sensors provide observations across wavelengths ranging from the visible through the infrared region of the spectrum with resolutions typically from 1 km (visible) to 4 km (infrared) at the sub-satellite point. However, the resolution becomes increasingly poor towards the edge of the disc due to the curvature of the Earth with no coverage of the polar regions. The images are normally acquired every 30 min, although sensors such as the Spinning Enhanced Visible and Infrared Imager (SEVIRI) on Meteosat Second Generation (MSG; Schmetz et al. 2002) have a nominal acquisition period of 15 min, with some sensors providing rapid-scan capabilities covering smaller areas as frequent as every 30 s (e.g., the GOES imager; Menzel and Purdom 1994). Although it may seem that GEO satellites have reached a consolidated sensor payload, new satellites are advancing our observing capabilities such as the EUMETSAT Meteosat Third Generation (MTG, https://www.eumetsat.int/meteosat-third-generation) now in orbit and the planned NOAA's Geostationary Extended Observations (GeoXO, Lindsey et al. 2025). Several mission concepts have been proposed for geostationary PMW sensors (e.g., Bizzarri et al. 2007; Lambrigtsen et al. 2022), however these would operate at high frequencies and only provide very limited global coverage at any one time.

#### 5.2.1.3 Specialist Orbit Missions

The use of specialist orbits, such as the Molynia or Tundra orbits, have been very limited to date. Of potential relevance to satellite precipitation measurements are the Russian Arctica-M N-series of satellites. These satellites operate in a Molynia orbit with a perigee of 1000 km and an apogee of 40,000 km with an inclination of 63.4° and an orbital period of about 12 h. Through using two satellites near continuous pseudo geostationary observations can be made the where the dwell time is the longest, in this case, over the Arctic. Each satellite is equipped with a 10-channel visible and infrared instrument with resolutions of about 1 km in the visible and 4 km in the infrared (see Asmus et al. 2021 for more details).

Another mission of interest is the proposed Canadian Arctic Observing Mission. Like the Arctica satellites, this will exploit a highly elliptical Tundra orbit to provide observations over the Arctic region. The Tundra orbit differs slightly from the Molynia orbit by providing a greater dwell time over the region of interest. One of the proposed mission architectures includes the visible/infrared Advanced Baseline Imager (ABI) that is currently employed on the US GEO satellites.

#### 5.2.1.4 Constellations

Where the data from multiple sensors contribute to a common geophysical parameter, they are often organised into a constellation to coordinate their observations and help improve temporal sampling or improve their retrievals through multispectral observations.

The current meteorological observing system, using LEO satellites in polar orbits, aims to provide data to the operational meteorological agencies. This constellation is built around satellites operating in three orbital planes, "early morning", "morning" and "early afternoon" (see Table 5.4) with satellites being provided by China, Europe and the US. Since the orbits provide data over the ascending and descending overpasses the constellation can provide global data every 4 h. Each satellite carries a VIS/IR sensor and PMW sounders: the vis/IR sensors have evolved over the last 50 years and provide a climate-scale data record, while the PMW

**Table 5.4** Current LEO polar-orbiting operational meteorological satellites

| Orbit | Satellite | Equatorial Crossing Time | Operating Agency | Instrumentation |
|---|---|---|---|---|
| Early morning | FY-3E | 0440 (desc) | CMA (CHINA) | VIS/IR/MWS |
| Morning | NOAA-19 | 0822 (desc) | NOAA (USA) | VIS/IR/MWS |
| | MetOp-B | 0931 (desc) | EUMETSAT (Europe) | VIS/IR/MWS |
| | MetOp-C | 0931 (desc) | EUMETSAT (Europe) | VIS/IR/MWS |
| | FY-3F | 1000 (desc) | CMA (China) | VIS/IR/MWS/MWI |
| Afternoon | NOAA-21 | 1325 (asc) | NOAA (USA) | VIS/IR/MWS |
| | NOAA-20 | 1325 (asc) | NOAA (USA) | VIS/IR/MWS |
| | SNPP | 1325 (asc) | NOAA (USA) | VIS/IR/MWS |

sounders are invaluable for the data they gather for input into numerical weather prediction (NWP) models.

Of note is while many of these satellites have a nominal equatorial time, most of the LEO satellite orbits drift during their lifetime, as shown in Fig. 5.15. For the early satellites such as NOAA-7 ("N7") with relatively short lifetimes, the amount of drift was small. However, for later missions such as NOAA-19 ("N19") its crossing time has change from about 1400 to 2300 UTC over its 20+ years lifetime. More recent satellites, such as the NOAA SNPP and NOAA-20 ("npp" and "N20") and the EUMETSAT MetOp series ("M1", "M2", and "M3") have station keeping capabilities ensuring that their overpass times are consistent across their data record.

Another element of the meteorological observing system are the geostationary satellites. As noted above, each of their satellite sensors can provide images over a full Earth disc. However, to provide global coverage a minimum of 5 operational satellites are needed (although these do not adequately cover the polar regions). The organisation of this GEO-constellation extends back to 1979, as shown in Fig. 5.16, and has been continuously updated/renewed up to the present. The current suite of satellites is shown in Table 5.5, with satellites operated by the United States, Europe, India, China and Japan.

The ability to combine the observations from these difference satellites is illustrated in Fig. 5.17 showing the 5 individual satellite images and the combined image. Such data is available with at least 1 h temporal sampling. The NOAA Climate Precipitation Center (CPC) routinely generate a 4 km 30-min global IR composite (Janowiak et al. 2001) alongside the GridSat product (Knapp and Wilkins

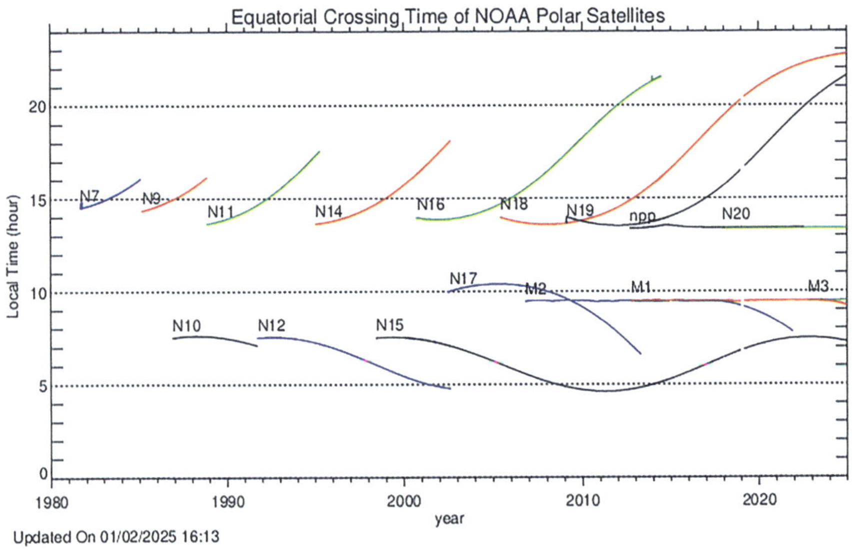


**Fig. 5.15** Equatorial Crossing Time (ECT) for NOAA Polar Satellites. (Source: NOAA-NESDIS Center for Satellite Applications and Research)

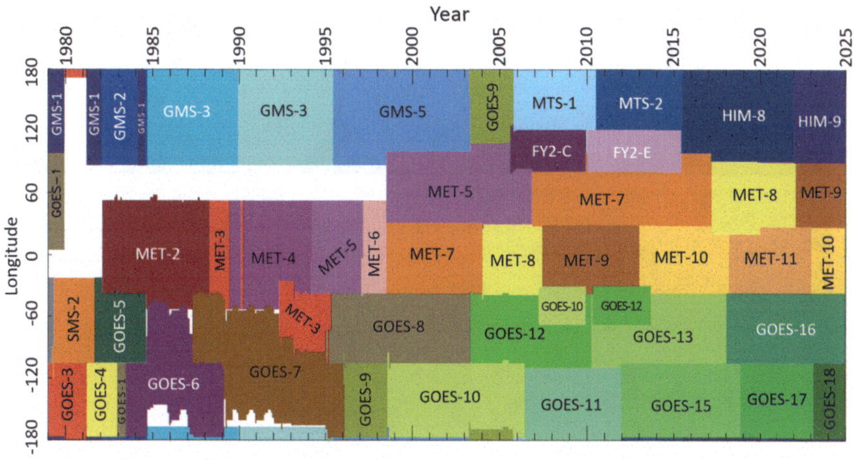


**Fig. 5.16** Timeline of geostationary satellites from 1979 to 2025 (Available at https://www.ncei.noaa.gov/products/satellite/international-satellite-cloud-climatology-b1 with additional information from WMO OSCAR database)

**Table 5.5** Current geostationary operational meteorological satellites. Italic text indicates those satellites that are in standby orbits to act as backups in case of failure

| Position | Satellite | Agency | Launched | Comment |
|---|---|---|---|---|
| 137.0°W | GOES-18 | NOAA (USA) | 01/03/2022 | Operational GOES-W (from 04/01/2023) |
| *108.0°W* | *GOES-14* | *NOAA (USA)* | *27/06/2009* | *Standby* |
| *104.7°W* | *GOES-17* | *NOAA (USA)* | *01/03/2018* | *Standby* |
| 75.2°W | GOES-16 | NOAA (USA) | 19/11/2016 | Operational GOES-E (from 18/12/2017) |
| 0.0°W | Meteosat-10 | EUMETSAT (Europe) | 05/07/2012 | Operational Europe/ Africa coverage |
| *9.5°E* | *Meteosat-11* | *EUMETSAT (Europe)* | *15/07/2015* | *Standby* |
| 45.5°E | Meteosat-9 | EUMETSAT (Europe) | 21/12/2005 | Indian Ocean Data Coverage (IODC) |
| 82.0°E | INSAT-3DS | IMD (India) | 17/02/2024 | India |
| 105.0°E | FY-4B | CMA (China) | 25/07/2013 | Asia |
| 140.7°E | Himawari | JMA (Japan) | 02/11/2016 | Asia/Japan/Australia |

2018). A new initiative has been started to reprocess all available data back to 1979 to ensure a consistent data product from the GEO constellation sensors.

Of particular interest here is the precipitation virtual constellation: since global precipitation cannot be adequately measured by a single satellite, numerous precipitation-capable satellites are needed. The current Global Precipitation Measurement (GPM) mission incorporates observations from 10 to 12 precipitation-capable satellite sensors provided by international partner organisations (Skofronick-Jackson et al. 2017). To fully exploit available observations, these sensors are a mix

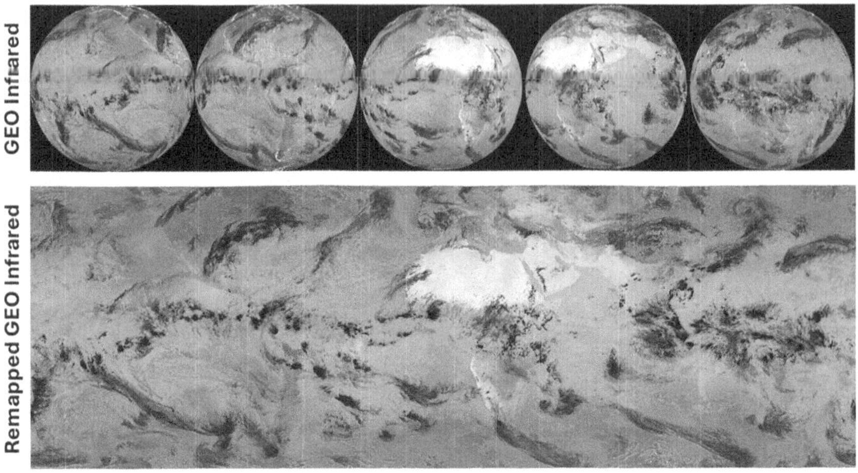


**Fig. 5.17** Infrared images from 5 GEO satellites (top) together with their remapped image (bottom) for 1200 UTC on 19 April 2025

of both PMW imagers and sounders: while the latter are primarily designed for retrieving temperature and humidity (Mo 1995) these instruments have proved invaluable for improving the temporal sampling of the constellation (Kidd et al. 2021).

The GPM mission consists of a Core Observatory (GPM-CO) and the constellation of satellites provided by international partners. The GPM-CO is a joint NASA and JAXA mission and was launched on 28 February 2014 (Hou et al. 2014) into a 65° inclination, 407 km altitude orbit, providing coverage from 68°N to 68°S. The GPM-CO carries two instruments: the Dual-frequency Precipitation Radar (DPR; Meneghini et al. 2021) comprises of Ku and Ka band radars (13.6, 35.5 GHz, respectively) to provide the 3-D structure of precipitation, and the GPM Microwave Imager (GMI) provides multi-frequency, wide-swath observations across 13 channels at 7 PMW frequencies. The combination of the DPR and GMI provides a reference standard for the calibration of observations from the partner sensors. The constellation includes the operational NOAA satellites (with the MHS and ATMS sensors), EUMETSAT (MHS) and the U.S. Department of Defense (DoD) SSMIS sensors. In addition, observations from JAXA's AMSR-2 and the joint Centre National d'Etudes Spatiales (CNES) and Indian Space Research Organisation (ISRO) Megha-Tropiques mission Sondeur Atmospherique du Profil d'Humidité Intertropicale par Radiométrie (SAPHIR) sensor are included (Table 5.6).

The GPM mission followed on from the very successful Tropical Rainfall Measuring Mission (TRMM; Kummerow et al. 1998, 2000) which carried the first space-borne Precipitation Radar (PR) together with the TRMM multi-channel Microwave Imager (TMI). Operating between November 1997 to April 2015 it provided observations between 35°N and 35°S and, like GPM, was in a non-sun synchronous allowing observations to be cross calibrated with other PMW radiometers.

**Table 5.6** Details of the GPM constellation satellites and sensors. Italic text indicates former satellites in the precipitation constellation

| Satellite | Agency | Sensor number | | Channels (GHz) | Retrieval resolution (km × km) |
|---|---|---|---|---|---|
| *AMW instruments* | | | | | |
| GPM | NASA/JAXA | DPR | | 13.6/35.5 | 5.4 × 5.4 |
| *TRMM* | *NASA/JAXA* | *PR* | ×*1* | *13.6* | *4.3 × 4.3* |
| *PMW imagers* | | | | | |
| GPM | NASA/JAXA | GMI | ×1 | 10.7–183.31 | 10.9 × 18.1 |
| DMSP F16,17,18,19 | US DoD | SSMIS | | 19.35–183.31 | 45 × 74 |
| GCOM-W1 | JAXA | AMSR2 | ×3 | 6.7–89.0 | 14 × 22 |
| *TRMM* | *NASA/JAXA* | *TMI* | ×*1* | *10.7–89.0* | *20.9 × 34.6* |
| *PMW sounders* | | | | | |
| NOAA-18,19, MetOp-A, B, C | NOAA/EUMETSAT | MHS | ×3 | 89.0–183.31 | 17.12 × 21.64 |
| NPP, NOAA-20,21 | NOAA | ATMS | ×2 | 23.0–183.31 | 16.51 × 16.22 |
| *Megha Tropiques* | *ISRO/CNES* | *SAPHIR* | ×*1* | *183.31 (× 6)* | *7.34 × 7.27* |

**Table 5.7** Details of the A-train satellites and sensors: acronyms can be found in the Appendices

| Satellite | Instruments | Observations | Products |
|---|---|---|---|
| Aqua | MODIS, AIRS/AMSU-A/HSB, AMS-E, CERES | VIS, IR, PMW (sounder and imager) | Land and ocean parameters, size information, temperature and moisture profiles |
| CloudSat | CPR | Cloud profiling radar | Cloud, liquid and ice profiles |
| CALIPSO | CALIOP, IIR | Lidar and IR | Cloud and aerosol profiles |
| PARASOL | POLDER | VIS/near IR polarimeter | Cirrus cloud optical properties |
| Aura | HIRDLS, MLS, TES, OMI | Limb sounding, IR and UV spectrometry | Trace gases, aerosol, cloud ice, ozone |

A full table can be found in Stephens et al. (2002)

While satellites within the GPM constellation operate independently other satellite constellations, or satellite convoys, operate with carefully choreographed orbits to ensure that the observations their sensors provide nearly coincident measurements. These convoys gain significant benefits from their multi-sensor observations. An example of such a convoy is the "Afternoon constellation" or A-train which was designed to bring together several different satellites and their sensors to provide complementary observations to evaluate cloud products and to develop new information about clouds and precipitation and their impact on the Earth's energy and radiation budget (Stephens et al. 2002). The A-train centres around the close formation-flying of five satellites, CloudSat, CALIPSO, Aqua, Aura and PARASOL (see Table 5.7). The A-train operated in a Sun-synchronous 705 km Polar orbit. The satellites orbits were actively maintained and routine manoeuvres coordinated to ensure no long-term changes to ensure consistency of the observations and their derived products. Time separations between the different satellites varied from about 15 s to just over 200 s, ensuring near co-incident observations. More information and results of the A-train and the CloudSat missions may be found in Stephens et al. (2002, 2008, 2018).

More recently the technological development of cubesats has led to several pioneer missions, such as the NASA/MIT Time-Resolved Observations of Precipitation structure and storm Intensity with a Constellation of Smallsats (TROPICS) mission (Blackwell et al. 2024). TROPICS was designed around six identical 3U satellites, each with a small, high frequency cross-track radiometer operating between 91.655 and 204.8 GHz. The satellites were designed to be placed in a low inclination orbit to retrieval profiles of temperature and humidity across the Tropics, particularly of tropical storms. An additional prototype was also built and became the initial satellite to be placed into a polar orbit in August 2021. Although two of the main six satellites were lost at launch, the remaining four were launched in May 2023. These sensors formed the basis of the first commercial passive microwave sensors managed by tomorrow.io.

## 5.2.2 Visible and Infrared Observations

Visible/Infrared sensors are carried on both GEO and LEO platforms and provide good resolution imagery at about 0.5–1 km for visible and 2–4 km for infrared. Several channels sense radiation in the visible part of the spectrum which may be re-combined to produce false-colour images. These channels primarily measure the reflected solar radiation, and thus generally provide only day-time imagery. However, more recent sensors are more sensitive to low light conditions and can operate at nighttime to sense small amounts of emitted light from man-made sources, such as urban areas, as well as reflected light under moonlit conditions. Visible images may be used to identify clouds with multi-spectral visible observations used for textural information, useful in classifying cloud types and microphysical structure. The thermal infrared channels measure emitted radiation from the Earth's surface and clouds (where present) and therefore can provide data both night and day. Amongst other things, they measure cloud top temperatures which forms the basis of many precipitation estimation techniques. Between the visible and thermal infrared lies the near and mid-infrared including wavelengths with atmospheric absorption lines due to water vapour. These channels are useful for determining the moisture content and profiles of the atmosphere. However, with both visible and infrared data the relationship between the cloud top characteristics and precipitation at the surface is generally poor.

### 5.2.2.1 Sensors

Many instruments have exploited the visible-infrared wavelengths since the beginning of satellite observations due to the technology available at the time. The first

Television and Infra-Red Observation Satellite (TIROS) was launched on 1 April 1960 with a Vidicon Camera System (VCS) capable of capturing visible images with a resolution of 0.3 km (narrow angle sensor). The Very High Resolution Radiometer (VHRR), launched 13 October 1973, provided both visible and thermal infrared observations at 0.9 km resolution. The Advanced Very High Resolution Radiometer (AVHRR; launched 28 March 1983) provided digital data over 4 channels, with a series of enhancements to channel selection to provide the modern-day AVHRR/3 with 6 channels and resolutions of about 1 km (note that channels 3A/3B operate in day/night mode). Details of the AVHRR/3 are shown in Table 5.8, together with the typical uses of each channel.

An example of the imagery of the AVHRR/3 instrument is shown in Fig. 5.18. The two visible channels, similar to what we see with our eyes, clearly show the areas of clouds as well as the land/sea areas. Channel 3, which in this case is in daytime mode (1.58–1.64 μm) reveals some subtle differences in the clouds, particularly between liquid and ice phase clouds. The two thermal infrared channels, 4 and 5, show the amount of emitted radiation: in this figure the darker areas are regions of low emissions (i.e., low temperatures) with lighter areas being those with more emissions (i.e., higher temperatures). The two thermal infrared channels are termed "split window" channels since although they both measure thermal emissions, the slight differences in the observed radiation can be used to distinguish thin high cloud (e.g., cirrus) from deeper, thicker cloud (such as cumulonimbus). The combination of the channels into a false colour composition (RGB = 124) allows a pseudo-colour image to be generated.

GEO satellites have carried and currently carry several different VIS/IR sensors with different capabilities. While the baseline observations have focused on providing imagery within the visible, mid-infrared/water vapour and thermal infrared bands, the current suite of sensors on these missions provide observations over numerous wavelengths from the visible through to thermal infrared to provide greater fidelity in the identification and retrieval of atmospheric components, including clouds and aerosols, and for precipitation retrievals derived from these cloud observations. Crucially, all these modern sensors have regular sampling strategies allowing images to be routinely generated at least every 15 min. These observations

**Table 5.8** Characteristics of the AVHHR/3 instrument covering the visible and infrared wavelengths

| Channel number | Resolution at nadir (km) | Wavelength (μm) | Typical use |
|---|---|---|---|
| 1 | 1.09 | 0.58–0.68 | Daytime cloud and surface mapping |
| 2 | 1.09 | 0.725–1.00 | Land-water boundaries |
| 3A | 1.09 | 1.58–1.64 | Snow and ice detection |
| 3B | 1.09 | 3.55–3.93 | Night cloud mapping, sea surface temperature |
| 4 | 1.09 | 10.30–11.30 | Night cloud mapping, sea surface temperature |
| 5 | 1.09 | 11.50–12.50 | Sea surface temperature |

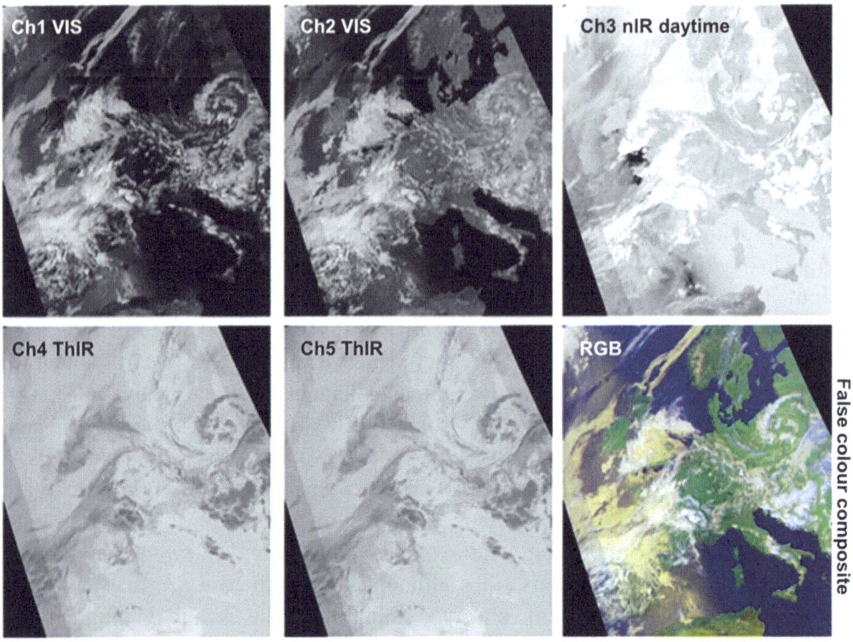


**Fig. 5.18** An example of the imagery from the NOAA-17 AVHRR/3 sensor for a case over Europe at on 20 July 2007, comprising of the two visible, near infrared (daylight) and two thermal infrared channels. A false colour composite is also included generated from a combination of these channels. (Processed from NOAA CLASS data)

are extremely important in providing consistent, long-term, frequent, regular, multi-channel low latency data for integration into multi-sensor, multi-satellite precipitation retrieval schemes. These modern sensors also have additional capabilities to provide rapid scan imaging which allow images every few seconds to be made: the latter being important for observing the development and progress of severe storms.

Two main types of sensors are currently utilised. The first is the Spinning Enhanced Visible Infra-Red Imager (SEVIRI) that is onboard the MSG series of satellites (see Fig. 5.19 and Table 5.9). The second type of sensor is the Advanced Baseline Imager (ABI) on the GEOS satellites, while the Japanese Himawari satellites carry an ABI-type sensor, the Advanced Himawari Imager (AHI) (see Table 5.10). The Chinese FY-4 GEO satellite series carry the Advanced Geostationary Radiation Imager (AGRI) with very similar channel selection as the ABI/AHI. All these sensors provide multi-spectral data at frequent and regular intervals to enable real time monitoring of weather systems.

#### 5.2.2.2 VIS/IR Algorithms for Precipitation

The availability of VIS/IR observations with good spatial and temporal resolution has allowed precipitation retrievals schemes to be developed. While the identification of clouds is relatively easy, providing a good estimate of the precipitation

**Fig. 5.19** Meteosat MSG full disc imagery for 24 November 2024. (Source: EUMETSAT)

falling from these clouds is more difficult. Although precipitation can be inferred from visible imagery since the brightness of a cloud is dependent upon its thickness, and thick clouds tend to be associated with surface rainfall, the relationship is rather poor. The texture of cloud tops derived from visible imagery can provide information on the types of cloud and hence, whether they are likely to be precipitating or not. Multi-spectral near infrared observations also provide information on the properties of cloud top particles, including their size and whether they are liquid or ice. However, VIS and near-IR imagery is only available during daylight. Nevertheless, the Clouds-Aerosols-Precipitation Satellite Analysis Tool (CAPSAT; Lensky and Rosenfeld 2008) using daytime and nighttime relationships for cloud scene classification, which can potentially be used for rain area delineation based on cloud microphysics. Since thermal IR imagery measures emitted radiation, it is available day and night. Taller, colder clouds tend to be associated with heavier rainfall, so by

**Table 5.9** Spectral bands of the Meteosat Second Generation SEVIRI instrument. The nominal resolution is 3 km except for the HRV channel which has a resolution 1 km

| Band (μm) | Central wavelength (μm) | Spectral interval (99% energy) (μm) | SNR or NEΔT @ specified input |
|---|---|---|---|
| HRV | N/A (broadband) | 0.6–0.9 | 4.3 @ 1% albedo |
| VIS 0.6 | 0.635 | 0.56–0.71 | 10.1 @ 1% albedo |
| VIS 0.8 | 0.81 | 0.74–0.88 | 7.28 @ 1% albedo |
| IR 1.6 | 1.64 | 1.50–1.78 | 3.0 @ 1% albedo |
| IR 3.9 | 3.92 | 3.48–4.36 | 0.35 K @ 300 K |
| WV 6.2 | 6.25 | 5.35–7.15 | 0.75 K @ 250 K |
| WV 7.3 | 7.35 | 6.85–7.85 | 0.75 K @ 250 K |
| IR 8.7 | 8.70 | 8.30–9.10 | 0.28 K @ 300 K |
| IR 9.7 | 9.66 | 9.38–9.94 | 1.50 K @ 255 K |
| IR 10.8 | 10.8 | 9.80–11.8 | 0.25 K @ 300 K |
| IR 12.0 | 12.0 | 11.0–13.0 | 0.37 K @ 300 K |
| IR 13.4 | 13.4 | 12.4–14.4 | 1.80 K @ 270 K |

**Table 5.10** Spectral bands of the Advanced Baseline Imager (ABI) as carried on the US GOES-16/17/18/19 satellites and the Advanced Himawari Image (AHI) as carried on the Himawari-8/9 satellites

| ABI central wavelength (μm) | ABI bandwidth (μm) | AHI central wavelength (μm) | AHI bandwidth (μm) | Resolution (s.s.p.) (km) |
|---|---|---|---|---|
| – | – | 0.455 | 0.05 | 1.0 |
| 0.47 | 0.04 | 0.510 | 0.02 | 1.0 |
| 0.64 | 0.1 | 0.645 | 0.03 | 0.5 |
| 0.86 | 0.04 | 0.860 | 0.02 | 1.0 |
| 1.38 | 0.03 | – | – | 2.0 |
| 1.61 | 0.06 | 1.610 | 0.02 | 1.0 / 2.0 |
| 2.26 | 0.05 | 2.260 | 0.02 | 2.0 |
| 3.90 | 0.20 | 3.850 | 0.22 | 2.0 |
| 6.15 | 0.90 | 6.250 | 0.37 | 2.0 |
| 7.00 | 0.40 | 6.950 | 0.12 | 2.0 |
| 7.40 | 0.20 | 7.350 | 0.17 | 2.0 |
| 8.50 | 0.40 | 8.600 | 0.12 | 2.0 |
| 9.70 | 0.20 | 9.630 | 0.18 | 2.0 |
| 10.30 | 0.50 | 10.450 | 0.30 | 2.0 |
| 11.20 | 0.80 | 11.200 | 0.20 | 2.0 |
| 12.30 | 1.00 | 11.350 | 0.30 | 2.0 |
| 13.30 | 0.60 | 13.300 | 0.20 | 2.0 |

observing cloud top temperatures, a measure of surface rainfall can be derived. However, although more direct than the visible, the cloud top temperature to surface rainfall is poor, and often changes through the life cycle of precipitation systems and between different climatological regions.

Despite the poor physical relationship, numerous techniques have been devised and used operationally. A good example of an IR-only techniques is the Global Precipitation Index (GPI; Arkin and Meisner 1987), which assigns a rain rate of 3 mm $h^{-1}$ to clouds tops colder than 235 K. Whilst developed for use over large time/spatial scales (e.g., monthly, 2.5° × 2.5°) it has proved useful at shorter/finer scales. The Auto Estimator (AE; Vicente et al. 1998) was developed primarily for hydrological and real-time applications using GEO IR observations to calculate precipitation using a power law regression calibrated by surface radar. Many other techniques have been developed (see Kidd et al. 2010) and despite their simplicity and problems, provide a benchmark against which other precipitation estimates may be compared.

To improve the cloud-top to surface precipitation relationship, retrieval schemes using observations from both the visible and infrared may be used. The ISCCP has a long-established methodology for categorising cloud using dual-channel observations (see Rossow and Schiffer 1991), as shown in Fig. 5.20. However, their

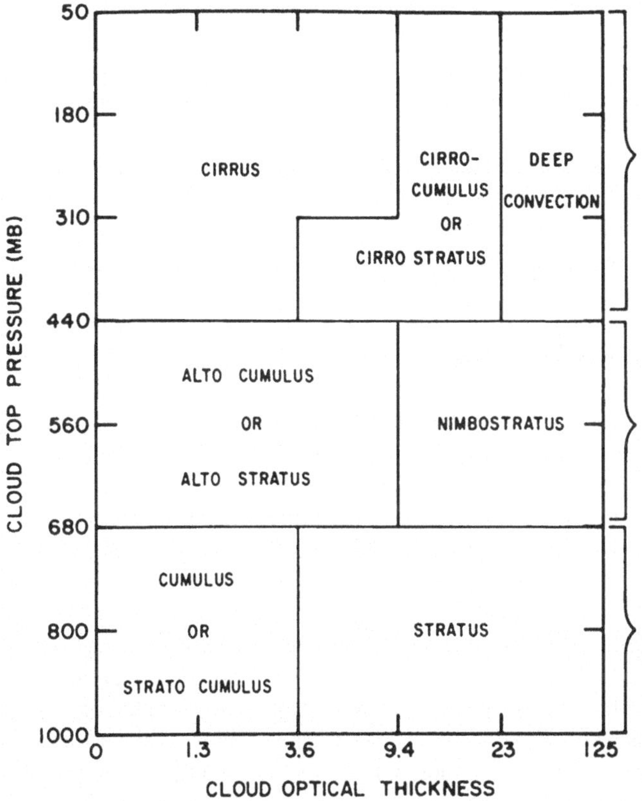


**Fig. 5.20** The ISSCP radiometric classification, based on WMO cloud types and general classes. The IR temperature is used as a surrogate for the cloud top pressure, while visible observations are used as a surrogate for the cloud optical thickness. At night the ISSCP classification only uses the IR to define low/medium/high clouds. (Source: NASA-GISS)

methodology imposes classes upon what is a continuous distribution in the observations. Desbois et al. (1982) and Yu et al. (1996) have shown that the distribution of the VIS vs IR observations is affected by substantial overlaps between the different categories.

The GOES Multi-Spectral Rainfall Algorithm (GMSRA; Ba and Gruber 2001) use five channels from GOES (VIS, near IR, water vapour, and two thermal IR channels) to extract cloud information, rain extent and amount depending upon the cloud/precipitation regime. Techniques will typically use information to delineate the precipitation based upon the cloud properties, then use another scheme to generate a precipitation intensity. These relationships often are split between convective and stratiform regions, not least due to the significant differences in the cloud-precipitation relationships, using water/ice phase discrimination of the cloud tops and/or the cloud top temperatures. Most techniques rely upon some calibration source, often using surface-based radar networks to provide the "true" precipitation intensity. One approach by Roebeling and Holleman (2009) uses the MSG SEVIRI observations to detect precipitating clouds using derived Cloud Water Path (CWP), particle effective radius, and cloud thermodynamic phase, then base a precipitation estimate on the CWP and cloud top height.

Multispectral techniques have also used neural networks to improve the delineation of precipitation. Work by Behrangi et al. (2009a) used the classification features of a Self-Organizing Feature Map (SOFM) together with probability matching techniques to separate single or multispectral data into rain/no rain classes: during daytime, the visible 0.65 μm channel can improve the delineation particularly when combined with other channels. At nighttime, two IR channels offer significant improvements over a single IR channel.

### 5.2.3 Passive Microwave

Within the microwave region of the spectrum the atmosphere is relatively transparent: at shorter wavelengths (higher frequencies) closer to the thermal infrared region, water vapour is the main attenuator of radiation emitted from the Earth's surface, and the main contributor of re-emitted radiation. However, at the longer wavelengths (lower frequencies), excluding the absorption lines, the atmosphere becomes more transparent with the main source of attenuation being precipitation-sized particles. The radiation received by passive microwave (PMW) sensors are termed brightness temperatures (Tbs) which are equal to the temperature of the object multiplied by its emissivity such that Tbs are the temperature that a black-body would have to have to generate that amount of radiation. The microwave region of the spectrum used for precipitation observations extends from about 6 GHz to above 200 GHz, although new sensors extent this to above 600 GHz. The most important channels lie within 10–230 GHz due to their more direct relationships with the precipitation-sized particles. Within the microwave regions there are several notable atmospheric absorption bands, including the oxygen bands between 50–60 GHz and around 118 GHz, and water vapour bands at 23.8 and 183.31 GHz. While these bands are primarily used by PMW sounding instruments for temperature and humidity retrievals, they can also be used for precipitation measurements, particularly bands around the 183.31 GHz absorption line.

PMW radiometers rely upon the upwelling radiation from the Earth's surface that is largely unaffected by the atmosphere and the presence of cloud, particularly at the lower frequencies (below 37 GHz). At these lower frequencies liquid particles of sufficient size (as in the case of precipitation) increase the upwelling radiation due to emission from the liquid droplets since most are generally smaller than 6 mm. At higher frequencies ice particles cause scattering of the upwelling radiation, resulting in a decrease in the at-sensor received radiation. However, there is some cross-over: for example, in cases of large hail a scattering signal can be found in the lower frequency observations, even as low as 10 GHz. While the higher frequencies are responsive to ice, the amount of ice can be used to infer surface precipitation more directly than is possible from infrared observations. The PMW observations therefore provide a more direct means of measuring precipitation than other observations. However, the received radiation also has a surface background component which is particularly important at the lower (more transparent) frequencies. The main factor in the background emissivity is water which has a low emissivity: the oceans have low emissions while land areas have higher emissions. This impacts the identification of precipitation over these different surfaces since the emissions from raindrops over land surfaces may be indistinguishable from the land surface background. However, the decrease in radiation at higher frequencies resulting from scattering by ice particles is discernible over land.

It should be noted that due to limited emitted radiation in the microwave region, the spatial resolution of an observation is related to the size of the antenna and inversely related to frequency of observation: observations at 10 GHz generally have very poor resolution (e.g., 100 km) while those made at high frequencies have good resolution (e.g., 5 km or better). Larger sensors carry antennas up to 2 m allowing reasonable resolution even at the low frequencies. The antennas on Cubesats must be much smaller (10 cm) restricting their use to high frequencies (89 GHz and above) with only moderate resolutions (16–25 km).

An example of PMW imagery is shown in Fig. 5.21 and shows the vertically polarised channels from the GPM GMI sensor ranging from 10 through to 183 GHz. The changing contrast between the land and sea background is clear: at low frequencies the ocean background is dark (low emissions) while the land surface is brighter (higher emissions). As the frequency of observation increases there is a greater contribution from the atmospheric water vapour, such that at 183 GHz there is little difference between the sea and land background: only where there is very little water vapour can the surface background be seen at the 183 GHz channel. In this example, at 10 GHz there is little atmospheric attenuation, although significant Radio Frequency Interference (RFI) can be seen. At 18 GHz some emission signal from raindrops can be seen around Sicily and the Adriatic: this signal increases at 23.8 and 37 GHz. At the higher frequencies (89 GHz and above) the scattering signal from ice particles can be observed over central Italy, as can scattering from surface snow over the Alps. The transition from the emission signal to the scattering signal is exemplified in Fig. 5.22, which shows PMW imagery of a hurricane. In this

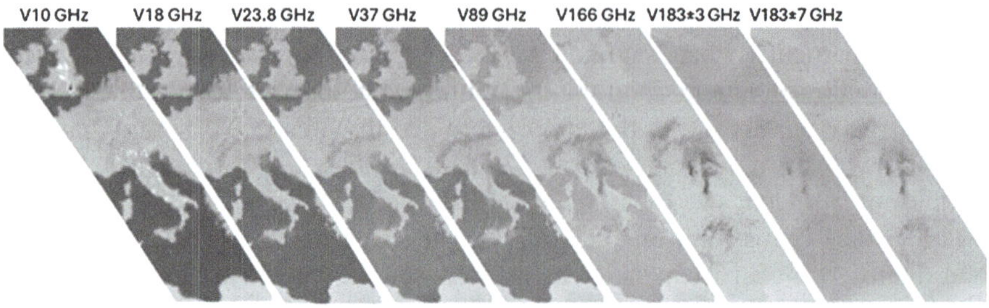


**Fig. 5.21** Images of a partial orbit of the GMI radiometer for vertically polarised channels. The bright spots at 10 GHz are emissions from device at the surface resulting in radio frequency interference. As the frequency of observation increases, so does the effect of water vapour in the atmosphere, thus obscuring the differences between the land and sea

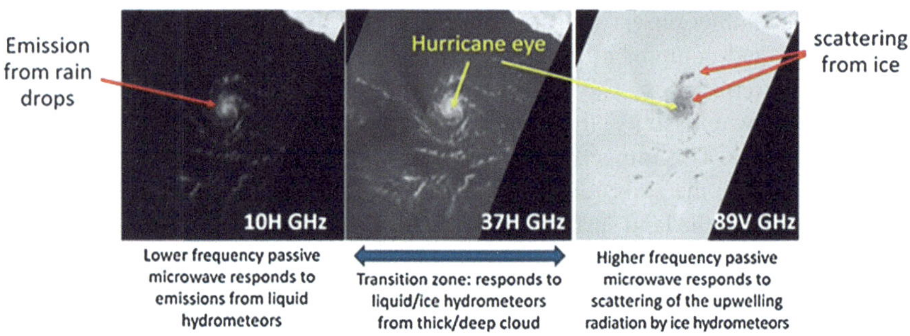


**Fig. 5.22** Passive microwave imagery of Hurricane Amanda, the first eastern Pacific hurricane of 2014, captured by the GMI on 24 May 2014 at 2330 UTC

case an emission signal is evident in the 10 GHz channel due to the significant amounts of liquid water. A close examination shows that the 10 GHz emission signal is not always co-located with the high-frequency scattering signal.

These examples show that PMW observations provide a wealth of information such that measurements of the precipitation intensity and type may be retrieved through the use of multi-spectral observations.

#### 5.2.3.1 Passive Microwave Sensors

Since observations in the microwave region of the spectrum are more directly related to the precipitation-sized particles in the atmosphere, many PMW sensors have been designed to operate in this region. These sensors can be categorised by their frequency of operation and by their scanning mechanism. PMW imagers generally observe the upwelling radiation within the window channels where the clear atmosphere is transparent whereas PMW sounders observe with the atmospheric absorption bands to extract vertical profiles of temperature and humidity. PMW images are generally conically scanning instruments where the Earth Incidence Angle (EIA) is constant, while PMW sounders are cross-track sensors with the centre of scan typically at the sub-satellite point. While the imagers benefit from constant resolution and atmospheric path, sounders have variable resolution that decreases away from the sub-satellite track, variable EIA, and variable atmospheric path (see Fig. 5.23). It should be noted that some conically-scanning "imagers" also provide observations at the atmospheric absorption channels, notably the 183.31 GHz channels. Although the exact frequencies used by each radiometer varies, they can be grouped into a number of broad ranges, as shown in Table 5.11.

The first practical PMW sensors with precipitation capabilities were the Electrically Scanning Microwave Radiometers (ESMR) on Nimbus-5 (ESMR-5), launched in 1972 and on Nimbus-6 (ESMR-6), launched in 1975 (Wilheit 1972). ESMR-5 operated at 19.35 GHz. At this frequency, since oceans have low emissivity, regions of low atmospheric water have low Tbs, with increasing Tbs with increasing liquid water content, and precipitation. Savage and Weinman (1975) demonstrated that while these Tbs could be related to cloud and precipitation other variables, such as sea surface temperature and water vapour, affect the Tbs such that a single channel alone is insufficient to provide accurate retrievals of precipitation. Nevertheless, in moderate to heavy intensities the precipitation signal becomes dominant (Wilheit et al. 1977). The ESMR-6 sensor operated at the higher frequency of 37 GHz and with dual polarization. This higher frequency is more sensitive to atmospheric water, resulting in a stronger cloud/precipitation signal over the oceans. The 37 GHz was also found to respond to scattering by ice particles, although differentiating these from cold background surfaces was problematic. The Nimbus-7 Scanning Multichannel Microwave Radiometer (SMMR; 1978–1987) represented a major advance due to operating at 6.6, 10.7, 18.0, 21.0 and 37.0 GHz (all dual polarised), with a nominal resolution of 27 km × 18 km at 37 GHz. The multi-frequency dual polarisation observations were ideal at retrieving a number of geophysical parameters, including precipitation.

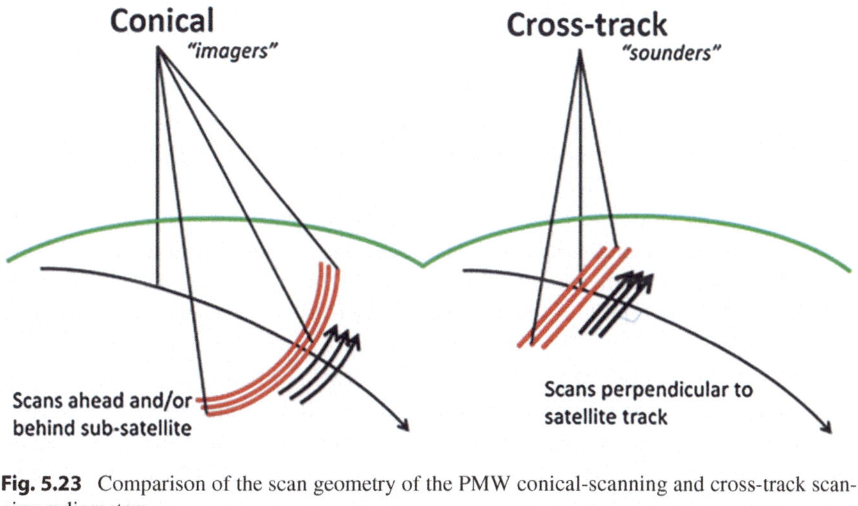


**Fig. 5.23** Comparison of the scan geometry of the PMW conical-scanning and cross-track scanning radiometers

**Table 5.11** Commonly used channels (GHz) for precipitation retrievals (left) together with other precipitation-capable sensors (right). The pairs of symbols are for use over land and ocean: available and used ●, available but not used ○, of limited use on its own ×

| | | Frequency (GHz) | | | | | | | | |
|---|---|---|---|---|---|---|---|---|---|---|
| | **Sensor** | ~6.9 | ~10.7 | ~19.0 | ~23.0 | ~37.0 | 50–60 | ~90 | ~150 | ~183 |
| Commonly used channels | TMI | | ●● | ●● | ●● | ●● | | ●● | | |
| | GMI | | ●● | ●● | ●● | ●● | | ●● | ●○ | ●○ |
| | AMSR-E | ○○ | ●● | ●● | ●● | ●● | | ●● | | |
| | AMSR-2 | ○○ | ●● | ●● | ●● | ●● | | ●● | | |
| | SSM/I | | | ●● | ●● | ●● | | ●● | | |
| | SSMIS | | | ●● | ●● | ●● | ○○ | ●● | ●○ | ●○ |
| | AMSU-B | | | | | | | ●● | ●● | ●● |
| | MHS | | | | | | | ●● | ●● | ●● |
| | ATMS | | | | | | | ●● | ●● | ●● |
| | SAPHIR | | | | | | | | | ●● |
| Unused/ underused channels | ESMR-5 | | | ○× | | | | | | |
| | ESMR-6 | | | | | ○× | | | | |
| | SMMR | ○× | ○× | ○× | ○× | ○× | | | | |
| | MSU | | | | | | ×× | | | |
| | SSM/T1 | | | | | | ×× | | | |
| | SSM/T2 | | | | | | | ○○ | ○○ | ○○ |
| | AMSU-A | | | | ○× | ○× | ○× | ○× | | |
| | MSMR | ○× | ○× | ○× | ○× | | | | | |
| | TMR | | | ○× | ○× | ○× | | | | |
| | JMR | | | ○× | ○× | ○× | | | | |
| | AMR | | | ○× | ○× | ○× | | | | |
| | MWR | | | | ○× | ○× | | | | |

In 1987 the US DoD Defense Meteorological Satellite Program (DMSP) series started with a series of passive microwave sensors, including the Special Sensor Microwave/Temperature (SSM/T1 and SSM/T2 that operated at 50–60 and 89–183 GHz respectively), and the Special Sensor Microwave/Imager (SSM/I), operating at 19.35, 23.235, 37.0 and 85.0 GHz. The SSM/I provided observations up to 85.0 GHz making precipitation retrievals over land feasible. The SSM/I was the focus of NASA's WetNet project (Barrett et al. 1994) that nurtured the development of many precipitation retrieval schemes. These sensors would later be developed into a single sensor, the Special Sensor Microwave—Imager/Sounder (SSMIS) from 2006 onwards, with a channel range from 19.35 to 183.31 GHz. The new US DoD Weather Service Follow-on Microwave (WSF-M) sensor operates from 10.85 to 89.0 GHz.

The Tropical Rainfall Measuring Mission (TRMM) Microwave Imager (TMI), launched in 1997, provided observations from 10.65 GHz through to 85.5 GHz, covering the main frequencies used for precipitation retrievals. The non-Sun-synchronous low inclination orbit of TRMM (35°N–35°S) was designed to capture the full diurnal cycle over the course of a few months. The GPM Microwave Imager (GMI), launched on the core observatory in 2014, extends the TMI channels to higher frequencies at

165.5 GHz (dual polarised) and 183.31 ± 3 and 183.31 ± 7 GHz. The 165.5 GHz channels are extremely useful for the identification and retrieval of snowfall. Importantly, the GMI is extremely well calibrated and is designed to be used as a reference standard for other PMW radiometers.

Alongside the US missions, several PMW imager missions have been launched and operated by the Japan Aerospace Exploration Agency (JAXA). These included the Advanced Microwave Scanning Radiometer (AMSR) sensors. The current AMSR-2 sensor operates at frequencies from 6.925 GHz through to 89.0 GHz providing spatial resolutions as fine as 3.1 km × 5.4 km at 89.0 GHz. The Chinese have PMW capabilities with their FY-3 satellite series with the Micro-Wave Radiation Imager (MWRI) that operates from 10.65 to 183.31 GHz and their Micro-Wave Humidity Sounder (MWHS) operating from 89.0 to 183.31 GHz. The ISRO/CNES Megha-Tropiques' SAPHIR sensor, with 6 channels around 183 GHz, has also contributed to precipitation measurements. Several of the single beam radiometers have also flown on ocean-focused satellites missions, such as the TOPEX and JASON satellites, tasked with providing atmospheric corrections to their altimeters.

While early PMW sounding observations by the Microwave Sounding Unit (MSU) at 50.3–57.95 GHz exploited the oxygen-band, these were by themselves not useful for precipitation. However, this sensor evolved into the Advanced Microwave Sounding Unit-A (AMSU-A) and -B (AMSU-B). AMSU-A operated from 23.8 to 89 GHz, while the AMSU-B operated from 89.0 to 183.31 GHz. The Microwave Humidity Sounder (MHS) provided similar observational capabilities as AMSU-B. The capabilities of the AMSU-A and AMSU-B/MHS sensors were combined into the Advanced Technology Microwave Sounder (ATMS) sensor flown on the NOAA Joint Polar Satellite System (JPSS).

Future missions include the EUMESAT European Polar Satellite—Second Generation (EPS-SG) that will carry the MicroWave Sounder (MWS) operating from 23.8 to 229 GHz, the MicroWave Imager (MWI) operating from 18.7 to 183.31 GHz, and the Ice Cloud Imager (ICI) that will operate from 183.31 to 664 GHz. The EPS-SG sensor channels will allow a full range of precipitation and related parameters to be observed and retrieved. JAXA's AMSR-3 sensor (launched 28 June 2025) extends the capabilities of the AMSR-2 sensor by providing additional low frequency channels to mitigate radio frequency interference, and by adding the 183 GHz high frequency channels.

#### 5.2.3.2 Algorithms

The evaluation of early PMW observations attempted to relate the measured radiances to our physical understanding of the processes and thus provide a measure of the precipitation. The ESMR-5 observations at 19.35 GHz showed that over the oceans increases in the brightness temperatures (Tbs) were the result of emissions from the drops of liquid water in the atmosphere, i.e. precipitation. While the 37 GHz observations on ESMR-6 provided increased sensitivity, Savage and Weinman (1975) demonstrated that while these Tbs could be related to cloud and precipitation other variables, such as sea surface temperature and water vapour, affect the Tbs such that a single channel alone is insufficient to provide accurate retrievals of precipitation, particular at low rain rates. However, the 37 GHz was also found to respond to scattering by ice particles, although differentiating these from cold background surfaces was problematic.

The Nimbus-7 SMMR instrument provided the first multi-frequency dual polarization observations and therefore was ideal for retrieving several geophysical parameters, including precipitation. Alishouse et al. (1990) used the SMMR 37 GHz to estimate precipitation of hurricanes using surface radar for calibration, achieving correlations of about 0.7. Rodgers and Siddalingaiah (1983) used the 37 GHz observations to classify rainfall over land surfaces, using the dual polarization to distinguish between low Tbs over wet land versus low Tbs caused by the scattering by precipitation. Other approaches generated quantitative measures of precipitation over land using the Tb differences between two frequencies, the basis of which is that both frequencies would response to changes in the background, while scattering caused by precipitation would be greater at 37 than at 18 GHz. A common theme through the early retrieval work was the "beam filling effect", a problem where only part of the observed field of view was filled with precipitation. However, Spencer (1986) showed that using the polarization difference could overcome this, thus greatly improving the potential precipitation retrievals. More importantly, the availability of dual frequency multi-frequency observations together with early radiative transfer work, allowed the development of physically based retrieval schemes to match theoretically calculated retrievals with the observations. Olson (1989) concentrated upon the parameters that had the greatest impact on the microwave radiances to devise a scheme to best match the observed radiance with the simulated radiances.

The launch of the US DMSP-F08 satellite with the first SSM/I sensor in 1987 was the next major milestone in the utilization of passive microwave for precipitation retrievals. This sensor provided passive microwave observations at 19.35, 22.235, 37.0 and 85.0 GHz with dual polarization at each frequency except the 22.235 GHz. The inclusion of the 85.0 GHz channel meant that the sensitivity to scattering by ice particles increases significantly and made retrievals over land practicable. NASA, realizing the potential of SSM/I observations, set up WetNet a group of US scientists with international participation. The group were provided with identical PC-based workstations and access to near real time SSM/I data to foster a distributed scientific collaborative network. The WetNet project included three Precipitation Intercomparison Projects (PIP-1,2,3) which ran alongside the GPCP Algorithm Intercomparison Programme (AIP-1,2,3). Both sets of intercomparisons used surface reference data for validation, ranging from fine scale instantaneous regional to global comparisons. The outcome of these intercomparisons was essentially that no single algorithm proved superior, with empirically based techniques performing as well as (or better than) physically based schemes, although the passive microwave schemes bettered the performance of VIS/IR techniques. A major drawback with physical retrieval schemes at this time was the excessive time needed to compute the retrievals: these used inverse radiative transfer modelling, so for each observation an ensemble of possible solution had to be calculated and compared. Some techniques circumvented this problem by generating model simulations beforehand to establish relationships between the precipitation and expected Tbs (e.g., Ferraro et al. 1994, 1998). Empirical techniques relied upon relatively simple relationships calibrated against surface reference data to generate estimates.

Development of physically based retrieval schemes continued particularly through the work of Mugnai et al. (1990), Smith et al. (1992) and Wilheit et al. (1994). This was in part due to significant advances in computing power, although offset somewhat by more detailed modelling capabilities. In addition, since the PMW observations were affected by many different parameters and precipitation types providing a sufficiently large sample from which to select the "correct" result was challenging. This led to the development of Bayesian schemes where the probabilities of precipitation systems were accounted for with the generation of previously (a priori) modelled databases (e.g., Kummerow and Giglio 1994). The main drawback at the time was that these modelled databases were limited to Cloud Resolving Model (CRM) cases and therefore were not always representative. Such schemes required an initial rain/no-rain test to remove artefacts and improve computational performance (Kummerow et al. 2001).

The launch of TRMM in 1997 with the TMI and PR greatly improved the representativeness of these physically based Bayesian schemes through linking the observed Tbs with co-temporal, co-located vertical profiles of precipitation as obtained from the PR (Kummerow et al. 2011). With the availability of TMI-PR comparisons, these physically-based Bayesian schemes were extended to include other PMW sensors, such as AMSR-E, and would be expanded further to provide estimates from the SSM/I, and SSMIS with the Goddard Profiling Scheme (GPROF; Kummerow et al. 2015), which was adopted as the official retrieval scheme for the GPM mission with further updates (see Randel et al. 2020). GPM provided the GMI sensor with the inclusion of the high-frequency 183 GHz channels, while the DPR provided dual frequency precipitation radar measurements.

One of the goals of the GPM mission was to improve the temporal sampling of global precipitation measurements using a wide range of sensors. It was realised that PMW imagers, like GMI, SSMIS and AMSR-2 could not provide the desired temporal sampling: observations from PMW sounding instruments were also needed. Early research on precipitation retrievals from sounding instruments were undertaken by Spencer (1993) who used the 50 GHz channels from the Microwave Sounding Unit (MSU) to generate coarse resolution climatological maps of precipitation over the oceans. The launch of NOAA-15 in 1998 with the AMSU-B sensor provided cross-track observations from 89 to 183 GHz with 16 km resolution at nadir. Ferraro et al. (2000) exploited the scattering properties of the 89 and 150 GHz channel, calibrated against surface observations, to generate precipitation retrievals, while Laviola and Levizzani (2011) utilised the 183 GHz scattering properties to identify and quantity precipitation. The GPM at-launch version of GPROF for the sounding instruments relied upon a modelled database for the high frequency channels and scan position corrections to account for the cross-track scan (Kidd et al. 2016), although this was replaced with a more observational database once sufficient GMI-DPR comparisons were available. While GPROF was applied to observations from the MHS and ATMS instruments, observations from SAPHIR were processed by the PRPS scheme utilizing an observational database. More recently the PRPS scheme has been applied to observations from the TROPICS (Kidd et al. 2022).

The GPROF scheme is used to provide precipitation retrievals from the current GPM passive microwave sensors, namely GMI, AMSR2, SSMIS, MHS and ATMS, while it has been used for retrievals from "legacy sensors", SSM/I, AMSU-B and AMSR-E. Note however that (i) GPROF retrievals to date are based upon single-sensor observations (with auxiliary input data) and (ii) GPROF does not necessarily utilize all the available channels where available. There is therefore scope to fully utilize observations made by other sensors, and to exploit observations from presently unused channels.

### 5.2.4 Active Microwave

Active microwave sensors (radar) provide the most direct measure of precipitation from satellite systems. This results from the backscatter of their emitted radiation by the precipitation-size particles, and their ability to provide the vertical distribution of these particles through the atmosphere. While the principles of measuring precipitation using satellite-based radars are similar to ground-based radars, there two important differences. First, satellite-based radars operate at higher frequencies, and second, ground clutter is advantageous to retrieving precipitation. Such sensors are, however, only available on low Earth orbiting missions, and the swath of such precipitation radars is limited to about 250 km. Cloud profiling radars that offer some precipitation capabilities are however, restricted to a single beam.

The practicalities of radars operating at higher frequencies finds that the sensitivity is much improved, allowing a greater identification and discrimination of cloud and precipitation sized particles. Ground-based weather radars generally operate in the C-band, or about 5.6 GHz. Satellite-based precipitation radars operate at Ku- and Ka-bands (13.6 and 35.0 GHz, respectively), while cloud radars operate at 94 GHz. However, at higher frequencies there is more atmosphere attenuation, particularly because of moisture or small particles, thus more processing is required to account for this. At higher frequencies multiple scattering also becomes very important due to the increase in scattering from the smaller particles. Higher frequencies also suffer from increased attenuation from precipitation, particularly at and above moderate intensities: at high intensities attenuation may be extreme with little of the radar emitted energy making it through to the lower layers near the surface. Trying to account for and correct for this attenuation is a key factor in precipitation radar measurements: by analysing the reflection from the ground surface, which is always very strong, an estimate of the path integrated attenuation may be calculated and used to correct the precipitation measurements through the vertical profile. However, the strong surface reflectance also limits how close to the surface these radars can sense precipitation: this is about 1 km at nadir but will vary depending upon the vertical resolution of the radar.

#### 5.2.4.1 Sensors

The first precipitation radar was flown on the Tropical Rainfall Measuring Mission (TRMM) that operated from 1997 to 2015 (Kummerow et al. 1998). TRMM's Precipitation Radar (PR) operated in a non–Sun-synchronous orbit at about 350 km above the surface. The PR was a Ku-band radar operating at 13.6 GHz with a sensitivity of 17 dBZ, equivalent to about 0.7 mm $h^{-1}$. The radar provided vertical measurements of the backscatter across 80 range bins from the surface to 20 km with 250 m vertical resolution, 4.3 km spatial resolution and a swath of 215 km.

Building upon the success of the PR, a Dual-frequency Precipitation Radar (DPR) was included on the Global Precipitation Measurement (GPM) mission (Hou et al. 2014). The DPR was developed by the Japan Aerospace Exploration Agency (JAXA) and National Institute of Information and Communications Technology (NICT). The DPR consists of two radars, a Ku-band precipitation radar (KuPR) operating at 13.6 GHz and Ka-band radar (KaPR) operating at 35.5 GHz. In the present configuration both the KuPR and KaPR provide data with profiles across a 250 km swath and a spatial resolution of 5 km × 5 km. Early in the GPM mission the KaPR scan pattern had 25 positions that matched the central 25 footprints of the KuPR with the remaining 24 footprints filling the gaps in between the former 25 footprints. The KuPR has a sensitivity of 15.46 dBZ while the KaPR sensitivity is up to 13.71 dBZ: this relates to 0.5 mm $h^{-1}$ for a combine retrieval. Moreover, the dual frequency capability allows a first guess of the DSD to be calculated due to the different frequency-response of the droplets. However, the detectability of precipitation depends on factors such as the DSD, attenuation by clouds, background noise, etc., but the DPR is more sensitive than the PR.

CloudSat carried the first 94 GHz (W-band) cloud profiling radar (CPR) with the ability to observe cloud particles and detect precipitation. CloudSat, a collaboration between the NASA Jet Propulsion Laboratory (JPL), the Canadian Space Agency (CSA), and the U.S. Air Force was launched on 28 April 2006 and started operations on 2 June 2006. From an altitude of 689 km, the CPR has a resolution on the ground of about 1.4 km cross track and 1.7 km along track. The CPR acquires a vertical profile with 125 samples with 240 m spacing resulting in a nominal vertical resolution of 485 m. To detect the weak scattering signal from clouds at 94 GHz, an overriding requirement of the CPR was the minimum detectable signal, which was set at −28 dBZ. This sensitivity was achieved through the high-power amplifier (HPA) with a nominal 1.7 kW power level, and the composite graphite antenna with low surface roughness to provide a highly directional beam.

The ESA EarthCARE mission, launched 28 May 2024, flies at a relatively low altitude of 393 km, in a sun-synchronous 97° polar orbit, with a descending-node crossing time of 14:00 MLST. It carries, amongst other sensors, a Cloud Profiling Radar (CPR) operating at 94 GHz developed by JAXA and NICT. It is the first space-borne Doppler cloud radar that estimates vertical profiles and motion of the cloud, and is equipped with a large antenna/reflector with a diameter of 2.5 m. The vertical range of the instrument can be selected in one of three modes to cover 20, 18 or 16 km depending on the latitude of the observation, while the lowest range bin is set to −1 km to utilise the surface backscatter for attenuation corrections. Although the use of 94 GHz is the same as the CloudSat CPR, the sensitivity of the radar is much better due to its lower orbit and larger antenna aperture size. The minimum radar reflectivity of the EarthCARE CPR is −35 dBZ as defined at the top of the atmosphere (20 km). The Doppler velocity can be measured ±10 m $s^{-1}$, with an accuracy of at least 1.3 m $s^{-1}$. With a pulse width of 3.3 μs a vertical resolution of 500 m is achievable (same as the CloudSat CPR) although this is over-sampled at 100 m, which is more precise than that of CloudSat (250 m). The instantaneous footprint size of the EarthCARE CPR is about 750 m.

#### 5.2.4.2 Algorithms

The higher frequencies used by the satellite-based precipitation (and cloud) radars require particular attention to the addition attenuation from atmospheric constituents, although the processing steps for each satellite-based radar is broadly similar. For the 13.8 GHz TRMM PR these include (i) attenuation correction, which is based on a hybrid of the Hitschfeld-Bordan method (Hitschfeld and Bordan 1954) and a surface reference method; (ii) selection of the default drop size distribution mode on the basis that the attenuation estimates derived from the model and the independent estimates from the surface reference with the nonuniform beam-filling correction are consistent for rain over ocean, and (iii) correction for the nonuniform beam-filling effect which is analysed as a two-dimensional problem. However, uncertainties in individual parameters affect the attenuation correction and rain estimates: parameters such as the drop size distribution, freezing height and the phase of the precipitating particles (which affects their density and shape), the inhomogeneity of the distribution of precipitation within footprint, attenuation due to cloud liquid water and water vapour, uncertainty of the surface scattering cross section, and fluctuation of the radar echo signal. Notably, the drop size distribution models that are consistent for convective rain over ocean are not consistent over land, indicating a change in the size distributions associated with convective rain over land and ocean, respectively (Kummerow et al. 2001).

For the GPM DPR precipitation rate retrieval an algorithm similar to that of the TRMM PR was developed: a combination of the Histchfeld-Bordan's attenuation correction method and the dual-frequency ratio method is used. The relationship between the precipitation rate and mass-weighted mean diameter replaced the relationship between the attenuation and the effective radar reflectivity factor. For both the single-frequency (Ka or Ku) and dual-frequency (Ka and Ku) algorithms, the forward retrieval methodology is applied with an assumed precipitation rate to mass-weighted mean diameter adjustment factor and the results evaluated to select the best value. The advantage of having dual frequencies is that a surface reference technique can compare the surface returns and thus improve the attenuation assumptions. However, in the presence of deep convection, certain characteristics of the DPR dual-frequency reflectivity profiles cannot be explained by single scattering calculations, but they are explained by multiple-scattering theory. Such characteristics are evident due to the absence of peaks in the surface reflectivity and anomalously small reflectivity slopes in lower parts of the vertical profile. These effect the

DPR-based rainfall retrievals and the stratiform/convective classification algorithms when dealing with deep convection.

The improved detectability of precipitation by the dual-frequency precipitation radar (DPR) was tested using two minimum detectable reflectivities of 12 and 18 dBZ applied to 6 months of synthetic GPM DPR data: the precipitation occurrence and volume increase by 21.1% and 1.9%, respectively, 40°N–40°S. The ability of the DPR to detect light precipitation consisted of two distinct types: shallow precipitation over eastern parts of subtropical oceans where deep convection is typically suppressed and, that associated with lower parts of anvil clouds associated with organised precipitation systems. In practice it was found that some observed reflectivities had lower values that the official minimum detectable reflectivity, despite these being consistent with true precipitation signals, suggesting that the original sensitivity specification might be too conservative.

The CloudSat CPR operating at 94 GHz is strongly affected by the presence of precipitation, both by scattering and by attenuation. The unattenuated near surface reflectivity of the ocean surface is about 0 dBZ, so any higher reflectivity is almost certainly a result of precipitation (about 0.03 mm $h^{-1}$). Since the reflectivity for falling snow is less, model information is included to adjust the threshold reflectivity. The Path Integrated Attenuation (PIA) is also very sensitive to the presence of precipitation which causes scattering of the radiation stream. Over the oceans, calculation of the PIA can be very accurate, providing a minimum sensitivity to precipitation at about 0.02–0.05 mm $h^{-1}$. Intermediate regimes, identified as rain-probable and rain-possible, are also retrieved based upon the reflectivity. However, while the delineation of precipitation is very good, due to the operating frequency the retrieval of heavier precipitation is more challenging. The EarthCARE CPR, operating at the same frequency as the CloudSat CPR, has similar challenges, although it does benefit from the availability of Doppler measurements that can be used to extract additional parameters. Various CPR calibrations are conducted to retrieve calibrated radar reflectivity and Doppler velocity, which are used for higher-level product processing. The radar reflectivity is calculated using the noise level estimation and CPR transmit power estimation, while retrieval of the Doppler velocity compensates for contamination resulting from the satellite velocity. The CPR Level 1b product contains the radar reflectivity and vertical Doppler velocity.

### 5.2.5 Advanced Techniques

The recent advances in data storage and computing power have opened new possibilities for a range of advanced techniques. These have encompassed the use of large data sets from multiple satellite and surface observations, improved communications and processing speed that allows near real time precipitation retrievals and improving the accuracy and temporal and spatial resolution of precipitation products. Improved computing power has allowed the development of Machine Learning (ML) and Artificial Intelligence (AI) techniques to help tease out additional information from the available observations.

### 5.2.5.1 Merged/Morphed Schemes

Precipitation measurements from satellite sensors are primarily categorised into two main groups by type of observation. The first group make use of the regular and frequent observations provided by the GEO sensors, but since these only carry vis/IR instruments, such estimation schemes are relatively indirect. The second group utilise the more direct PMW estimation schemes but since these sensors are only carried on the LEO satellites, the observations are much less frequent and unevenly distributed across the diurnal cycle. To exploit the benefits of these two groups several techniques have been developed to utilise the regular sampling of the vis/IR with the directness of the PMW. Techniques that have been developed to exploit both the vis/IR and PMW observations are themselves categorised into two: those that use the PMW to calibrate the IR data and those that use the IR-derived cloud motion with the PMW precipitation estimates.

Techniques that adjust IR-based estimates or generate IR radiance-precipitation calibration curves using the PMW (or other reference data) are termed blended algorithms (e.g., Kidd et al. 2003). These techniques correct or adjust the information used to transform the IR cloud top temperature into a precipitation estimate. All IR-derived estimates essentially fall into this category since even a simple cloud-thresholding techniques, such as the GPI, use reference data to set the rain/no-rain threshold and rain rate conversion. More complex techniques, such as the TRMM Multi-Satellite Precipitation Algorithm (TMPA) (Huffman et al. 2007), ingests data from PMW sensors and GEO IR data adjusting for the different satellite retrievals before combing the data into the final precipitation product. Other techniques continuously update the PMW-IR relationship as new coincident observations are received: this ensures that changes in the meteorological regimes are properly represented in their final precipitation estimates. These techniques include the Passive Microwave InfraRed (PMIR) scheme (Kidd et al. 2003), the Microwave-adjusted IR Algorithm (MIRA; Todd et al. 2001), the Naval Research Laboratory scheme [Turk et al. 2000], the Microwave-Infrared Combined Rainfall Algorithm (Marzano et al. 2004), the Self-Calibrating Multivariate Precipitation scheme (Kuligowski 2002), and the Microwave/Infrared Rain Rate Algorithm (Miller et al. 2001). However, although techniques that use the PMW precipitation estimates to calibrate the IR data show great promise, they are ultimately limited by the indirectness of the IR observations to sense the at-surface rainfall.

While the best observations for global precipitation measurements are undoubtedly satellite-based precipitation radars, these have only been available since 1997, generally with only one in orbit at once and with very limited swath width. PMW observations, however, also provide a more direct measure of precipitation than the IR observations, albeit without the temporal sampling of the IR. While the IR observations of precipitation maybe direct, they can provide a reasonable measure of cloud movement: this property can be used to advect the PMW retrieved precipitation between the successive satellite overpasses with the direction/speed of the IR-derived motion. The increased access to satellite (and surface) data and in computing (storage and processing) power has enabled the evolution of practical motion-based (or advection) techniques that can be run operationally, i.e. in near real time. Many of these techniques incorporate a suite of modules to inter-calibrate the input data and to ensure consistency of the generated output product. These are morphing techniques since the try to change, or evolve, the precipitation features from one time to the next. Examples of techniques include the Climate Prediction Center Morphing (CMORPH) technique (Joyce et al. 2004), the Global Satellite Mapping of Precipitation (GSMaP; Kubota et al. 2007), the merged product of the EUMETSAT Satellite Applications Facility on Support to Operational Hydrology and Water Management (H-SAF; Mugnai et al. 2013) and the Integrated Multi-satellitE Retrievals for GPM (IMERG) product. The main drawback of these techniques is that the cloud motion may not accurately reflect the movement of the precipitation at the surface. Bellerby et al. (2009a, b) developed a multi-sensor satellite rainfall monitoring methodology called the Lagrangian Model based on a conceptual cloud development model driven by GEO imagery, locally updated using the PMW rainfall estimates. The technique uses a single-band thermal IR GEO data to characterise cloud motion, growth, and decay as input to a simple, linear, semi-Lagrangian, conceptual cloud mass balance model, with representations of convective and stratiform processes. The Integrated Multi-satellite Retrievals for the Global Precipitation Measurement (GPM) Mission (IMERG; Huffman et al. 2020) in its latest version draws on the output from a number of precipitation products, advanced merger schemes such as Kalman filtering and the use of NWP motion vectors (see Sect. 5.3 for a full description).

To address user needs for "the best available precipitation product" several techniques combine all available data sets, including those from various satellite observations, gauge data and, where available, surface radar. Such schemes need to account for errors inherent in each of the input data sets: the most common mechanism is using Kalman filters to minimise the differences between these data before generating the final product (Joyce et al. 2009). A widely used product is the Global Precipitation and Climatology Project (GPCP; Huffman et al. 1997) which incorporates surface gauges data where available, to produce a homogenous global precipitation product. Another product used, for example, NWP model verification, climate studies, and hydrological applications, is the Climate Prediction Center Merged Analysis of Precipitation (CMAP; Xie and Arkin 1997), which incorporates satellite IR along with gauge and reanalyses data from the National Centers for Environmental Prediction (NCEP) and NCAR.

Other data may be incorporated into retrieval schemes, such as lightning data, either from ground-based detection networks or from space sensors, such as the TRMM LIS the Optical Transient Detector (OTD). The inclusion of cloud-to-ground lightning data has been shown to reduce the RMSE of satellite precipitation products (Grecu et al. 2000). Lightning data has also been useful in the study of determining precipitation features, such as the occurrence and distribution of severe storms (see Cecil et al. 2005).

#### 5.2.5.2 Machine Learning

Machine learning, encompassing neural networks (NN, aka Artificial Neural Networks), now has widespread uses and applications throughout society and science. Although the basics of ML are not new, the improvements in their design coupled with computing power have spurred recent advances. NNs are essentially mechanisms to link the complexity of the interconnections between observations and the geophysical parameters to be measured (Fig. 5.24). An important advantage of NNs is their ability to model nonlinear functions in a finite number of steps and are therefore very appropriate for use in processing remote sensing data sets, particular for classification tasks.

The use of machine learning (ML) tools across Earth Observation has grown rapidly, not least since it can identify and extract relationships that might be otherwise difficult to discern through conventional means. Applying ML techniques can be used as a step towards ensuring optimal extraction of the information content (here, precipitation rate) from the observations. ML techniques such as Random Forests, Extreme Gradient Boost, Convolutional Networks, U-nets, Quintile Regressions, and similar, have been applied to link single or multiple dimension inputs (observations) to the exponentially distributed targets (precipitation rate).

Early studies, such as Hsu et al. (1997) used neural networks for precipitation estimation from IR data, with good correlations for monthly accumulated, degree-scale estimates, with estimates at finer temporal/spatial (0.1° × 0.1°, 30-min) scales being investigated by Tapiador et al. (2004). To take advantage of the superior retrieval ability of the PMW and the spatial and temporal resolution of geostationary satellites NNs for combining passive microwave (PMW) data with GOES IR observations have been developed (Bellerby et al. 2000; Sorooshian et al. 2000). Further development of multi-spectral, multi-sensor precipitation retrieval

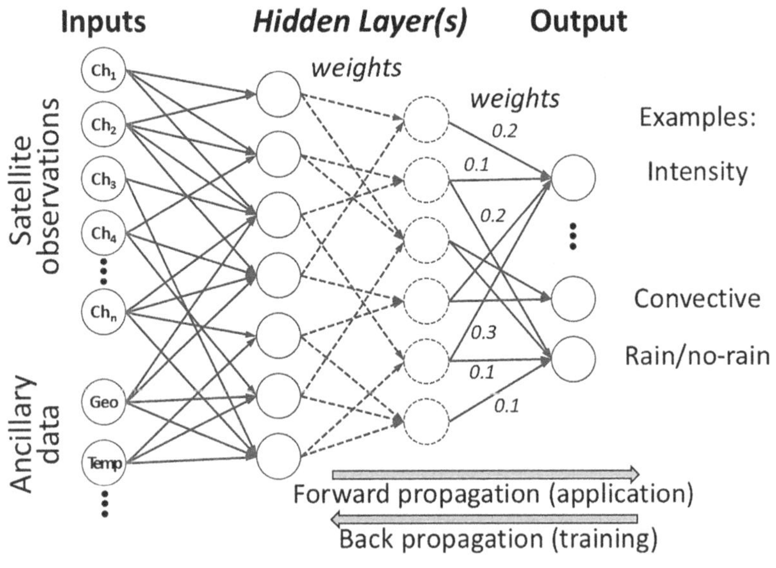


**Fig. 5.24** Example of a simple neural network with satellite observations used as in-put together with (optional) ancillary data. The hidden layer(s) represent the "neurons", or connections, whose weightings are determined through training (back propagation) with reference data

techniques for near real time estimation from GEO vis/IR using an adaptive ANN updated with PMW observations (as and when these become available) are designed to allow products to be used for a variety of studies including hydrological modelling, drought monitoring, soil moisture analysis, and flood forecasting (Behrangi et al. 2009b). More recently precipitation retrievals using convolutional neural networks (CNN) have become popular, even handling solid precipitation through combined PMW and IR data. ML has also been used in the prediction of precipitation: e.g., Pegion et al. (2022) studied daily precipitation anomalies over the southeastern US, while Yang et al. (2023) looked at the prediction of seasonal precipitation. The use of ML within numerical weather prediction has shown that it can garner better results for several atmospheric variables, but not necessarily for precipitation.

In many cases for precipitation retrieval, the ML models are constructed to operate in a two-step approach and applied in a consistent manner to all sensors (and combinations thereof) to construct the links between the available observations and corresponding surface precipitation rate. The first step, a classification model, is trained to delineate precipitating from non-precipitating observations. The second step then employs a regression model to estimate the actual precipitation intensity. The classification task considers the delineation of precipitation as a binary problem (0 = dry, 1 = precipitating) of retrieving the state of the sample given the observations used as the input. A Random Forest (RF) classifier and Convolutional Neural Network (CNN) are typically suitable for such as task. Given the simplicity of the ML architectures additional information may be needed, for example, for PMW observations since the mechanisms behind the observations over land and over water are different, the classification is usually done separately for ocean and land surfaces. To illustrate the ability of the RF classifier to delineate between dry and precipitating observations when provided with the GMI's 13 channel Tbs as input, a global (land + ocean) model, trained using a single year of GMI Tb co-located DPR precipitation intensities produced an 84% success rate in identifying non-raining samples, and 89% identifying raining samples. Once the precipitating observations are identified, other ML regression tasks may be employed to retrieve the precipitation intensity using, for example CNN architecture (with three convolutional and two dense layers, with ReLU activation and Adam optimization). The performance for retrieving precipitation rate is however very much dependent upon the initial training data, both in terms of quality of the data, but also the representativeness of the data.

ML retrievals are also able to provide a measure of uncertainty in the regression task (i.e. precipitation estimate). The steps in estimating uncertainty typically include generating a distribution of predictions such that the mean and variance of these multiple predictions will provide uncertainty estimates of retrieved values. Performing ensemble runs involves training multiple instances of the same ML model, each with slightly different initial weights or/and trained on different data subsets. This allows the calculation of the mean and variance across the ensemble predictions to be generated and provide a measure of uncertainty. While many sophisticated techniques exist and are readily available, simple approaches are often sufficient and easy to implement into the training framework.

Despite advances, using ML for precipitation estimation remains challenging. The extremely skewed distribution of instantaneous precipitation intensities towards zero makes any retrieval very sensitive. For example, small changes in the resolution of the observation, or slight changes in the assumptions on the model. Furthermore, the training data sets are a key part of the ML process: the quality and representative of such data is crucial. Many of the successes of applying ML to precipitation estimate has been for large scale, accumulated precipitation where the observation-precipitation relationships are more stable and are inherently less noisy. Nevertheless, NNs clearly hold significant potential, not least in simplifying the processing steps of retrieval schemes and therefore facilitating near real-time processing. However, issues remain about their "black box" operation and therefore the interpretability of their results: they may provide better, faster results, but they do not necessary reveal any insights into the physical linkages or processes that produce those results.

## 5.3 Precipitation Products

Precipitation data sets are derived from observations made at the Earth's surface, from satellites observations or from a combination of data sources. There is a rich selection of precipitation products from which users can choose, however, selecting one from the large number of available products can be a problem. The correct, or at least, the most appropriate product will depend upon the user requirements driven by their application(s). Factors governing the choice of precipitation products include the availability of the data on a local, regional or global scale, the longevity of the data record, the temporal and spatial resolution of the data products and ease of accessing and processing the data. For most users the more comprehensive datasets which combine data from multiple sources, including gauge data, are likely to be most suitable since these try to be as close to the true precipitation measurement as possible. These datasets are therefore typically "Level 3" products, or those that are multi-sourced and gridded, although the "Level 2" instantaneous, full resolution products are of interest to those who need finer scale data.

The Level 3 precipitation data products provide time-area averages on regular uniform latitude/longitude grid. Precipitation data covering defined geographic regions, such as a certain country or watershed, typically needs to be calculated from these gridded data, although some data providers do offer the ability to provide data for subsampled regions. The Level 3 data records generally cover time resolutions from daily to monthly, although fine temporal/spatial resolution data is available to address user requirements. Instantaneous scales can be found in Level 2 data records, while longer time resolutions (seasonal or annual) can be easily generated from the monthly data products. It should be noted that while many data products adhere to common data exchange format (e.g., Hierarchical Data Format, HDF,

Network Common Data Form, netCDF, etc), others do not, using a "flat binary" format: consequently, users may choose to use the former for ease of processing.

Each of these different precipitation products each represent estimates of the true value. Initial processing of the data, such as quality control of the raw data, followed by the computational procedures that are used to convert the observations, whether satellite or surface data, into the precipitation products have assumptions that can lead to potential errors. For surface data sets this may include, for example, the mapping of the point gauge data onto a regular Cartesian grid, or the remapping of the radial radar measurements onto a regular grid. For satellite data sets, there are inherent errors and associated with precipitation retrieval schemes, while combining and mapping the data with different native resolutions to a regular grid may introduce remapping errors. It is therefore clear that any of the precipitation products will have not only known errors, but also unknown errors.

## 5.3.1 Surface Precipitation Products

### 5.3.1.1 Gauge Data

The only observations that provide good spatial coverage and length of record are rain gauges, even with large gaps over the oceans and in sparsely populated regions. Gauge data represent the closest set of observations we have to the true precipitation measurements. While the various errors and uncertainties associated with the different gauge design (wind effects, evaporation, and side-wetting, etc) need to be considered in the analysis, perhaps the greatest challenge is the gathering, quality-control, and archiving of these data records from the wide range of international weather organisations. The WMO sets standards to ensure a degree of commonality, such as gauge design, siting of gauges, common temporal sampling (e.g., 3 hourly, daily, etc.) and sufficient data latency for user applications. However, there is a large range of user requirements, such as flood monitoring and forecasting that requires fine scale near real time data which contrasts with the requirements of (for example) climate applications which needs long term stability in data records. The number of gauges, as noted in Sect. 5.1, may be encouragingly high, but this number typically includes all the gauges that have existed and does not represent the number currently in operation, or those with a long-term data record. Indeed, for the 3-hourly GTS data fewer than half of these stations regularly report rainfall measurements.

A spatially and temporally consistent precipitation product generated from the point gauge data requires much care. While the gauge is deemed to represent the rainfall amount in its immediate vicinity (a few metres to preferably a few kilometres), at instantaneous scales this distance is typically very small (Habib et al. 2001) and is dependent upon the meteorology of the precipitation event and the local topography. Ideally, multiple gauges should be present within each grid box of the mapped precipitation product. Interpolation of gauge data should be used with great caution, particularly for instantaneous or short-duration events, and then only with interpolation procedures designed for precipitation. However, since gauge data used in precipitation products are often accumulated over longer periods (e.g., daily for monthly), the data may be deemed to be representative of the regional precipitation regime (Bell et al. 1990).

It should be noted that not all gauge observations may be publicly accessible. First, the availability of data from different countries/regions depends upon local administrations, with more than one agency often responsible for rainfall data. Additional and potentially large numbers of gauge observations are collected by commercial organisations (e.g. water companies), but these are generally restricted due to commercial reasons, may not be fully archived or achieve high levels of quality control needed for robust precipitation products.

Of great interest to many users are gauge-derived global precipitation products that form the de facto standard precipitation measurements. The benchmark for these global precipitation products is that of the Global Precipitation Climatology Centre (GPCC; Becker et al. 2013). About 180 institutions provide data to the GPCC from about 85,000 gauges at a daily scale. This data is carefully quality controlled for accuracy and consistency before being mapped onto a regular latitude/longitude grid. The distribution of these gauges globally varies greatly: over regions such as western Europe, United States, eastern South America, India, and the more populated regions of Australia many gauges exist. However, other regions have sparse gauge networks and, of course, over the oceans very few island-based gauges are available (Fig. 5.25a). This distribution therefore impacts the representativeness of the final product. The GPCC generates a series of global precipitation products, starting with a "first guess" $1^\circ \times 1^\circ$ monthly product based upon ~7500 SYNOP station reports, generated 5 days after the end of the month. A "full" data product, at $1^\circ \times 1^\circ$ daily and $0.5^\circ \times 0.5^\circ$, $1^\circ \times 1^\circ$, $2.5^\circ \times 2.5^\circ$monthly scales is generated from about 85,000 gauges 1–2 years after data collection (Figure 5.25b): this delay is necessary to gather the data and process it (Becker et al. 2013). Also available from the GPCC are anomalies based upon these monthly gauge products, allowing users (e.g., water resources, climate, etc) to easily assess changes in precipitation across the global land area (see Fig. 5.25c).

Other global gauge precipitation products are available that have been created using different procedures (e.g., quality control, mapping). It should be noted however, that these data products incorporate similar or the same gauge information as the GPCC product, rather than providing unique information from additional gauges. The Climate Prediction Center (CPC) Gauge-Based Analysis of Global Daily Precipitation (Xie et al. 2010) and the Global Historical Climatology Network (GHCN; Menne et al. 2012) provide daily gridded precipitation products although the number of available gauges varies considerably by region and/or year, with a maximum of just over 30,000 stations (the United States accounting for about half of that number). Snow depth data products are also generated by the GHCN from about 17,000 stations (virtually all in the United States). The Climate Research Unit at the University of East Anglia global precipitation dataset (Mitchell and Jones 2005) concentrates on providing a consistent long term historical precipitation records for climate applications.

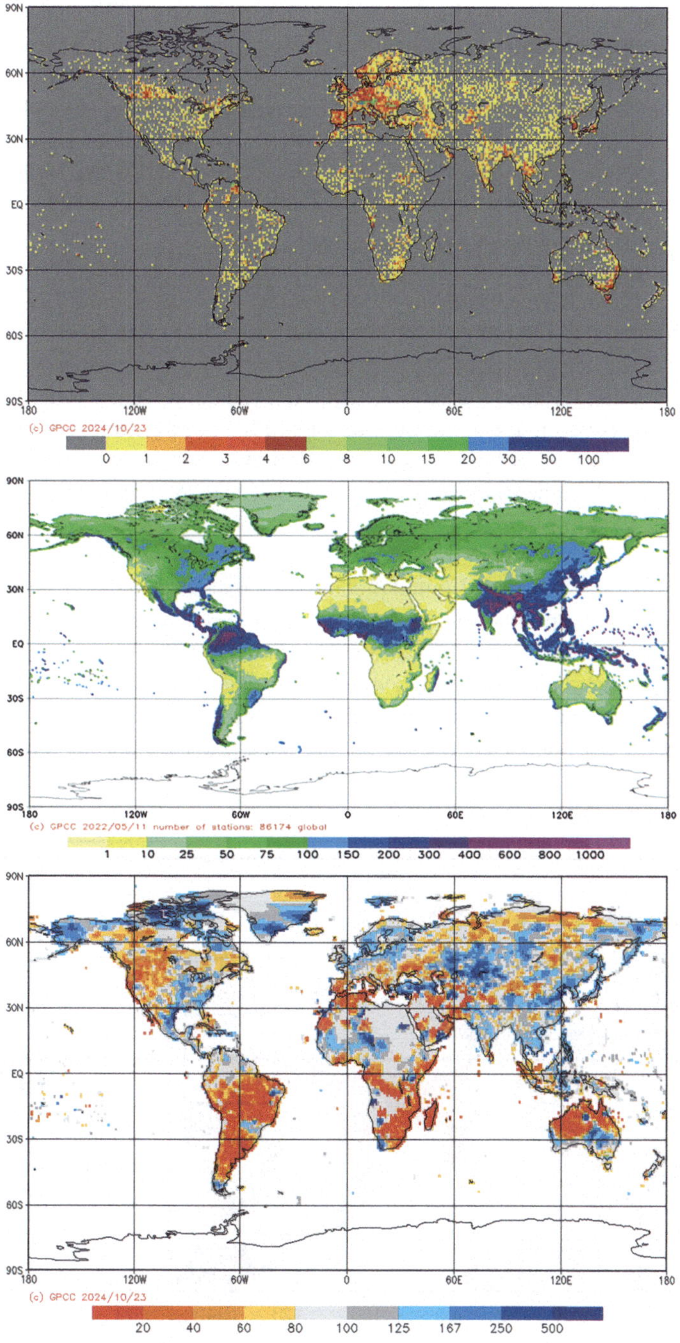


**Fig. 5.25** GPCC precipitation products for (**a**) July 2024 number of available stations per 1° × 1° grid box; (**b**) GPCC 0.25° × 0.25° precipitation product (mm/month) and; (**c**) % of normal (1951–2000) precipitation for July 2024. (Source: https://www.dwd.de/EN/ourservices/gpcc/gpcc.html, © The Authors under CC-BY-4.0 license)

Regional precipitation products are generated to exploit additional gauge data that is made available through local data exchange agreements, often providing data with limited geographical coverage and time periods. The largest regional scale network, the Asian Precipitation—Highly-Resolved Observational Data Integration Towards Evaluation of Water Resources (APHRODITE), produces precipitation products across four main regions including Russia, the Middle East, Monsoon Asia and Japan (see Fig. 5.26). While the data set only extends back to 1951, it is based upon the locally available dense gauge networks provided by contributing countries. Gridded daily products are generated at nominal resolutions of 0.25° × 0.25° and 0.5° × 0.5°, although over Japan this is 0.05° × 0.05°. A total of about 12,000 gauges are incorporated into the Eurasian product, and about 8000 gauges in the Monsoon Asia product.

Over the United States, the NOAA Climate Prediction Center (CPC) creates a gauge analysis covering the conterminous United States on a fine-resolution grid in near real time (see https://psl.noaa.gov/data/gridded/data.unified.daily.conus.html and Fig. 5.27). These products are available at 0.25 × 0.25 daily (and coarser) resolutions from 1948 through to the present.

Surface-based precipitation products over the ocean are scarce, not lease due to the paucity of data. However, limited precipitation data is available from island and atoll gauge data, shipborne measurements and inferred from shipborne observations. The Comprehensive Pacific Rainfall Database (PACRAIN) contains daily and monthly precipitation records from a variety of sources across the tropical region of the Pacific basin. The Pacific Islands Global Climate Observing System program is designed to fill in the gaps in observations and provide capacity building and

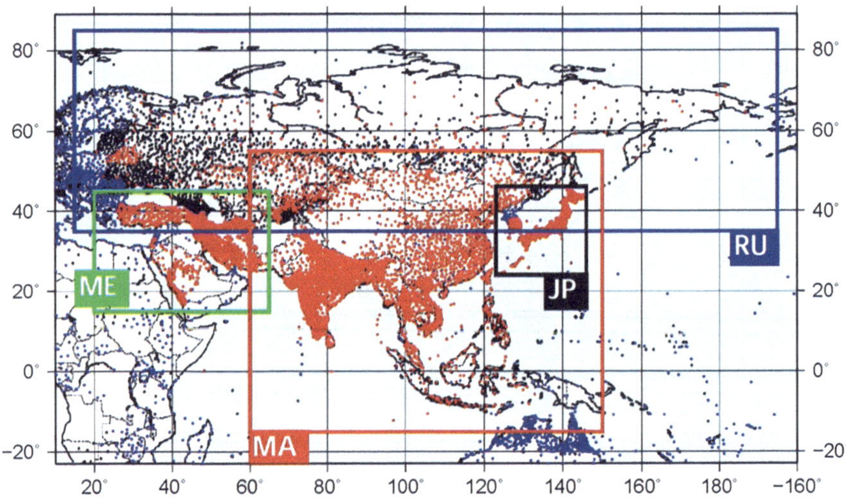


**Fig. 5.26** Regions of the APHRODITE precipitation products together with the gauge locations used in the product. Blue dots represent GTS-based data, black pre-compiled data, and red individual collections. (Public domain)

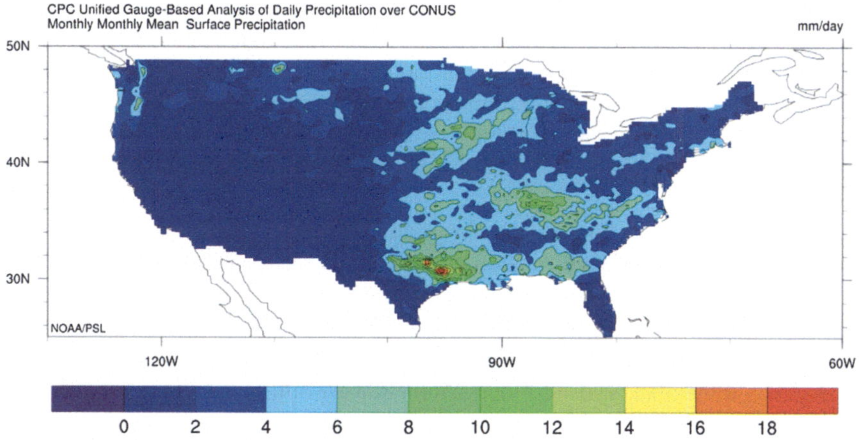


**Fig. 5.27** Example of the NOAA CPC unified gauge-based analysis of daily precipitation over CONUS for May 2024. (Data downloaded and plotted from https://psl.noaa.gov/data/gridded/data.unified.daily.conus.html)

technology transfer to Pacific Island Nations allowing the expansion of the network and increased collaboration with local meteorological agencies. The PACRAIN database includes quality control, observation and data entry standardisation: limited data records starting in 1847 are available and have been developed into a data set with daily observations from up to 710 sites (Greene et al. 2008), a third of which are from atoll sites which may be deemed representative of the open ocean.

Shipborne present weather observations from the Comprehensive Ocean–Atmosphere Data Set (COADS) have been analysed to allow the characteristics of oceanic precipitation to be determined (see Petty 1995). Up to 70 years of data is now available and has been studied to provide, for example, the frequency of precipitation over the oceans (Petty and Tran 2023). Although these shipborne data are visual observations which require assumptions to generate a precipitation product, they provide unique and essential information for the verification and validation of satellite products and model analyses. More quantitative information, albeit with limited coverage, has been generated through the OceanRAIN dataset that provides shipborne measurements of precipitation from research vessels operating worldwide (Klepp et al. 2018).

#### 5.3.1.2 Surface Radar Products

The coverage provided by operational radars on a global scale is somewhat limited, being concentrated primarily over the United States, Canada, Europe, western Russia, Japan, South Korea, Australia, New Zealand, India and parts of South America. Although these are regions where the density of gauge data is generally adequate and despite some repetition of coverage, radars can provide regular spatial measurements at fine time scales that fill a niche in the measurement of precipitation, at least on a local to regional scale. Data from radar networks that meets a wider range of user requirements is available primarily from the United State, Europe and Japan, although single site radars do provide high quality radar data that may be used for hydrological, satellite or model validation. While the value of high temporal and spatial resolution products are evident, it should be noted that even with meticulous quality control, artefacts may still be present in the data which may not be obvious at the fine scale but become evident when generating longer term accumulations.

Radar data over the United States forms an integral part of their weather service, particularly for providing information on severe whether such as tornadoes or tropical storms. It is perhaps no surprise therefore that data from their radar network is of high quality and easily obtainable. The National Weather Service's (NWS) 160 Weather Surveillance Radars—1988 Doppler (WSR-88D; Zhang et al. 2016) provide observations that feed into their Multi-Radar/Multi-Sensor System (MRMS; Kirstetter et al. 2020). The MRMS integrates data streams from these radars, surface and upper air observations, lightning detection systems, satellite observations, and forecast models, and generates several products including hail detection, quantitative precipitation estimates, etc. In particular, the generation of the products at fine temporal and spatial scales, 0.01° × 0.01° and 2 min, makes them extremely important for flash flood forecasting (e.g., Flash Flood Monitoring and Prediction System, FFMP, system) and river and water resources forecasting where timeliness is critical. Included with the precipitation product are quality indices and precipitation type flags. An example of the MRMS precipitation data is shown in Fig. 5.28,

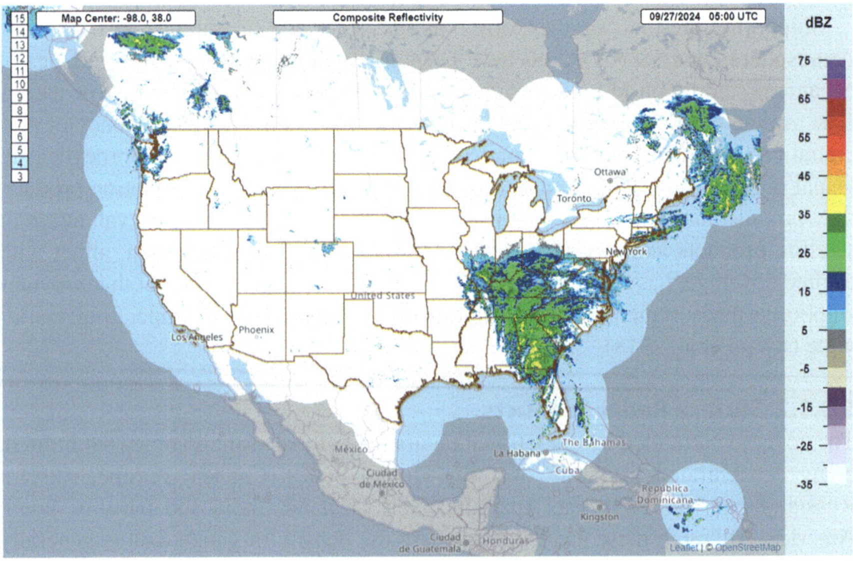


**Fig. 5.28** MRMS radar composite over the United States and southern Canada on 27 September 2024 at 0500 UTC. The large region of precipitation is from Hurricane Helene. (Data from https://mrms.nssl.noaa.gov/qvs/product_viewer/)

showing the precipitation associated with Hurricane Milton over the southeast of the United States.

Over Europe regional meteorological organisations generate radar-based precipitation products. For example, radar data over the United Kingdom is made available by the UK MetOffice through the Centre for Environmental Data Analysis (CEDA). A total of 16 dual polarised Doppler radars are composited to generate mapped precipitation products at 1 km × 1 km, 5-min resolution over the UK (Kitchen and Illingworth 2011). The processing data system applied are series of algorithms to reduce anomalous propagation errors, correct for variations in the vertical profiles, attenuation and for variations in the DSD (Harrison et al. 2009, 2012). A Europe-wide radar product is available through the EUMETNET Operational Programme for the Exchange of Weather Radar Information (OPERA; Huuskonen et al. 2014). A total of 200 (mostly C-band) radars from across Europe contribute to the product which generates mapped precipitation at a 2 km × 2 km 15-min resolution (Saltikoff et al. 2019). An example of the OPERA radar product is shown in Fig. 5.29.

Other regions of note include Japan, Australia, South Korea, India and some countries in South America. Japan operates the Radar-Automated Meteorological Data Acquisition System (AMeDAS) which ingests data from 20 ground-based weather radars and gauge information from 1320 automatic weather stations to generate a 2.5 km × 2.5 km 30-min product (Makihara et al. 1996). Over Australia 70 C- and S-band radars used to generate a nationwide precipitation product (Soderholm et al. 2022), while over South Korea radar precipitation products are generated from 26 radars (mostly C- and S-band) from three agencies (Lee et al. 2012).

### 5.3.2 Satellite Precipitation Products

The ability of satellite sensors to observe the whole globe allows truly global precipitation products to be created. In particular, the provision of these products over data-sparse regions, both land and ocean, makes their data indispensable for many global scale studies. Satellite precipitation estimates are available at different processing levels, from the individual satellite sensors at their native, instantaneous resolutions, through to multi-satellite, multi-sensor products that are calibrated by, or incorporate, surface reference data. The selection of an individual product will depend upon the user requirements: while certain data sets may provide fine scale precipitation data in near real time, their accuracy may not be as good as a multi-sensor product that is available later. For many of these satellite precipitation data sets the reader is directed to the GCMS IPWG web site which tabulates the available data, together with point of contacts (see https://www.eorc.jaxa.jp/IPWG/data/datasets.html).

#### 5.3.2.1 Visible/Infrared Products

Precipitation products generated from the visible and infrared observations provide good temporal and spatial resolution estimates. Visible observations rely upon the

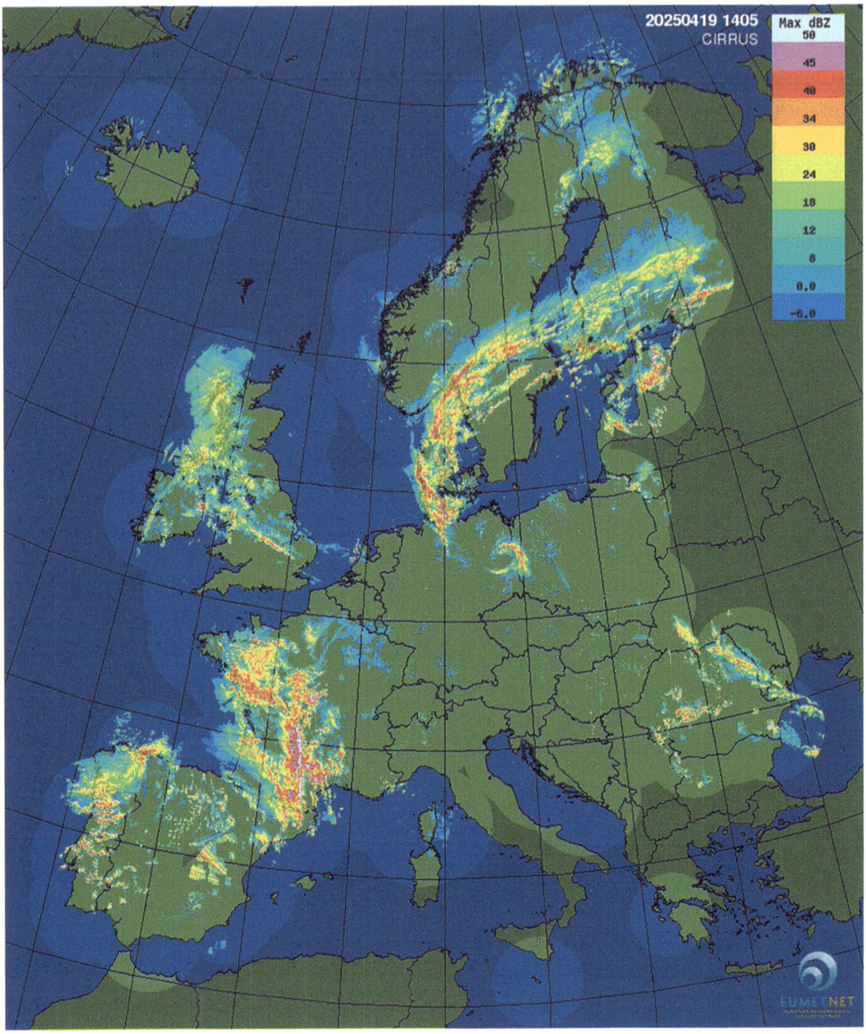


**Fig. 5.29** An example of a EUMETNET OPERA radar composite product at 1735 UTC on 10 October 2024. (Available from https://www.eumetnet.eu/observations/opera-radar-animation/)

reflected solar radiation, consequently, these can only be usefully employed during daylight, and often only to within a few hours of dawn/dusk due to changes in the illumination conditions. Many of the visible/infrared products therefore rely upon infrared observations that measure the thermally emitted radiation from the surface and clouds. The latter allows the cloud top temperatures to be measured and their relationships with the surface precipitation to be utilised. While combined visible/infrared techniques exist, these are really two techniques using visible/infrared observations during the daytime and an infrared-only observations at night. Most visible/infrared retrieval products are based upon the observations made by the

GEO sensors since, despite the relatively poor cloud top to surface precipitation relationship, the temporal and spatial sampling compensates to a certain degree. Machine learning techniques have been used with a good degree of success to extract additional information from the multi-spectral observations provided by the newer GEO sensors (e.g., SEVIRI, ABI/AHI), although these too rely upon the generally indirect cloud top to surface precipitation relationship.

Perhaps the best known and most used single sensor precipitation product is the Global Precipitation Index (GPI; Arkin and Meisner 1987). In fact, the GPI is a single channel technique relying purely upon thresholded thermal infrared observations to produce a precipitation estimates: any IR temperature colder than 235 K is deemed to be precipitating at 3 mm $h^{-1}$. While the instantaneous estimates are clearly poor, the relationship is surprisingly robust when used over larger temporal/spatial scales (e.g., daily, 2.5° × 2.5°). While the GPI products are available online, it is sufficiently simple that most users can easily generate it directly from (calibrated) IR imagery. Other GPI-like products are also available as "adjusted" GPI products: since there is no direct IR to precipitation relationship each IR algorithm is calibrated so that the rain/no-rain threshold and the rain rates vary through comparison with reference data sets. These variables may be calibrated/applied regionally and/or globally and over time spans from hours to seasonal.

Global precipitation products generated by more complex schemes are available that overcome some of the drawbacks of simple threshold schemes like the GPI. The Precipitation Estimation from Remotely Sensed Information using Artificial Neural Networks (PERSIANN) Cloud Classification System (CCS) products are routinely available through the University of Irvine, CA (see https://chrsdata.eng.uci.edu/ and Hong et al. 2004). This product is based upon the extraction and categorization of cloud-patches based on cloud height, areal extent, and texture using GEO observations. The conversion of the observations to rain rates is done using predefined specific rain-rate and brightness temperature curves. Products are available in near real time at hourly 0.04° × 0.04° resolution.

The Self-Calibrating Multivariate Precipitation Retrieval (SCaMPR) products (see https://www.star.nesdis.noaa.gov/smcd/emb/ff/SCaMPR.php) use observations from sensors onboard the 5 operational GEO satellites. Five infrared bands are used, closest to 6.0, 7.0, 8.5, 11.0, and 12.0 µm, with the exact wavelengths depending on the particular sensor. In addition, the texture of the 11 µm band is used to help define convective cells. The Climate Prediction Center (CPC) combined microwave (MWCOMB) data set is used as the calibrator for the product with adjustments for evaporation in dry atmospheres. Although initially developed for use over the US, the SCaMPR products are now available as a global product every 10 min with a nominal (sub-satellite) spatial resolution of 2 km × 2 km.

The Center of Hydrometeorology and Remote Sensing (CHRS) at the University of California Irvine (UCIrvine) generates several products through its PERSIANN framework (https://chrsdata.eng.uci.edu/; Hong et al. 2004). These are summarised in Table 5.12. The operational PERSIANN product uses a neural network function to classify and estimate precipitation using geostationary infrared data with adaptive training allowing the network parameters to be updated regionally and over

**Table 5.12** Summary of PERSIANN precipitation products produced by the Center of Hydrometeorology and Remote Sensing at the University of California Irvine

| Scheme | Spatial resolution | Temporal resolution | Coverage | Data range |
|---|---|---|---|---|
| PERSIANN | 0.25° × 0.25° | Hourly | 60°N–60°S | 2000–present |
| PERSIANN-CCS | 0.04° × 0.04° | Hourly | 60°N–60°S | 2003–present |
| PERSIANN-CDR | 0.25° × 0.25° | Daily | 60°N–60°S | 1983–present |
| PDIR-now | 0.04° × 0.04° | Hourly | 60°N–60°S | 2000–present |
| PERSIANN-CCS-CDR | 0.04° × 0.04° | 3-hourly | 60°N–60°S | 1983–present |

time. The precipitation product is generated at a resolution of 0.25° × 0.25° daily, covering 60°S to 60°N. A real-time global high resolution 0.04° × 0.04° hourly PERSIANN-CCS precipitation product is available from 2003 to the present. This product generates a precipitation estimate by identifying cloud patches based upon their texture, geometry, evolution and cloud top height: these factors are then used to determine the infrared-intensity relationship to assign precipitation intensities. Based upon PERSIANN, the Dynamic Infrared Rain Rate near real-time (PDIR-Now) has been developed to exploit the high temporal sampling of the GEO IR sensors to produce 0.04° × 0.04° hourly precipitation within 15–60 min of data collection, necessary for hydrological applications, such as flood forecasting and mapping (https://irain.eng.uci.edu). The product makes use of a dynamic IR-precipitation intensity relationship based upon climatology.

A 40+ year precipitation product is available in the form of the PERSIANN-Climate Data Record (PERSIANN-CDR). This scheme generates daily rainfall estimates at 0.25° × 0.25° resolution over 60°S to 60°N. The aim of this product is to address the need for a consistent, long-term, high-resolution (quasi) global precipitation dataset for studying the changes and trends in daily precipitation and extreme precipitation events. The product is generated using the infrared data from the GridSat remapped GEO sensors and adjusted to the Global Precipitation Climatology Project (GPCP) monthly product (see below) to ensure consistency throughout the data record. Alongside the PERSIANN-CDR product, is a CDR version of the PERSIANN-CCS product: the PERSIANN-CCS-CDR. This product is available since 1983 at a resolution of 0.04° × 0.04° 3-hourly and incorporates the GridSat and CPC-4 km IR data together with bias adjustment using the GPCP data set.

Regional precipitation products are also available and offer several advantages, such as being tuned to the local meteorology and climatology, often at a higher resolution than global estimates and often available in very near real time. For example, the Tropical Applications of Meteorology using SATellite data (TAMSAT; Maidment et al. 2020) provides daily, pentad, monthly and seasonal precipitation estimates for the whole of Africa at a resolution of 4 km × 4 km (see Fig. 5.30). The technique uses a linear cloud-top temperature to rainfall relationship which is calibrated locally with rain gauge data: near real time gauge data currently allows operational products to be generated and delivered for several applications, such as climate monitoring and risk assessment.

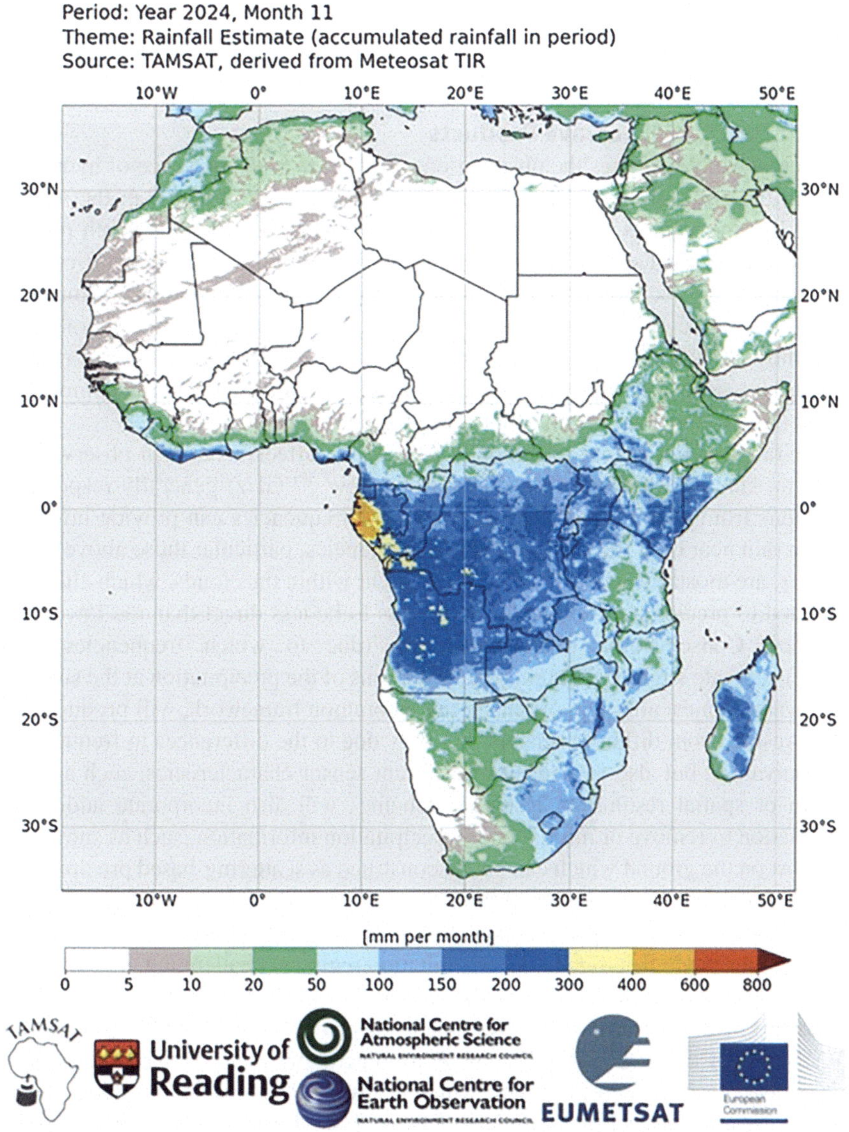


**Fig. 5.30** Example of the University of Reading TAMSAT product for November 2024. (See https://research.reading.ac.uk/tamsat/latest-maps/)

The main advantage of the visible/infrared products is the availability of frequent (~15 min), good resolution (1–4 km) observations that are available, particularly from GEO sensors. These products can achieve reasonable results even with relatively simple techniques, and additional information on cloud top properties/microphysics can be extracted. The disadvantages are that the visible imagery is limited to daylight and the cloud-top to surface precipitation relationship is indirect: not all clouds precipitate, and not all precipitating clouds are identified, consequently many products require reference data for calibration.

#### 5.3.2.2 Passive Microwave Products

At microwave frequencies the observations are related to the presence of hydrometeors in the atmosphere and are therefore physically more direct than the visible and/or IR observations. However, the sampling by PMW sensors is much poorer, with each individual sensor providing typically no more than two observations per day at any given location on Earth. Such sparse sampling requires use of multiple satellites to achieve fine-scale temporal coverage. Due to the poor temporal sampling most PMW precipitation products are either provided at Level 2 (full resolution instantaneous) or Level 3 (gridded, accumulated), where the accumulation period (e.g., monthly) overcomes the sparse temporal sampling.

The range of PMW sensors and their frequencies affects what their observations relate to: since the lower frequencies (at or below 37 GHz) generally respond to emissions from liquid water, sensors with these frequencies can provide information on rain near or at the surface. Higher frequencies, particular those above about 90 GHz, are mostly responsive to the ice content within the clouds, which although is related to precipitation to the surface, tends to be less direct than the lower frequencies. Consequently, different sensors (due to which frequencies they sesnse) generate slightly different representations of the precipitation at the surface. Retrieval schemes, although often sharing a common framework, will produce different results from different sensors, not only due to the differences in frequencies of observation, but also as a result of different sensor characteristics, such as scan pattern or spatial resolution. Retrieval schemes will also incorporate additional information to remove or mitigate non-precipitation information, such as snow, ice, and frost on the ground which can be misconstrued as scattering-based precipitation signals. Finally, it should be noted that all precipitation schemes and their products find snow/ice-covered surfaces, coastal regions (due to land/water mix) and complex terrain (due to heterogeneous precipitation regimes) challenging.

Central to the GPM mission is the retrieval of precipitation using a common framework. Building upon the development of precipitation retrievals scheme for TRMM, GPM uses the GPROF scheme to generate retrievals for all the passive microwave sensors except SAPHIR, which uses the PRPS scheme. Products generated by these techniques are available as Level 2 and Level 3 data sets through the GPM Precipitation Processing System (PPS) data repository (https://gpm.nasa.gov/data) or through the Goddard Earth Sciences Data and Information Services Center (GES DISC; https://disc.gsfc.nasa.gov/). The Level 2 data products provide instantaneous precipitation intensities at the retrieval resolution of the GPROF sensor (see Table 5.6). While the GMI and AMSR2 retrievals are at medium resolution of 10.9 km × 18.1 km and 14 km × 22 km respectively, the SSMIS retrievals are much coarser at 45 km × 74 km. Care is therefore needed in comparing the retrievals from different sensors even if using the same retrieval scheme. Level 3 gridded data is also available from both PPS and GES DISC, although as noted above, care is needed to ensure that the products derived from the different satellites is sufficient to provide a representative sample of the precipitation over a given time. An example of the GPROF product for the GMI sensor is shown in Fig. 5.31 below, together with the DPR overpass. The example is of the landfall of Hurricane Ian at 1741 UTC on 29 August 2021 and clearly shows the differences in the resolution between the GPROF retrieval (10.9 km × 18.1 km) and the DPR product (5.4 km × 5.4 km), particularly over eye wall and adjacent rain bands.

The EUMETSAT Hydrological Satellite Application Facility (H SAF) provides a range of precipitation products at Level 2 and Level 3 at https://hsaf.meteoam.it/Products/ProductsList?type=precipitation (Mugnai et al. 2013; Sanò et al. 2015). These include the Level 2 products for the SSMIS (H01), MHS (H02B), ATMS (H18) and a combined Level 3 gridded instantaneous precipitation product (H68). The SSMIS product is generated in near real time using a physically based Bayesian scheme using an a priori database populated with cloud/dynamic-radiation database derived from a cloud resolving model. For the MHS and ATMS products the Bayesian scheme is replaced with one based upon a neural network approach. Figure 5.32 shows the H SAF product generated for the SSMIS, MHS and ATMS for a single overpass (mapped onto the Meteosat projection). Meanwhile, the EUMETSAT Climate Monitoring Satellite Application Facility (CM SAF) provides a climate-scale precipitation product as part of the Hamburg Ocean Atmosphere Parameters and Fluxes from Satellite Data (HOAPS) with a resolution of monthly 0.5° × 0.5°.

An ensemble of Level 2 precipitation products from the conically scanning passive microwave sensors was generated under a NASA Making Earth System Data Records for Use in Research Environments (MEaSUREs) project (see https://www.earthdata.nasa.gov/about/competitive-programs/measures). The sensors include the AMSR-series, TMI, GMI SMMR and the SSM/I and SSMIS series, covering the period 1978–2020. Each data product file contains the retrievals from 15 different techniques, along with a quality score for each technique. All these retrievals are at the native resolution of each of the sensors to facilitate the intercomparison of the products: this permits some of the fundamental characteristics of each techniques precipitation retrievals to be compared. Figure 5.33 shows the distribution of

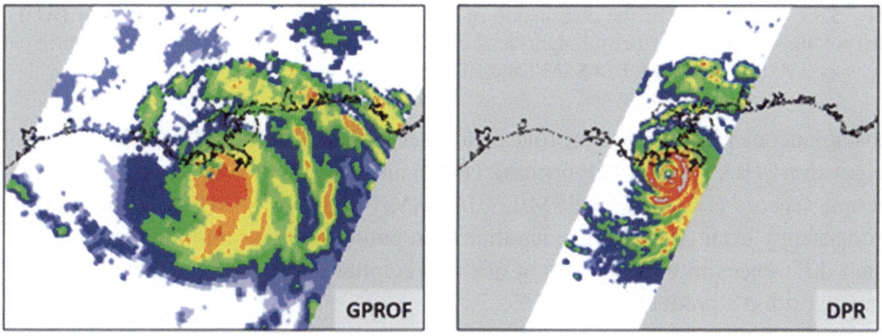


**Fig. 5.31** An example of the GPROF-GMI Level 2 product for Hurricane Ida over the Mississippi delta at 1741 UTC, 29 August 2021 together with the DPR product

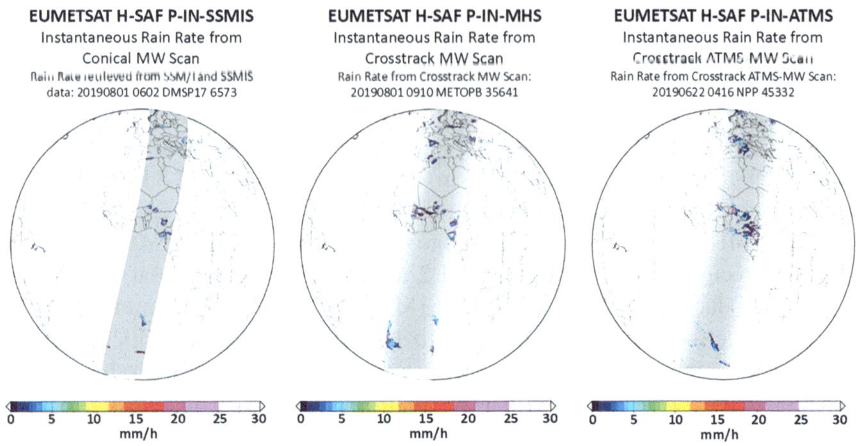


**Fig. 5.32** Examples of the EUMETSAT H SAF Level 2 products for the SSMIS (F17, 0602 UTC, 1 June 2019), the MHS (MetOp-B, 0910 UTC, 1 June 2019) and the ATMS (NPP, 0416 UTC, 22 June 2019). (Source: EUMETSAT H-SAF)

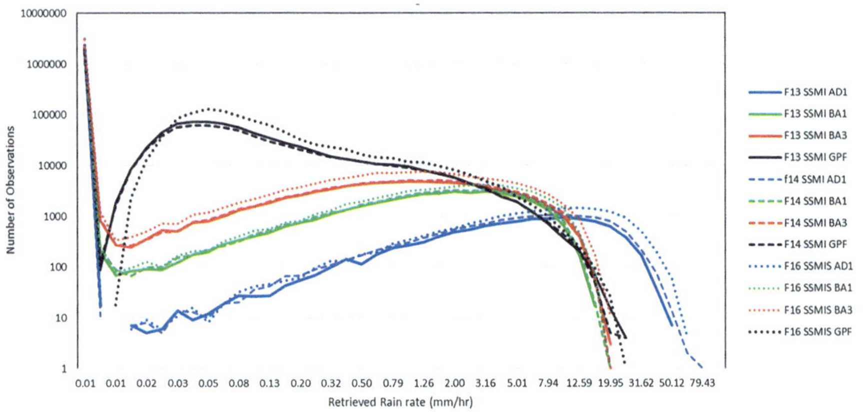


**Fig. 5.33** Comparison of the distribution of instantaneous precipitation for the Adler (AD1), Bristol algorithm1 (BA1) Bristol algorithm2 (BA2) and GPROF (GPF), and for three different sensors (DMSP F13 SSM/I, F14 SSM/I and F16 SSMIS)

instantaneous precipitation for four difference techniques, the Adler (AD1), Bristol algorithm1 (BA1) Bristol algorithm2 (BA2) and GPROF (GPF), and for three different sensors (DMSP F13 SSM/I, F14 SSM/I and F16 SSMIS). While there is consistency from one sensor to another for a particular algorithm, there are significant differences in the distribution of the precipitation intensities between each of the algorithm's products.

### 5.3.2.3 Satellite Active Microwave/Radar Products

The most direct measure of precipitation from space from precipitation radars. However, since only two precipitation radars have operated (the TRMM PR and

GPM DPR), each with relatively narrow swaths, the amount of at-surface precipitation information available from them is extremely limited. While the CloudSat CPR also provides precipitation products, the CPR was a single beam radar designed primarily for cloud studies, rather than precipitation retrievals. Nevertheless, data products from both the precipitation radars and cloud radars can provide useful information either for instantaneous case studies (where overpasses are available) or where multiple observations can elucidate the characteristics of precipitation over the longer-term.

Precipitation radar products are available for the TRMM PR and GPM DPR from the NASA PPS (see https://gpm.nasa.gov/data). Products are available at Level 2 (instantaneous, native resolution) for each satellite orbit. The Level 2 products are designed to determine the general characteristics of the precipitation, correct for attenuation, and estimate profiles of the precipitation water content, or rainfall rate. For the DPR the particle size distributions in the phase of the precipitation (i.e., liquid or frozen) are also provided. Note that due to greater sensitivity, the DPR will provide better estimates of rainfall and snowfall rates than those of the TRMM PR data. The PR data is available with a 4.3 km × 4.3 km spatial resolution and 250 m vertical resolution, while the DPR data is available with a 5.4 km × 5.4 km spatial resolution and 125 m vertical resolution. Level 3 DPR precipitation products are also available, derived from the Level 2 DPR results. The high-resolution 0.25° × 0.25° product includes the mean precipitation, standard deviation, and number of observations, while the low-resolution 5.0° × 5.0° product includes the mean precipitation, standard deviation, number of observations and a histogram of the distribution of the precipitation intensities. For both the Level 2 and Level 3 data the PR is available from December 1997–March 2015 and the DPR is available from April 2014–present.

The CloudSat CPR produces three precipitation products. The first is the 2C-PRECIP-COLUMN (see https://www.cloudsat.cira.colostate.edu/data-products/2c-precip-column; Haynes et al. 2009). Due to the greater attenuation experienced by the CPR, the retrieval algorithm makes use of the path integrated attenuation to determine if surface precipitation is present, and if so, the intensity. The former is calculated over all surfaces, the intensity only over the oceans. The 2C-PRECIP-COLUMN product provides input data (precipitation flag, freezing level, precipitation type and PIA) to the 2C-RAIN-PROFILE product (https://www.cloudsat.cira.colostate.edu/data-products/2c-rain-profile; L'Ecuyer et al. 2002) which infers vertical profiles of liquid precipitation and ice water content together with the associated surface precipitation intensity. Due to the sensitivity of the CPR, estimates of falling snow are possible and are represented by the 2C-SNOW-PROFILE product (https://www.cloudsat.cira.colostate.edu/data-products/2c-snow-profile; Wood et al. 2013, 2014). This product provides estimates of the vertical profiles of snowfall intensity together with the snow particle size distribution and snow water content, as well as providing an estimates of snowfall intensity at the (near) surface. Parameters and their uncertainties are then used to estimate uncertainties for the snowfall rates and snow water contents. Although all these products are only available along the ground track of the satellite, they provide useful long-term measurements that can be used to inform and/or verify other satellite-derived observations. Figure 5.34 shows the mean occurrence of snowfall from 2006 to 2010 as derived from the 2C-SNOW-PROFILE product.

## 5.3.3 Merged Multisource Products

The goal of the merged or multisource products is to provide improved precipitation estimates over that of any single-sensor or single band product. To do this a product may incorporate information from multiple sensors to reduce the errors and uncertainties inherent within a retrieval scheme or include multiple products to improve the temporal and/or spatial resolution of the precipitation estimate. Ultimately, combination schemes seek to take advantage of the strengths of individual products while minimising their weaknesses. However, the traceability of the error and uncertainties within such combined products becomes complicated since these also incorporate errors from the individual-sensor estimates.

### 5.3.3.1 Multi-Sensor

A good example of a multi-sensor product is the GPM Combined Radar-Radiometer Algorithm that exploits the co-temporal, co-located observations of passive (GMI) and active (DPR) microwave sensors to provide accurate, high-resolution estimates of the vertical distribution of precipitation to the surface. This high-quality data is important not only for user applications, such as studying precipitation storm structures, but also for providing a single common reference dataset that can be used to "cross-calibrate" precipitation estimates from the passive microwave radiometers in the GPM constellation. The latter is crucial for ensuring a consistent, high time-resolution precipitation record for climate and model validation applications. The

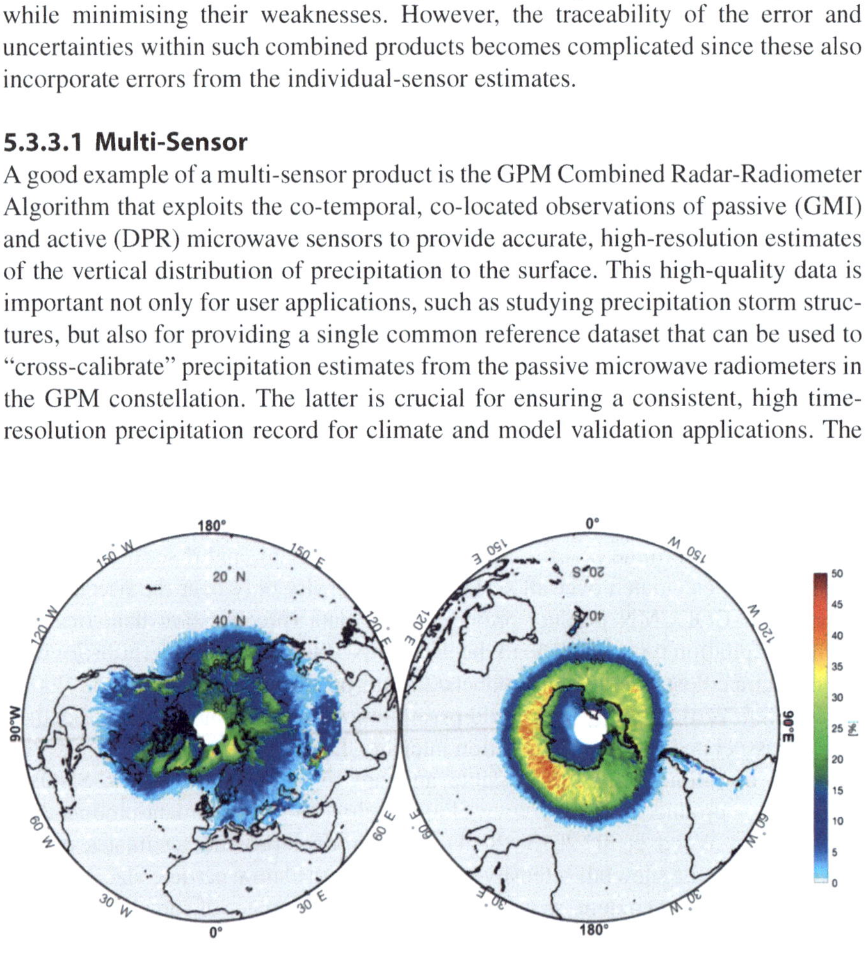


Fig. 5.34 CloudSat frequency of occurrence of frozen precipitation (snowfall) from 2006 to 2010. (Milani and Kidd 2023. © The Authors under CC-BY-4.0 license)

combined technique uses the PMW observations to provide more information on the atmospheric attenuation which is then fed into the precipitation radar retrieval scheme, thus providing an improved product. The current GPM combined product (https://gpm.nasa.gov/data/directory) is available in near real-time (<3 h), or within 40 h for the research product (ensuring highest quality control), and is provided with a horizontal resolution of 5 km × 5 km and a vertical resolution of 125 m covering the DPR-swath. A similar product is available for the TMI-PR combination, but with 250 m vertical resolution, extending the data record back to 1997.

#### 5.3.3.2 IR-PMW Techniques

The Climate Prediction Center morphing (CMORPH) product uses GEO IR-derived motion vectors to propagate the PMW precipitation estimates generated from the sensors on LEO satellites. The shape and intensity of the precipitation features are modified (morphed) in between the PMW observations by a time-weighted linear interpolation. The CMORPH product is available in near real time since 2002 with a 30-min 0.08° × 0.08°. CMORPH includes improved snow-screening process to help screen out anomalously high PMW rainfall due to ice or snow cover. A revised version of CMORPH, developed in 2007, takes advantage of a Kalman filter (KF)-based approach to the morphing of the PMW precipitation field: precipitation estimates are calculated by the forward and backward propagation of the LEO PMW-derived precipitation at the beginning/end times. If the gap between these two PMW observations is longer than 90 min, an IR-based precipitation estimate is used. The CMORPH product was reprocessed in 2017 to provide bias-corrected precipitation estimates to form a 0.08° × 0.08° 30-min CDR precipitation analysis back to January 1998. This reprocessing allows a single version of the Level 2 PMW precipitation product to be used ensuring consistency throughout the data record, while the CPC daily gauge analysis is used to bias-correct the final product.

The GPM mission provided an impetus to exploit the precipitation estimates generated from the GPM's constellation of PMW sensors combined the frequent infrared data GEO IR observations. The Integrated Multi-satellitE Retrievals from GPM (IMERG) is designed to produce fine time- and space-scale global precipitation estimates spanning the TRMM and GPM eras (1997–present) by inter-calibrating, merging, and interpolating all available precipitation estimates from PMW satellite sensors, together with microwave-calibrated infrared (IR) satellite estimates, precipitation gauge analyses, and other precipitation estimators. Inter-calibrated satellite precipitation estimates, and other data (including monthly surface gauge analyses) are utilised to generate half-hourly, 0.1° × 0.1° precipitation product, globally (see Fig. 5.35).

IMERG utilises precipitation estimates generated from the GPROF retrieval scheme as the main input data. The sensors include the conically scanning SSM/I, SSMIS and AMSR-series, together with the cross track ATMS, MHS and SAPHIR sensors. Since each type of sensor has different characteristics, inter-calibration is performed to ensure consistency between them. To provide the necessary temporal sampling, GEO-IR data are fed into the PERSIANN-CCS scheme and the CMORPH-Kalman Filter to create half-hourly estimates where no PMW estimate is

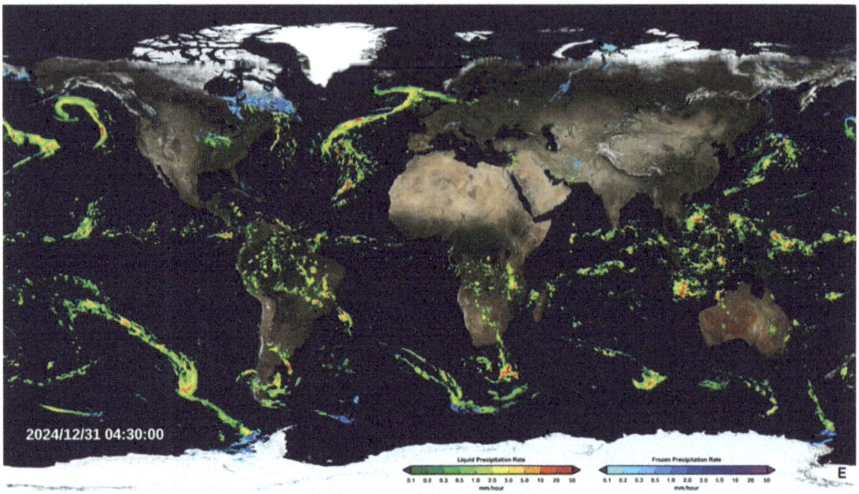


**Fig. 5.35** Visualisation example of the IMERG precipitation product for 0430 UTC on 31 December 2024. Note the inclusion of snowfall estimates (in blue shades) to provide a truly global precipitation product. (Source: NASA)

available. Note that the motion vectors are provided by the vertically integrated water vapour derived from models. To control any overall bias in the final product, the GPCC precipitation gauge analysis is used. Precipitation phase is determined using surface type, surface pressure, surface temperature, and surface humidity (see Sims and Liu 2015). The final precipitation product also incorporates ancillary data, namely surface type, snow/ice surface extent and surface temperature, relative humidity, and surface pressure. Three products are available, depending upon the latency (data collection to availability). An "Early" product is available 4 h after the observation time where time is important, for applications such as flash flood analysis. A "Late" product is generated 14 h after the observation time when most of the satellite estimates are available, for applications such as crop forecasting. The Final product is made available 3.5 months after observation time when all satellite data is available with the necessary quality control checks and calibration made against surface reference data. Retrospective processing has extended the data set back to start of the TRMM mission in 1997.

The Japanese GPM product, the Global Satellite Mapping of Precipitation (GSMaP) is provided by JAXA to provide hourly global precipitation at a resolution of $0.1° \times 0.1°$. Based upon early work with TRMM, GSMaP has been adapted and improved with the observations available from the GPM. Several products are generated, comprising of a standard product, a near real-time product, a real-time product and a reanalysis product (see Table 5.13). Each product, except for the real-time product, also has a gauge-adjusted version.

The core algorithms underlying the GSMaP products are based on the PMW precipitation retrieval algorithm (Aonashi et al. 2009), the PMW–IR combined

**Table 5.13** Summary of the products available for the GSMaP

| GSMaP Product | Abbreviation | Resolution | Latency | Update interval |
|---|---|---|---|---|
| Standard | GSMaP_MVK | 0.1 × 0.1 hourly | 3 days | 1 h |
| Standard gauge adjusted | GSMaP_Gauge | | | |
| Near real time | GSMaP_NRT | | 4 h | |
| Near real time gauge adjusted | GSMaP_Gauge_NRT | | | |
| Real time | GSMaP_NOW | | 0 h | 0.5 h |
| Reanalysis | GSMaP_RNL | | Occasional reprocessing of past time periods | |
| Reanalysis gauge adjusted | GSMaP_Gauge_RNL | | | |

After Kubota et al. (2020), reprinted under permission from Springer Nature

algorithm, and gauge-adjustment algorithm. GSMaP also accounts for orographic precipitation, snowfall estimation and changes in the sensors over the data record. Ancillary data used as operational inputs include the JMA global analysis (GANAL) and forecast (FCST) data set (6-hourly, 0.5° resolution) to atmospheric conditions for the look-up tables used by the PMW algorithms. The real-time product, GSMaP_NOW, is generated over the regions covered by the GEO IR observations (i.e., Himawari, Meteosat and GOES) with these real time products being made available 30 min after the observation time and updated every 30 min, with a 30-min forward extrapolation by applying cloud motion vectors. The near real-time product is available about 4 h after the observation time, while the standard product is available after 3 days to allow the full set of input data to be used. Other variants of the GSMaP precipitation product include the GSMaP reanalysis products (GSMaP_RNL and GSMaP_RNL_Gauge) which are calculated using the Japanese 55-year reanalysis (JRA55), and the RIKEN GSMaP Nowcast (GSMaP_RNC) which provides precipitation forecasts 6 h in advance using hourly-updated global precipitation data based on satellite observation. These products can be visualised and downloaded from https://sharaku.eorc.jaxa.jp/GSMaP/ (see Fig. 5.36).

The Global Precipitation Climatology Project (GPCP) precipitation product is well crafted and calibrated to suit a range of climatological applications. The GPCP was formed in 1986 as an international effort under the auspices of the World Climate Research Program (WCRP) with an initial focus on data access and distribution. The GPCP also initiated the Global Precipitation Climatology Center (GPCC) to collect and process global rain gauge data. The precipitation product of the GPCP is considered a climate data record, providing a consistent, globally complete analysis of precipitation through the integration of various data sets. Up to version 2.3 the precipitation product was generated at 2.5° × 2.5° monthly time scales which allowed earlier data sets to be incorporated. The latest GPCP product, Version 3.2 Precipitation Analysis, provides globally complete analyses of surface precipitation with a spatial resolution of 0.5° × 0.5° for monthly time scales beginning 1983, and daily time scales from June 2000 to the present. This version, as in

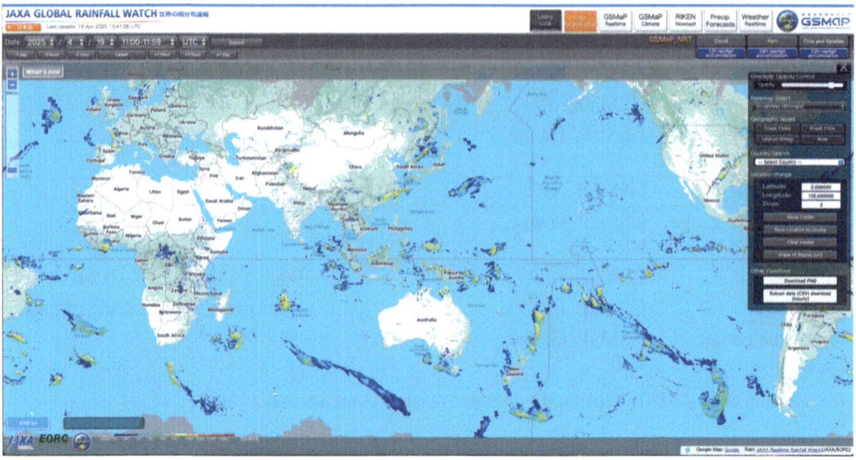


**Fig. 5.36** An example of the JAXA GSMaP dashboard at https://sharaku.eorc.jaxa.jp/GSMaP/ displaying the near real time (subsetted) precipitation map. (Source: JAXA)

previous GPCP versions, incorporates precipitation estimates from LEO microwave data, GEO IR data, sounder-based estimates, and surface rain gauge data exploiting the strengths of various inputs while aiming for temporal and spatial homogeneity. Version 3.2 also incorporates the latest techniques based upon improved intercalibrations between sensors, as well as new fine-scale multi-satellite estimates, allowing characteristics of precipitation to be better defined such as the liquid precipitation fraction (to define rain vs snow) for both the monthly and daily products, with the provision of a quality index in the monthly product. A full description of the new version 3.2 product is provided by Huffman et al. (2023). Figure 5.37 illustrates the Version 3.2 Precipitation Analysis for 2007.

The development of a comprehensively sourced precipitation product is the Multi-Source Weighted-Ensemble Precipitation (MSWEP) product. This global precipitation product provides $0.1^\circ \times 0.1^\circ$ 3-hourly data since 1979 within about 3 h from real-time. The product merges surface, satellite, and reanalysis data to generate a high-quality combined precipitation (Beck et al. 2017a, b). Gauge data used in the analysis is carefully screened to remove false zeroes (often used instead of "missing") and to adjust the reporting times to reduce mismatches with the other data inputs. Details of the scheme can be found in Beck et al. (2019) together with the range of input data and the weighting schemes used to combine the data, with the data available from https://www.gloh2o.org/mswep/.

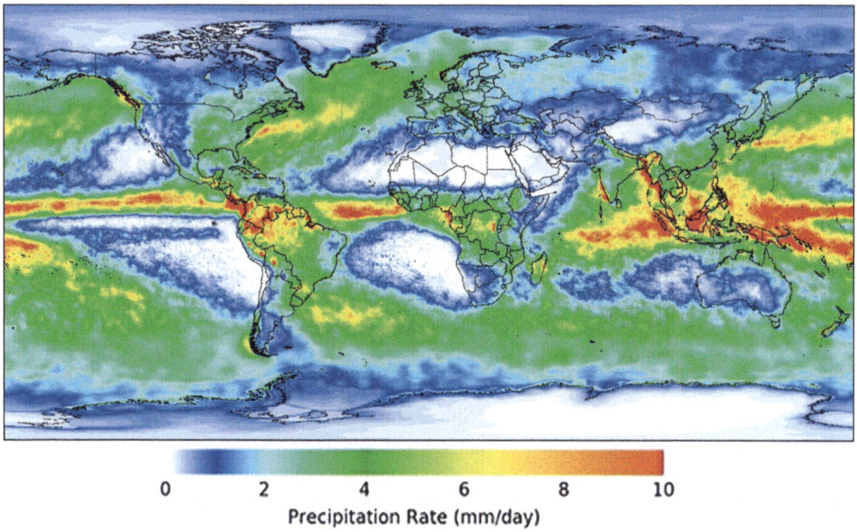


**Fig. 5.37** GPCP Version 3.2 Precipitation Analysis product of mean daily precipitation for 2007. Estimates are based on satellite data from the international constellation of IR, passive microwave, and other sensors selected for homogeneity (a Climate Data Records [CDR] approach), merged with analyses of surface precipitation gauge data contributed by numerous operational and research organizations around the world

## 5.4 Applications

Precipitation measurements are used in a wide variety of user applications. This section offers a small sample of the range and types of applications that surface, and satellite precipitation measurements are used in.

Table 5.14 provides examples of applications that use precipitation data obtained from gauge, radar and satellite observations. The data used in each applications depends largely upon the suitability (spatial and temporal resolution), the availability and the latency of the data sets.

### 5.4.1 Nowcasting

The nowcasting of precipitation allows the short-term prediction of upcoming precipitation events through changes in the position and intensity of rainfall. In particular, the nowcasting of precipitation, such as heavy rain or thunderstorms, is particularly important for extending the lead times in early warning for disaster preparedness, typically within the 5–6 h period. However, precipitation nowcasting solely through traditional short-term numerical models is problematic due to the complexity of precipitation systems, particularly at very fine scales (e.g., 1 km/5 min).

**Table 5.14** Examples of thematic areas, topics and applications where satellite precipitation estimates are being used for situational awareness and decision-making

| Area | Topic | Application |
|---|---|---|
| Extreme events and disasters | Flooding | Incorporation in hydrologic and routing models for flood estimation, streamflow, and inundation |
| | Landslides | Nowcasting of potential landslide activity, rainfall intensity, and duration characteristics for landslide occurrence |
| | Tropical cyclones | Improved characterization of tropical cyclone track, intensity, and 3D visualization |
| | Reinsurance | Definition of extreme precipitation thresholds to support microinsurance |
| | Wildfires | Management and situational awareness of rainfall accumulation in affected areas |
| | Disaster response | Situational awareness of extreme precipitation in potentially affected areas |
| Water resources and agriculture | Drought | Evaluation of precipitation anomalies leveraging extended temporal record |
| | Water resource management | Assessment of freshwater inputs to basins and reservoirs to quantify water fluxes |
| | Famine early warning | Inclusion of satellite precipitation in crop forecast modelling |
| | Food security | Assessing water crises that may impact water balance in vulnerable areas |
| Weather and climate modelling | Numerical weather prediction | Assimilation of Level 1 brightness temperatures within NWP modelling for initializing model runs |
| | Land surface modelling | Data assimilation into land surface models to estimate environmental variables |
| | Climate variability and change | Verification and validation of climate modelling precipitation outputs |
| Public health and ecology | Disease tracking | Tracking precipitation anomalies with environmental conditions for vector or waterborne diseases |
| | Animal migration | Monitor changes in precipitation that are associated with animal migration patterns |

Kirschbaum et al. (2017). 

The primary source of data for the nowcasting of precipitation is typically weather radar due to the regular and frequent observations (1–2 min), coupled with good spatial resolution (ca. 1 km). Techniques used range from the simple advection of precipitation to more complex schemes which incorporate addition information. The underlying premise is that past successive radar data (images) can be used to establish the movement of the precipitation which can then be used to extrapolate the future location of the precipitation. Such schemes are often used for real time flood management where nowcasting and short-term forecasting is crucial (Winterrath et al. 2012). Since the movement and dynamics of precipitation is often poorly captured by simple advection schemes, more advanced techniques have been developed to account for the speed of precipitation/weather systems, homogeneity of the precipitation (which affects identification of features used for tracking) and the dynamic nature of precipitation (e.g., initiation, growth, decay). In these techniques, not only is the movement of the precipitation considered, but also the amount of precipitation which may be enhanced/reduced through consideration of external factors (e.g., relief) and model data to help with the dynamics of the weather system. The Short-Term Ensemble System (STEPS; Pierce et al. 2012; Liguori et al. 2012) uses nowcasts for precipitation rates and accumulation based upon weather radar, extrapolated model analysis of forecast precipitation, as well as the uncertainties within the radar precipitation field (e.g., ZR relationships). The main factor in the success of these hybrid schemes is the variability in the meteorological conditions, with their benefits being best for the first hour. These schemes are generally more skilful than either the radar or model nowcasts used individually (Ballard et al. 2012).

The nowcasting of orographic precipitation is extremely important since heavy precipitation over small (<200 km$^2$) steeply sided basins often leads to flash floods and debris flows, typically with a 1–2 h lead time which is below the spatial/temporal resolution of models. Panziera et al. (2011) developed a scheme using radar data for the initial precipitation field, coupled with model information. The radar data provided 3-dimensional precipitation information, while the model provided low, medium, cross and upper air flow allowing the orographic forcing to be accounted for. Such schemes also allow the development of ensemble short-term forecasts: Nakakita et al. (2012) in their error ensemble approach showed improved 1 h accuracy in mountainous regions. Other approaches have used schemes such as the McGill Algorithm for Precipitation forecasting using Lagrangian Extrapolation (MAPLE) for nowcasting over complex terrain (Mandapaka et al. 2012). Machine learning (ML) models have the potential to allow complex precipitation nowcasting tasks to be performed in an efficient way through training on large datasets. Piran et al. (2024) studied the Soyang Dam basin in South Korea using weather radar to train their models and showed some improvements in short-term prediction of precipitation.

However, despite the range of schemes developed for the nowcasting of precipitation, the skills of such short-term forecasts are very much limited to a few hours with correlations typically of about 0.66 after 1 h, and 0.5 after 2 h. Ultimately, the skill is very much dependent upon the meteorological and background conditions, as well as the spatial resolution of the schemes. Nevertheless, for hazard warnings (e.g., floods, landslides, etc) the nowcasting and short-term forecasting of precipitation is extremely valuable.

### 5.4.2 Hydrology—Floods and Landslides

There are many hydrological applications that utilise precipitation data sets (Gebremichael and Hossain 2010). These include applications that require the data in (near) real time to enable the precipitation data to be incorporated into hydrological models to provide timely flood forecasting as well as for post-event analysis to allow lessons to be learnt from hydrologically significant events. Longer term precipitation data is of great value for water resource management, allowing the monitoring of water storage in catchments and reservoirs and the analysis of fine scale precipitation intensities over a climate-scale record allows intensity-duration statistics to be derived for use in recurrence analysis.

For near real time flood monitoring and mapping radar data is ideal due to its temporal and spatial sampling, where this is available. For flood prediction anywhere around the globe, satellite-based precipitation mapping is most appropriate with many near real time products available. However, to improve our understanding of the issues associated with flood dynamics, the spatial-temporal characteristics of the data are of great importance, as is the representativeness of the precipitation data. While gauge data provides the most direct measures of precipitation, the representativeness of gauges is a key element in determining suitability for flood monitoring and prediction. This is especially true where the gauge density is poor compared with the spatial variability of the precipitation, and particularly so in regions of varied topography with small drainage basins. While much work has looked at the use of satellite-based precipitation measurements in flood forecasting in particular, and hydrological applications in general, the spatio-temporal issues between the hydrological scales and precipitation scales remains problematic (Behrangi et al. 2011; Maggioni et al. 2013). For example, the hydrological community often incorporates satellite precipitation data for hydrological modelling since over poorly gauged or large river basins such data are often the only low-latency precipitation information that allows sufficient lead time for river flow prediction (Maggioni and Massari 2018).

The problem of relating the variability of the precipitation and the complexity of the hydrological system to the final river flow remains challenging (see Krajewski et al. 2019). However, access to near-real-time precipitation estimates on a global scale is valuable for regional assessments of current or potential flood events. For example, the TMPA precipitation estimates are used in the Global Flood Monitoring System (GFMS; http://flood.umd.edu/) with the goal of providing near-real-time flood estimates and forecasts between 50°N and 50°S (Wu et al. 2014). The GFMS couples a land surface model with a river tracing model to generate thresholds for flood detection. GFMS identifies potential flood areas in real time and provides temporal histories of flooding conditions at each pixel, including estimates of flood detection and severity (water depth above the threshold) and streamflow (and its flood threshold) (Wu et al. 2012, 2014). The products from the GFMS are designed to provide users with situational awareness of potential flooding conditions and associated triggering variables. The TMPA precipitation product has also been used to aid global precipitation forecasts from the Goddard Earth Observing System, version 5 (GEOS-5; Molod et al. 2015), extending hydrological calculations out to 5 days. This global model provides 0.125° resolution output, although finer-resolution (1-km) global streamflow, surface water storage, and inundation mapping estimates are available in real time. The current model takes advantage of improved resolution and accuracy of the IMERG precipitation product.

An example of the utility of a system such as GFMS is shown below in Fig. 5.38. In December 2014 Malaysia experienced very heavy rainfall that resulted in

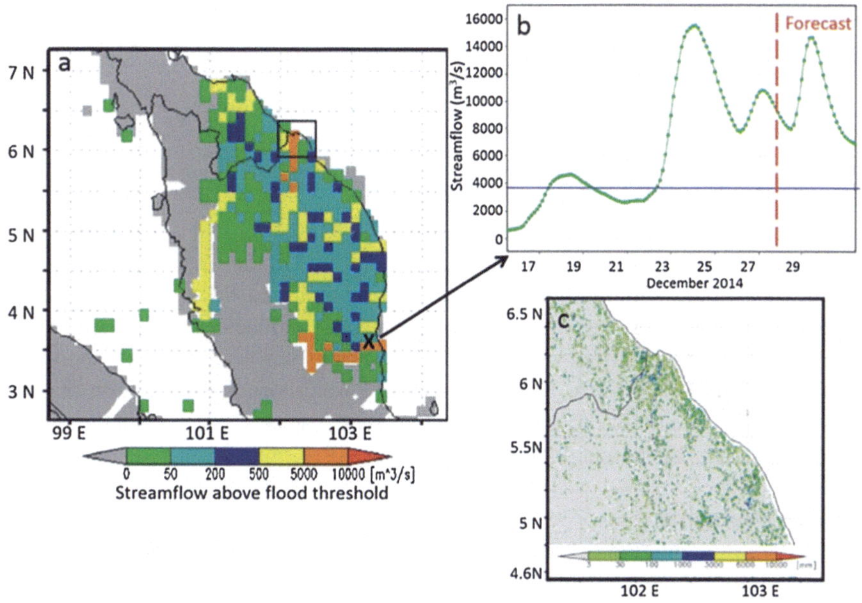


**Fig. 5.38** A case study utilising the GFMS tool applied to flooding across Malaysia in December 2014: (**a**) map of streamflow above flood threshold on 27 December, (**b**) river discharge history (for the "X" in (**a**)) from 27 to 31 December (left of the vertical red dashed line is the GFMS output, to the right is the modelled output), and (**c**) a high resolution 1-km flood inundation estimate (black box (a)). (Kirschbaum et al. 2017. © American Meteorological Society. Used with permission)

significant flooding, leading to 17 fatalities and USD $284 million in damage. The GFMS product (Fig. 5.38a) shows the streamflow above the flood threshold on 27th December. Continuous monitoring and modelling of the rivers allows the flow to be estimated, as shown in Fig. 5.38b for one of these rivers. Historical flood inundation information together with finer scale flood products (1 km) helps to identify the particular localities that are (or will be) affected by the flooding (Fig. 5.38c for the small box in Fig. 5.38a). Flood monitoring and prediction systems using satellite precipitation data, such as the GFMS, have an important role in supporting the rapid response to affected inundated areas and communities, particularly in developing countries with limited surface data, helping emergency responders and decision-makers.

A combination of antecedent conditions combined with periods of sustained rainfall may also induce landslides and debris flows. Near real-time GPM precipitation products are routinely used in the Landslide Hazard Assessment for Situational Awareness (LHASA) model (Kirschbaum and Stanley 2018). Through combining background fields of landslide susceptibility based upon slope angles, soil typography, etc. with the precipitation, potential landslide hotspots can be identified and monitored. The International Centre for Water Hazard and Risk Management (ICHARM) uses the JAXA GSMaP near-real-time data in the Integrated Flood

Analysis System (IFAS) in Asian regions such as Pakistan. JAXA has also collaborated with the Philippine Institute of Volcanology and Seismology (PHIVOLCS) studying the use of GSMaP for landslide warnings.

## 5.4.3 Droughts and Disease

The influence of precipitation on the environment around us, and our reliance upon the fresh water that precipitation brings, is exemplified in the form of droughts and disease. While too much rain can induce flooding and landslides, too little rainfall can result in drought where water stress can seriously reduce crop yields leading to famine. Little or no rainfall over various time scales can lead to water shortages impacting water availability. Many diseases are waterborne: precipitation plays a key role in the availability of water that allows such diseases to spread and multiply, again impacting the human population.

### 5.4.3.1 Droughts

The lack of precipitation causes drought impacting food security. The variability in precipitation caused by changes in weather patterns, such as the El Niño Southern Oscillation (ENSO), has a significant effect in many of the food-producing regions in Africa and Central and South America. Satellite observations of precipitation, due to their global view, provide a critical part of hydroclimate monitoring for organizations that track food and water security and for studying large-scale changes in the rainfall patterns and their teleconnections. Satellite precipitation products are used in schemes, such as the Famine Early Warning System Network (FEWS Net; Funk et al. 2019) which uses precipitation information to determine the start of the growing season and generate drought metrics (Kirschbaum et al. 2017). A consequence of drought is increased crop stress leading to susceptibility to pests and disease and reduced crop yield, resulting in food insecurity (Dinku et al. 2010), although when the rain falls is also crucial: the timing of the onset of rainfall, which determines the start of season, is important for the early estimation of growing season outcomes like crop yield loss (Brown and de Beurs 2008). The Water Requirement Satisfaction Index (WRSI; Verdin and Klaver 2002) uses the start of the season in conjunction with historic crop yields to develop drought vulnerability models over southern Africa and western Sahelian Africa (Jayanthi et al. 2014) and Ethiopia (Senay and Verdin 2003). Using satellite-derived global precipitation data and population density to derive the precipitation per head of population, a metric of stress can be calculated (Shepherd et al. 2016).

Satellite-derived precipitation data is also used to assess the precipitation in fire-prone regions, particularly along the Western seaboard of the United States and western Canada. In these regions the snowfall during the winter forms an important, natural storage of fresh water. The lack of accumulated snow during the winter months can seriously affect the water availability during the following summer months when dry conditions can enhance the likelihood of forest fires (see Field 2020). Global Fire Weather Database (GFWED) includes satellite data that are used to both detect the active fire locations using visible and infrared observations and are combined with the satellite precipitation data and other weather data to monitor and predict the spread of a wildfire.

#### 5.4.3.2 Disease

The transmission of vector-borne disease in human and animal populations is a function of coupled ecological and social factors. Precipitation, and its variability, influences the underlying dynamics of disease since rainfall alters the routes through which diseases are transmitted. While excessive precipitation can increase the potential for transmission across a wider area, drought can concentrate transmission into smaller localities, such as around water sources. Precipitation also affects soil moisture and the formation of puddles or ephemeral ponds which provide breeding sites for insects and bacteria.

Satellite precipitation data contributes to the understanding, monitoring, and prediction of vector-borne disease risk, particularly in regions with poor in situ precipitation measurements. Timely information on precipitation may be used as a precursor indicator of risk since rainfall influences the habitat for insects allowing the risk of disease transmission to be predicted by several weeks. This in turn provides health officials with information to make the necessary decisions. Satellite precipitation estimates have been used to study and predict malaria (Midekisa et al. 2012; Lauderdale et al. 2014; Zinszer et al. 2015). Valle et al. (2013) and Zaitchik et al. (2014) used precipitation estimates from TRMM and GPM as input to a Land Data Assimilation System (LDAS) to generate estimates of soil moisture variability and ponding potential, resulting in a 1-km resolution risk assessment for malaria-prone regions in Peru and Ecuador. Similar use of satellite data and LDAS showed that mosquito numbers declined during periods of heavy rainfall due to a loss of breeding sites. Crucially, these studies showed that both total precipitation and soil moisture anomalies are significant predictors of malaria cases with several weeks' lead time.

The usefulness of satellite precipitation to vector-borne disease analysis has been studied by many including Beck et al. (2000), Machault et al. (2011) and Tourre et al. (2009), with diseases that are linked to water availability including schistosomiasis (Xue et al. 2011), Rift Valley fever (Guilloteau et al. 2014), and black plague (MacMillan et al. 2012; Moore et al. 2012; Monaghan et al. 2012). Longer term variations in precipitation can also be related to precipitation anomalies: for example, cholera is linked to the availability of water and its quality (De Magny et al. 2012; Eisenberg et al. 2013; Jutla et al. 2013). Relating precipitation anomalies and cholera anomalies to climate variations, such as ENSO allows an assessment of the seasonal likelihood of outbreaks to be made (Moore et al. 2017).

### 5.4.4 Weather and Climate

The observation and measurement of precipitation extremes, both intense events and lack of precipitation, is central to science and society, both from a meteorological and a climate perspective. The World Climate Research Program (WCRP) established a Grand Challenge on Weather and Climate Extremes. The availability of multiple satellite-based datasets, reanalyses from multiple centres and ground-based gridded datasets allows comparisons of how well precipitation is portrayed and represented. A main challenge is the scale dependency of precipitation: for example, extreme or intense precipitation can arise from either very short duration events or from long(er) duration moderate intensity events.

Over land areas with good gauge density, documentation of precipitation events is generally good, at least at the daily scale. This information becomes less comprehensive for finer temporal scales due to the instrumentation available. Use of daily data should also be treated with caution since an "extreme" event may straddle 2 days, such that the 2 days in themselves might not be deemed to constitute an extreme event. Of course, satellite observations and model reanalysis may also be used over land, providing additional information and coverage where gauge coverage is poor. However, it should be noted that differences in intense precipitation across these different products can be significant (Alexander et al. 2020): satellite precipitation products showed potential of climate scale (30–40 years) analyses of extremes alongside gauge data, model reanalyses less so. Studies into these extremes and their trends generally show that the location of the extremes in time and space were in agreement, although the amounts differed (Bador et al. 2020). In terms of trends, Donat et al. (2019) found that the totals and extremes had increased in humid regions since the 1950s. However, care is needed due to the heterogeneity of many of the data sets, particularly through the satellite data record (i.e., 1980s onwards) and in the derived precipitation data sets (Sun et al. 2017).

Over the oceans data from conventional sources is scarce, typically restricted to networks of buoys (Wu and Wang 2019), rain gauges or radars on atolls or islands (Greene et al. 2008; Henderson et al. 2017) or ship-based observations or disdrometer measurements (Klepp et al. 2018). In particular, the examination of extremes and trends in precipitation are difficult due to the paucity of data. Satellite products through their global coverage have the potential to provide useful information, although De Meyer and Roca (2021) noted that satellite base products showed significant discrepancies, particularly between the multi-sensor estimates and those from single sensor estimates. Similarly, model reanalysis showed significant differences in the extreme instantaneous precipitation compared with the OceanRAIN data sets Burdanowitz et al. (2019).

Measuring, monitoring and documenting the changes in the characteristics of precipitation because of climate change has become increasingly important. Crucially, the length of precipitation events and of droughts are likely to change, impacting agriculture, hydrology, and water resources (Trenberth et al. 2003). Changes in global temperature have shown a clear increase due to climate change: changes in global precipitation have been found to be widespread with significant increases, although with less coherent changes (Alexander et al. 2006). Using observations and global climate models, Donat et al. (2016) showed increases in extreme daily precipitation over both wet and dry regions over the last six decades, very much in line with the Clausius-Clapeyron constraint (Pall et al. 2006).

Over both land and ocean, it is important to understand the limitations of the different data sets that are being utilised. For example, in studying extremes it should be noted that the PERSIANN precipitation product is truncated above 120 mm day$^{-1}$. Satellite sensors observe different things related to precipitation, rather than precipitation directly, consequently, many products may not capture warm rain processes properly since many rely upon the scattering by ice to provide estimates over land. In addition, satellite precipitation estimates over coastal regions, and complex terrain may be inconsistent, while most satellite-based estimates have some variation in performance with season. Crucially, as Herold et al. (2016) concluded, the spatial scales of the observations, models and satellite data are a significant factor in discrepancies between the different precipitation data sites.

The direct assimilation of precipitation into numerical prediction models is problematic, not least since precipitation as a parameter is very variable and many precipitation measurements are not necessarily comparable with the input characteristics of datasets for models. Nevertheless, precipitation-capable observations are extensively used within the modelling community, particularly for severe weather events. Numerical weather prediction centres worldwide rely on microwave and IR-based observations to improve weather prediction and forecasting of tropical cyclone track and intensity, not least to reduce the forecast track uncertainty and improve lead times to help disaster preparedness. The NOAA/NESDIS/Center for Satellite Applications and Research (STAR) and the NASA Global Modeling and Assimilation Office (GMAO) assimilates GMI passive microwave observations into the global data assimilation systems used to initialize NWP forecast models. In STAR, clear-sky, ocean-only brightness temperatures from the GMI are assimilated to improve the analysis and forecast skill in the NOAA Global Data Assimilation System/Global Forecast System (GDAS/GFS; Wang et al. 2013). In a study of Hurricane Julio during the first 2 weeks of August 2014 two assimilation experiments were performed, the first (control) experiment included all conventional and satellite observations operationally assimilated in the GDAS/GFS, with the second experiment (DAGMI) added GMI observations to the configuration of the control experiment. The DAGMI run showed less uncertainty, more consistency cycle to cycle, and better agreement with the official track that there would be no direct impact from Hurricane Julio on Hawaii.

For climate modelling activities, the inclusion of precipitation data is helping to improve models predicting the Earth's future climate. Climate models help capture the state of the atmosphere, land, ocean, and ice and the processes and interactions among them. Improving the elements of the climate models, such as model parametrization, data assimilation and validation, help to evaluate and improve climate models. Satellite observations of precipitation (and indirect measurements of latent heating) can provide useful information that can be compared with model outputs. Many satellite precipitation products are now available with a sufficiently long record that seasonal/regional climatologies, can be generated with reasonable confidence. Comparisons of the satellite observed precipitation with the climatology allows the precipitation anomalies to be assessed. These satellite products can be used to evaluate the results of the climate models and reanalysis products with the purpose of determining whether the climate model can accurately represent what satellites are observing and allow improvements in the parameterization of the models.

Figure 5.39 shows a comparison of the GISS-E3 GCM seasonal climatology (upper) for January and July with the GPM-IMERG climatology (below) for the same months. While the GISS-E3 climate model generally simulates the patterns observed by GPM quite well, discrepancies are noticeable both in terms of amount and intensity. In January the extent and amount of the precipitation across the south-eastern Pacific Ocean is much reduced in the GISS-E3 output, while in July the amounts and distribution in the GISS-E3 output are much less than that portrayed by the IMERG product. Using satellite products such as the GPM-IMERG product as a benchmark help to identify where improvements in the climate models are needed and thus help to make future climate predictions more reliable.

### 5.4.5 Water, Energy, Food Nexus

The modern world is increasingly interconnected, not just in terms of physical communications, but in terms of the impacts of the changes that are occurring in the natural world. The increases in human population coupled with changes in our climate puts significant pressure on the Earth's ecosystems. Interconnections between the different elements of the Earth system and our society are complex and the

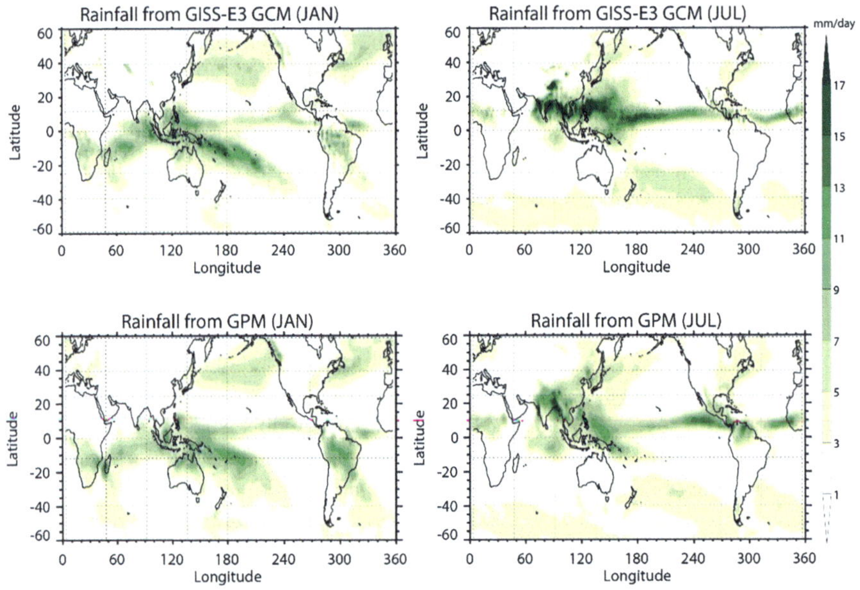


**Fig. 5.39** Comparison of the modelled GISS-E3 GCM seasonal climatology (upper) and the satellite-based GPM-IMERG climatology (below) for January (left) and July (right). (Source: NASA)

realisation of the limits to our natural resources presents a significant challenge. For example, food production requires water and energy, water extraction and distribution require energy and energy production requires water (see Fig. 5.40). Changes in the cost of energy impacts the availability and delivery of water and this also impacts the price of food. Providing a sustainable future involves understanding and managing many different aspects of nature and society.

Satellite data sets provide a wealth of data, which, together with other data, are key in unravelling the links between water, energy and food. The United Nations 2030 Agenda for Sustainable Development plan includes 17 Sustainable Development Goals (SDG) to provide a global blueprint for dignity, peace, and prosperity for people and our planet. Alongside this the World Resources Institute (WRI), a non-governmental organization (NGO) ingests data, including satellite data, into their interactive data platforms. They work with more than 60 countries worldwide to help develop sustainable solutions to critical problems, focusing on seven essential inter-related issues: climate, energy, food, forests, water, cities, and the ocean. Providing an integrated approach to global development, using data and analysis, technical tools and active engagement with decision-makers to provide the information needed to make informed decisions about the environment and the well-being of people across the globe. For example, water insecurity is a growing problem worldwide, thus using multiple data sources to identify water-related high-risk areas helps to manage freshwater resources. Similarly, decision making to anticipate monitor and prepare for natural disasters such as landslides and forest fires has potential to save lives.

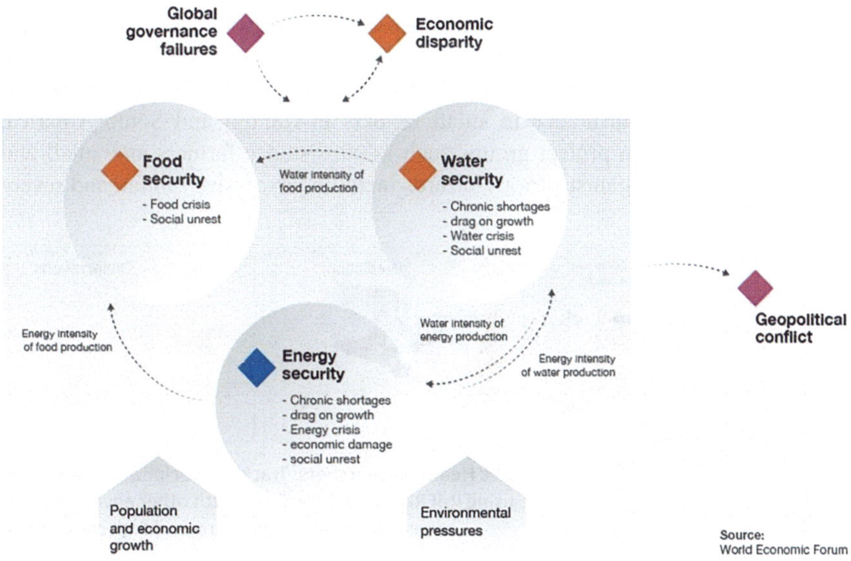


**Fig. 5.40** Implications of the Energy Water Food Network (EWFN). (Source: The World Economic Forum 2011, public domain)

As noted above, the strong links between the climate and disease are well established. Using satellite data, and in particular precipitation data to monitor and map the distribution and amounts of precipitation can help in the prediction of outbreaks of not only diseases. Pests, such as locusts, also pose significant risks to food production: by understanding their lifecycle action can be taken to mitigate, or even prevent, significant damage. Figure 5.41 illustrates how satellite data monitoring temperature and rainfall may be used to evaluated water insecurity and subsequently cholera risk. Maps of areas posing a high cholera risk can be generated and resources deployed to those areas, thus reducing the impact of any outbreak of the disease.

### 5.4.6 Commerce

Satellite data sets are increasingly used in the commercial world, examples of which include insurance, water resources and agri-business. Early examples of exploiting satellite precipitation data sets relate to crop forecasting, not least for assessing crop yields and therefore the cost of food commodities in the trading markets. Insurance companies also use precipitation data sets to evaluate the return periods of storms as a well as provide information on the impact of severe weather events in determining damage payouts.

The impacts of natural hazards events, such as extreme precipitation or drought, disproportionately affect the developing world where individuals and farmers are often not insured and who live and work in conditions that leave them vulnerable to natural disasters. This may lead to economic and environmental challenges if preventive or mitigating measures are not taken in time. However, many of those in the developing world are not able to afford insurance. However, "community" insurance schemes have been established to fill this gap, such as the Microinsurance Catastrophe Risk Organization (MiCRO; Fig. 5.42), supported various organizations, provides local insurance to small farmers in Central and South America. MiCRO is designed to protect groups such as subsistence farmers and small and micro-entrepreneurs against various hazards including excessive rainfall and severe

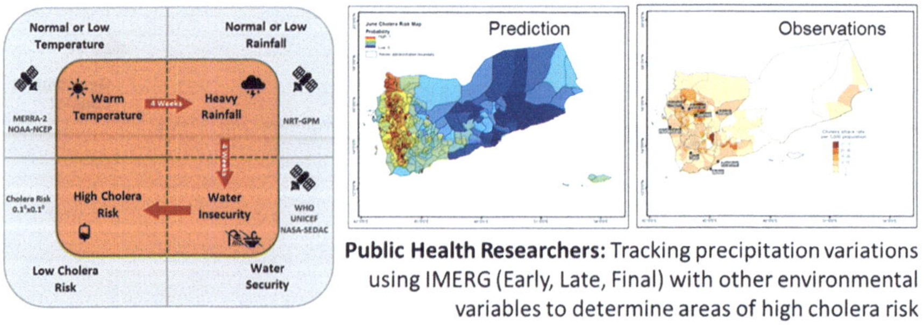


**Fig. 5.41** Environmental conditions impacting the formation of a cholera outbreak, together with a prediction of risk derived using satellite data, together with the observed number of cases. (Source: NASA)

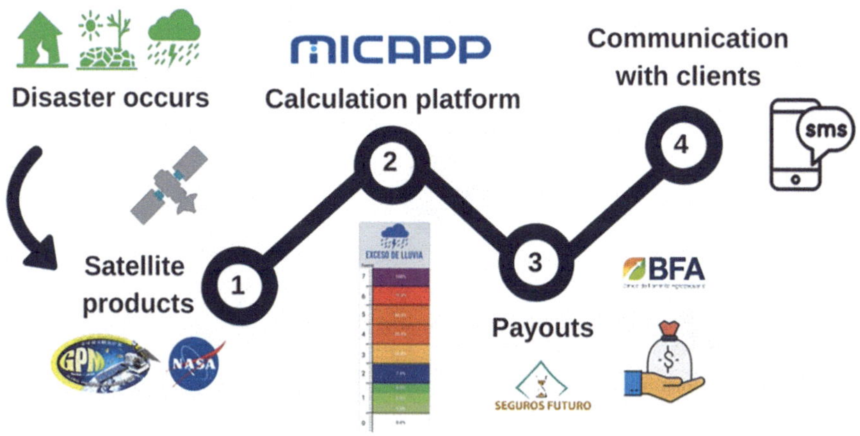


**Fig. 5.42** The payout process of MiCRO depicting actions taken when a disaster strikes: NASA satellite data (1) is used to establish payout triggers through a calculation platform (2). When a disaster occurs and depending on its severity, payment is determined (3) and clients are informed via text (4). (Source: NASA)

drought. An index, i.e. rainfall, is calculated from satellite precipitation products: when this index is triggered, a payout is made directly to the clients, avoiding the lengthy and costly loss adjustments process.

On the other side of the world, satellite precipitation data is helping farmers in India to better manage their scarce water resources due to the growing competition over finite water resources. Many of these farmers own small plots and depend on groundwater, channels and rainwater for irrigation. The lack of reliable information on the weather and subsequent water availability often results in inefficient use of the water they have. Knowing when and how much to water their crops can help conserve critical freshwater resources. The University of Washington Civil and Environmental Engineering (UW CEE) has developed an irrigation advisory system to help small farmers receive irrigation information directly to their mobile phones. The Provision of Advisory for Necessary Irrigation (PANI) system (Hossain et al., 2017) uses satellite precipitation data combined with ground sensors from each plot of land (e.g. temperature, humidity and wind speed). This data is then used to generate irrigation advisories customised to each plot of land, which is communicated to farmers through text messages (see Fig. 5.43).

From late 2018 through mid-2019 one hundred and fifty farmers in northern India tested the PANI system and reported that 85% of them benefitted from changing their irrigation habits, resulting in reducing their expenditure and water usage, while improving their crop productivity.

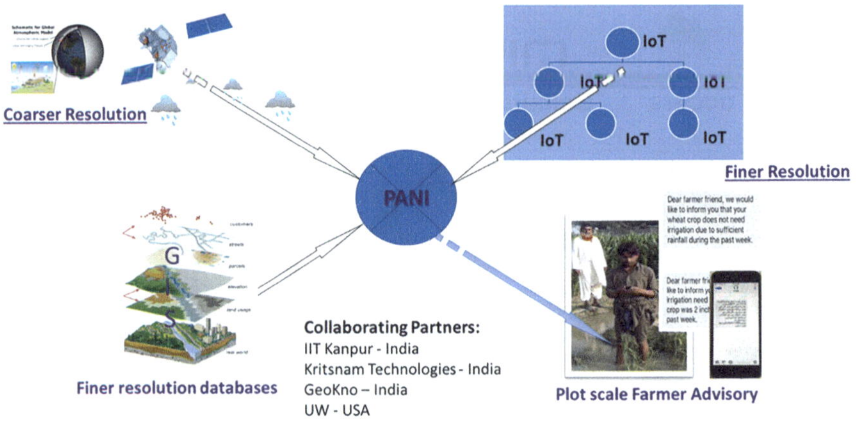


**Fig. 5.43** The PANI concept, combining satellite and weather model data (upper left) with finer resolution geographic data (lower left) and personalised farm data (upper right) to generate irrigation advisories in real-time via cell phones. (Source: NASA and F. Hossain)

## 5.5 What We Know and What We Need to Understand

A range of observing systems measure precipitation. However, any precipitation measurement contains some errors and uncertainties related to how directly each instrument observes the precipitation. Quantifying these errors and uncertainties is difficult however, and different user communities often have different expectations of them. While surface gauges represent the most direct measure of precipitation, their point sampling is subject to representativeness errors, particular in regions of spatially varied precipitation such as over mountainous and coastal regions. While modern gauges often provide good temporal sampling in near real time, historical gauge records are generally at daily or even monthly resolution. Weather radar provides good temporal sampling with spatial coverage. However, radar measurements are less direct than gauges since they rely upon the backscatter of radiation from the precipitation particles: this backscatter to intensity relationship varies across (and often within) different meteorological situations. Furthermore, both gauges and weather radar are very much limited to land areas, with few measurements possible over the vast expanses of the oceans.

Satellite sensors have the capability of providing truly global observations. Observations in the visible and infrared part of the spectrum, while providing good spatial and temporal sampling, only provide information on cloud tops, precipitation being inferred from these cloud top characteristics. Passive microwave observations are more direct since they respond to the precipitation particles (hydrometeors) within the atmospheric column, albeit with relatively poor spatial and temporal sampling. In addition, these passive microwave observations need to be converted into estimates of precipitation: this is dependent upon the channels and spatial resolution used in the observation as well as the retrieval scheme and its assumptions. While precipitation radar provides the most direct spaceborne measures of precipitation at reasonable resolution, the number of such radars is extremely limited and thus the temporal sampling is extremely poor. Furthermore, satellite-based precipitation radars also suffer from the same issues that weather radars do in determining the backscatter to precipitation intensity relationships. For passive microwave and precipitation radar observations another issue is their temporal sampling: these sensors provide "snapshots" of the precipitation at different times of day. In some regions, particularly the Tropics, there is a significant diurnal cycle in the precipitation which impacts the satellite-based estimates made over a 24-h period.

Table 5.15 summarises the temporal and spatial sampling of both surface and satellite-based observing systems for precipitation, while Fig. 5.44 illustrates the different systems. Note that each type of instrument measures precipitation in slightly different ways, consequently precipitation measurements from them will be slightly different.

Satellite precipitation products have been the subject of many validation (or at least verification) studies using surface data as the reference standard. Such studies are important to both the algorithm developer, to ensure that the retrieval scheme is operating as expected, and to the users to evaluate the accuracy of the precipitation product (Kidd et al. 2018). The validation usually uses surface reference data derived from high quality gauge and radar network using a range of scales, from the instantaneous to monthly, and from footprint-scale to 2.5° scale, to address their usefulness across a range of applications. Since the point measurements of gauges are not necessarily directly compatible with the spatial sampling of satellite products, validation using gauge data is usually performed at relatively coarse resolution over areas with good gauge densities. At the instantaneous scale, weather radar data is often used since these data can be more easily resampled to match the observations and spatial scale of the satellite products.

Large scale intercomparisons of a range of satellite precipitation products have been conducted through specific projects, such as the NASA Precipitation Intercomparison Projects (Adler et al. 2001) designed to evaluate the SSM/I PMW retrieval schemes, and the GEWEX organised the Algorithm Intercomparison Program that compared a range of retrieval products, including those from VIS/IR

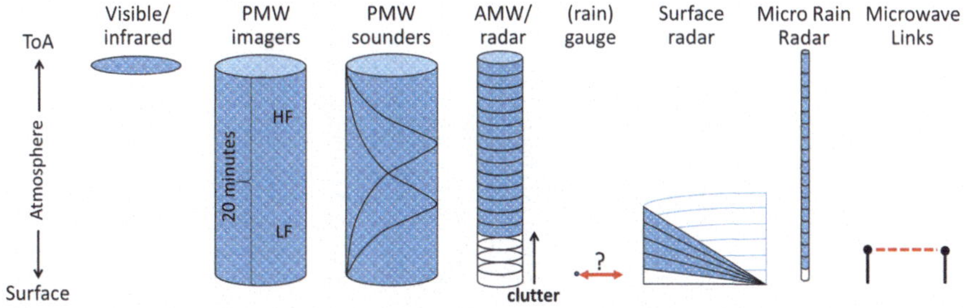


**Fig. 5.44** Visualisation of precipitation observations by satellite sensors covering visible/infrared, passive microwave imagers, passive microwave sounders and precipitation radar, together with surface-based measurements from gauges, weather radar, micro rain radars and microwave links

**Table 5.15** Comparison of different types of precipitation measurements

| | Instrument | Temporal | Spatial | Notes |
|---|---|---|---|---|
| Surface | Gauges: accumulation | Variable | Point | Temporal scale dependent upon observation frequency |
| | Gauges:Tipping Bucket | Quantised | Point | Quantisation of bucket (0.1 or 0.2 mm or 1/100") and data logger |
| | Disdrometers | Instantaneous | Point | Individual drop measurements |
| | Micro rain radar | Instantaneous | Point | 30 vertical levels |
| | Weather radar | Instantaneous | Radial | Radial measurements of dBZ converted to a Cartesian grid |
| Satellite | Visible imagery | Instantaneous | 1–4 km | Intermittent (LEO)/15 min sampling (GEO) of cloud tops |
| | Infrared imagery | Instantaneous | 1–4 km | Intermittent (LEO)/15 min sampling (GEO) of cloud tops |
| | Passive Microwave Imagers | Column | 5–25 km | Intermittent sampling (LEO)<br>Resolution = frequency dependent |
| | Passive Microwave Sounders | Column | 16–48 km | Intermittent sampling (LEO)<br>Resolution = frequency/scan position dependent |
| | Active Microwave (radar) | Instantaneous | 5 km | 80 vertical levels; intermittent sampling (LEO) |

observations, PMW and combined schemes (Ebert et al. 1996). While many of these intercomparisons showed that no single precipitation product was "best", the PMW schemes generally performed better than the visible/infrared products. The International Precipitation Working Group (IPWG; Levizzani et al. 2018) has provided some continuous near real time assessment of precipitation over several validation regions (Ebert et al. 2007; Kidd et al. 2012, 2021) with daily, 0.25° × 0.25° resolution, using surface gauge and/or radar data. More specific validation is required at finer scale to better understand precipitation processes. The GPM mission incorporates dedicated ground validation campaigns designed to collect detailed information from multiple sensors carried on multiple platforms, including satellites, aircraft and at the surface (Hou et al. 2014; Skofronick-Jackson et al. 2015; Petersen et al. 2020). These campaigns, carried out over specific regions, provide a comprehensive set of observations enabling a better understanding of how the precipitation is observed by the satellite sensors, as well as detailed characteristics of the precipitation itself (Jensen et al. 2016; Houze et al. 2015, 2017). A comparison between the different types of intercomparison studies is shown in Table 5.16.

Long-term validation of instantaneous satellite precipitation products using high quality radar and gauge data over Europe and the southeast US showed good agreement between the satellite and surface products (Kidd et al. 2018). The precipitation radar provided highest overall correlation and skill score, followed by the GMI-based product and then other sensors. Large scale (global/monthly) validation analysis by Beck et al. (2017a, b) showed that the performance of precipitation products varied over different regions. Typically, the "best" products are the multi-satellite, multi-source products that incorporate all available observations. However, such products often contain the same gauge data that are used in the validation and would therefore be expected to perform well, at least over regions where there is good gauge data coverage.

**Table 5.16** Comparison of typical Ground Validation programs and the IPWG validation effort

| Criteria | Ground validation program | IPWG validation |
|---|---|---|
| Purpose of validation | Physical understanding of precipitation processes and systems | Assessment of precipitation products for developers and user communities |
| Type of validation | Priority on physical, but also statistical metrics | Focus on descriptive indicators and statistical metrics |
| Sources of validation data | Specifically collected for, or by project investigators | Contributed by validation groups from established sources |
| Source of precipitation products | Specific products provided by developers | Developers provide products directly to validation groups |
| Types of validation data | Diverse, multi-tier specialist instrumentation (gauge, radar, aircraft), in specific locations | Gauges and/or radars, usually part of a national network over specific regions of interest |
| Types of precipitation products | Full resolution/instantaneous, single-sensor or model products | Time/area averaged model or blended satellite products |

The conclusion of these intercomparison/validation exercises is that no single precipitation product stands out as being superior: the performance of each product has its own subtleties and nuances depending on region, time scales or spatial resolutions. Nevertheless, the validation exercises should continue possibly with the aid of automated algorithms (e.g., Arulraj et al. 2024) to produce long time statistical appraisals. Ultimately, the user should determine the suitability of a particular product to meet the requirements of their application.

## References

Alexander, L.V., Zhang, X., Peterson, T.C., Caesar, J., Gleason, B., Klein Tank, A.M.G., Haylock, M., Collins, D., Trewin, B., Rahimzadeh, F., Tagipour, A., Rupa Kumar, K., Revadekar, J., Griffiths, G., Vincent, L., Stephenson, D.B., Burn, J., Aguilar, E., Brunet, M., Taylor, M., New, M., Zhai, P., Rusticucci, M., Vazquez-Aguirre, J.L.: Global observed changes in daily climate extremes of temperature and precipitation. J. Geophys. Res. Atmos. **111**, D05109 (2006). https://doi.org/10.1029/2005JD006290

Alexander, L.V., Bador, M., Roca, R., Contractor, S., Donat, M.G., Nguyen, P.L.: Intercomparison of annual precipitation indices and extremes over global land areas from in situ, space-based and reanalysis products. Environ. Res. Lett. **15**, 055002 (2020). https://doi.org/10.1088/1748-9326/ab79e2

Alishouse, J.C., Ferraro, R.R., Fiore, J.V.: Inference of oceanic rainfall properties from the Nimbus 7 SMMR. J. Appl. Meteorol. Climatol. **29**, 551–560 (1990). https://doi.org/10.1175/1520-0450(1990)029<0551:IOORPF>2.0.CO;2

Aonashi, K., Awaka, J., Hirose, M., Kozu, T., Kubota, T., Liu, G., Shige, S., Kida, S., Seto, S., Takahashi, N., Takayabu, Y.N.: GSMaP passive microwave precipitation retrieval algorithm: algorithm description and validation. J. Meteorol. Soc. Jpn. **87A**, 119–136 (2009). https://doi.org/10.2151/jmsj.87A.119

Arkin, P.A., Meisner, B.N.: The relationship between large-scale convective rainfall and cold cloud over the Western Hemisphere during 1982–84. Mon. Weather Rev. **115**(1), 51–74 (1987). https://doi.org/10.1175/1520-0493(1987)115<0051:TRBLSC>2.0.CO;2

Arulraj, M., Petković, V., Wen, S., Ferraro, R.R., Meng, H.: NPreciSe - an automated satellite precipitation product assessment tool. Sci. Data. **11**, 1042 (2024). https://doi.org/10.1038/s41597-024-03877-x

Asmus, V.V., Milekhin, O.E., Kramareva, L.S., Khailov, M.N., Shumakov, I.A.: Arktika-M: the world's first highly elliptical orbit hydrometeorological space system. Russ. Meteorol. Hydrol. **46**, 805–816 (2021). https://doi.org/10.3103/S1068373921120013

Atlas, D., Ulbrich, C.W.: Path- and area-integrated rainfall measurement by microwave attenuation in the 1–3 cm band. J. Appl. Meteorol. Climatol. **16**(12), 1322–1331 (1977). https://doi.org/10.1175/1520-0450(1977)016<1322:PAAIRM>2.0.CO;2

Ba, M., Gruber, A.: GOES Multispectral Rainfall Algorithm (GMSRA). J. Appl. Meteorol. Climatol. **40**(8), 1500–1514 (2001). https://doi.org/10.1175/1520-0450(2001)040<1500:GMRAG>2.0.CO;2

Bador, M., Boé, J., Terray, L., Alexander, L.V., Baker, A., Bellucci, A., Haarsma, R., Koenigk, T., Moine, M.-P., Lohmann, K., Putrasahan, D.A., Roberts, C., Roberts, M., Scoccimarro, E., Schiemann, R., Seddon, J., Senan, R., Valcke, S., Vanniere, B.: Impact of higher spatial atmospheric resolution on precipitation extremes over land in global climate models. J. Geophys. Res. Atmos. **125**(13), e2019JD032184 (2020). https://doi.org/10.1029/2019JD032184

Ballard, S.P., Li, Z., Simonin, D., Buttery, H., Charlton-Perez, C., Gaussiat, N., Hawkness-Smith, L.: Use of radar data in NWP-based nowcasting in the Met Office. In: Moore, R.J., Cole, S.J., Illingworth, A.J. (eds.) Weather Radar and Hydrology, vol. 351, pp. 336–341. IAHS Publ. (2012)

Barrett, E.C., Adler, R.F., Arpe, K., Bauer, R., Berg, W., Chang, A., Ferraro, R., Ferriday, J., Goodman, S., Hong, Y., Janowiak, J., Kidd, C., Kniveton, D., Morrissey, M., Olson, W., Petty, G., Rudolf, B., Shibata, A., Smith, E., Spencer, R.: The first WetNet Precipitation Intercomparison Project: interpretation of results. Remote Sens. Rev. **11**, 303–373 (1994). https://doi.org/10.1080/02757259409532268

Beck, L.R., Lobitz, B.M., Wood, B.L.: Remote sensing and human health: new sensors and new opportunities. Emerg. Infect. Dis. **6**, 217–227 (2000). https://doi.org/10.3201/eid0603.000301

Beck, H.E., van Dijk, A.I.J.M., Levizzani, V., Schellekens, J., Miralles, D.G., Martens, B., de Roo, A.: MSWEP: 3-hourly 0.25° global gridded precipitation (1979–2015) by merging gauge, satellite, and reanalysis data. Hydrol. Earth Syst. Sci. **21**, 589–615 (2017a). https://doi.org/10.5194/hess-21-589-2017

Beck, H.E., Vergopolan, N., Pan, M., Levizzani, V., van Dijk, A.I.J.M., Weedon, G.P., Brocca, L., Pappenberger, F., Huffman, G.J., Wood, E.F.: Global-scale evaluation of 22 precipitation datasets using gauge observations and hydrological modeling. Hydrol. Earth Syst. Sci. **21**, 6201–6217 (2017b). https://doi.org/10.5194/hess-21-6201-2017

Beck, H.E., Wood, E.F., Pan, M., Fisher, C.K., Miralles, D.M., van Dijk, A.I.J.M., McVicar, T.R., Adler, R.F.: MSWEP V2 global 3-hourly 0.1° precipitation methodology and quantitative assessment. Bull. Am. Meteorol. Soc. **100**(3), 473–500 (2019). https://doi.org/10.1175/BAMS-D-17-0138.1

Becker, A., Finger, P., Meyer-Christoffer, A., Rudolf, B., Schamm, K., Schneider, U., Ziese, M.: A description of the global land-surface precipitation data products of the Global Precipitation Climatology Centre with sample applications including centennial (trend) analysis from 1901–present. Earth Syst. Sci. Data. **5**, 71–99 (2013). https://doi.org/10.5194/essd-5-71-2013

Behrangi, A., Hsu, K.-L., Imam, B., Sorooshian, S., Huffman, G.J., Kuligowski, R.J.: PERSIANN-MSA: a precipitation estimation method from satellite-based multispectral analysis. J. Hydrometeorol. **10**(6), 1414–1429 (2009a). https://doi.org/10.1175/2009JHM1139.1

Behrangi, A., Hsu, K.-L., Imam, B., Sorooshian, S., Kuligowski, R.: Evaluating the utility of multispectral information in delineating the areal extent of precipitation. J. Hydrometeorol. **10**, 684–700 (2009b). https://doi.org/10.1175/2009JHM1077.1

Behrangi, A., Khakbaz, B., Jaw, T.C., AghaKouchak, A., Hsu, K., Sorooshian, S.: Hydrologic evaluation of satellite precipitation products over a mid-size basin. J. Hydrol. **397**, 225–237 (2011). https://doi.org/10.1016/j.jhydrol.2010.11.043

Bell, T.L., Abdullah, A., Martin, R.L., North, G.R.: Sampling errors for satellite-derived tropical rainfall: Monte Carlo study using a space-time stochastic model. J. Geophys. Res. Atmos. **95**, 2195–2205 (1990). https://doi.org/10.1029/JD095iD03p02195

Bellerby, T., Todd, M., Kniveton, D., Kidd, C.: Rainfall estimation from a combination of TRMM Precipitation Radar and GOES multispectral satellite imagery through the use of an artificial neural network. J. Appl. Meteorol. Climatol. **39**(12), 2115–2128 (2000). https://doi.org/10.1175/1520-0450(2001)040<2115:REFACO>2.0.CO;2

Bellerby, T., Hsu, K.-L., Sorooshian, S.: LMODEL: a satellite precipitation methodology using cloud development modeling. Part I: algorithm construction and calibration. J. Hydrometeorol. **10**, 1081–1095 (2009a). https://doi.org/10.1175/2009JHM1091.1

Bellerby, T., Hsu, K.-L., Sorooshian, S.: LMODEL: a satellite precipitation methodology using cloud development modeling. Part II: validation. J. Hydrometeorol. **10**, 1096–1108 (2009b). https://doi.org/10.1175/2009JHM1092.1

Bizzarri, B., Gasiewski, A.J., Staelin, D.H.: Observing rain by millimetre-submillimetre wave sounding from geostationary orbit. In: Levizzani, V., Bauer, P., Turk, F.J. (eds.) Measuring Precipitation from Space Advances in Global Change Research, vol. 28, pp. 675–692. Springer Nature (2007). https://doi.org/10.1007/978-1-4020-5835-6_50

Blackwell, W., Cunningham, A., Diliberto, M., Donnelly, S., Kidd, C., Kim, M.J., Leslie, R., Milstein, A., Perras, G., Pieper, M., Prince, J., Zorn, N.: Results from the NASA tropics mission after one year in orbit. In: IGARSS 2024-IEEE International Geoscience and Remote Sensing Symposium, pp. 6689–6692 (2024). https://doi.org/10.1109/IGARSS53475.2024.10641472

Brown, M.E., de Beurs, K.M.: Evaluation of multi-sensor semi-arid crop season parameters based on NDVI and rainfall. Remote Sens. Environ. **112**(5), 2261–2271 (2008). https://doi.org/10.1016/j.rse.2007.10.008

Bumke, K., Clemens, M., Grassl, H., Pang, S., Peters, G., Seltmann, J.E.E., Siebenborn, T., Wagner, A.: Accurate areal precipitation measurements over the land and sea (APOLAS). BALTEX Newsl. **6**, 9–13 (2004) Available at https://oceanrep.geomar.de/id/eprint/1807/1/Newsletter6.pdf

Burdanowitz, J., Buehler, S.A., Bakan, S., Klepp, C.: The sensitivity of oceanic precipitation to sea surface temperature. Atmos. Chem. Phys. **19**(14), 9241–9252 (2019). https://doi.org/10.5194/acp-19-9241-2019

Cecil, D.J., Goodman, S.J., Boccippio, D.J., Zipser, E.J., Nesbitt, S.W.: Three years of TRMM precipitation features. Part I: radar, radiometric, and lightning characteristics. Mon. Weather Rev. **133**, 543–566 (2005). https://doi.org/10.1175/MWR-2876.1

De Magny, G.C., Thiaw, W., Kumar, V., Manga, N.M., Diop, B.M., Gueye, L., Kamara, M., Roche, B., Murtugudde, R., Colwell, R.R.: Cholera outbreak in Senegal in 2005: was climate a factor? PLoS One. **7**, e44577 (2012). https://doi.org/10.1371/journal.pone.0044577

De Meyer, V., Roca, R.: Thermodynamic scaling of extreme daily precipitation over the tropical ocean from satellite observations. J. Meteorol. Soc. Jpn. **99**(2), 423–436 (2021). https://doi.org/10.2151/jmsj.2021-020

Desbois, M., Seze, G., Szejwach, G.: Automatic classification of clouds on Meteosat imagery: application to high-level clouds. J. Appl. Meteorol. Climatol. **21**, 401–412 (1982). https://doi.org/10.1175/1520-0450(1982)021<0401:ACOCOM>2.0.CO;2

Dinku, T., Ceccato, P., Cressman, K., Connor, S.J.: Evaluating detection skills of satellite rainfall estimates over desert locust recession regions. J. Appl. Meteorol. Climatol. **49**(6), 1322–1332 (2010). https://doi.org/10.1175/2010JAMC2281.1

Donat, M.G., Lowry, A.L., Alexander, L.V., O'Gorman, P.A., Maher, N.: More extreme precipitation in the world's dry and wet regions. Nat. Clim. Chang. **6**, 508–513 (2016). https://doi.org/10.1038/nclimate2941

Donat, M.G., Angélil, O., Ukkola, A.M.: Intensification of precipitation extremes in the world's humid and water-limited regions. Environ. Res. Lett. **14**, 065003 (2019). https://doi.org/10.1088/1748-9326/ab1c8e

Duchon, C.E., Essenberg, G.R.: Comparative rainfall observations from pit and above ground rain gauges with and without wind shields. Water Resour. Res. **37**, 3253–3263 (2001). https://doi.org/10.1029/2001WR000541

Ebert, E.E., Manton, M.J., Arkin, P.A., Allam, R.J., Holpin, G.E., Gruber, A.: Results from the GPCP Algorithm Intercomparison Programme. Bull. Am. Meteorol. Soc. **77**(12), 2875–2888 (1996). https://doi.org/10.1175/1520-0477(1996)077<2875:RFTGAI>2.0.CO;2

Ebert, E.E., Janowiak, J.E., Kidd, C.: Comparison of near-real-time precipitation estimates from satellite observations and numerical models. Bull. Am. Meteorol. Soc. **88**(1), 47–64 (2007). https://doi.org/10.1175/BAMS-88-1-47

Eisenberg, M.C., Kujbida, G., Tuite, A.R., Fisman, D.N., Tien, J.H.: Examining rainfall and cholera dynamics in Haiti using statistical and dynamic modeling approaches. Epidemics. **5**, 197–207 (2013). https://doi.org/10.1016/j.epidem.2013.09.004

Ferraro, R.R., Grody, N.C., Marks, G.F.: Effects of surface conditions on rain identification using the SSM/I. Remote Sens. Rev. **11**, 195–210 (1994). https://doi.org/10.1080/02757259409532265

Ferraro, R.R., Smith, E.A., Berg, W., Huffman, G.J.: A screening methodology for passive microwave precipitation retrieval algorithms. J. Atmos. Sci. **55**, 1583–1600 (1998). https://doi.org/10.1175/1520-0469(1998)055<1583:ASMFPM>2.0.CO;2

Ferraro, R.R., Weng, F., Grody, N.C., Zhao, L.: Precipitation characteristics over land from the NOAA-15 AMSU sensor. Geophys. Res. Lett. **27**, 2669–2672 (2000). https://doi.org/10.1029/2000GL011665

Field, R.D.: Using satellite estimates of precipitation for fire danger rating. In: Levizzani, V., Kidd, C., Kirschbaum, D.B., Kummerow, C.D., Nakamura, K., Turk, F.J. (eds.) Satellite Precipitation Measurement Advances in Global Change Research, vol. 69, pp. 1131–1154. Springer Nature (2020). https://doi.org/10.1007/978-3-030-35798-6

Funk, C., Shukla, S., Thiaw, W.M., Rowland, J., Hoell, A., McNally, A., Husak, G., Novella, N., Budde, M., Peters-Lidard, C., Adoum, A., Galu, G., Korecha, D., Magadzire, T., Rodriguez, M., Robjhon, M., Bekele, E., Arsenault, K., Peterson, P., Harrison, L., Fuhrman, S., Davenport, F., Landsfeld, M., Pedreros, D., Jacob, J.P., Reynolds, C., Becker-Reshef, I., Verdin, J.: Recognizing the Famine Early Warning Systems Network: over 30 years of drought early warning science advances and partnerships promoting global food security. Bull. Am. Meteorol. Soc. **100**(6), 1011–1027 (2019). https://doi.org/10.1175/BAMS-D-17-0233.1

Gebremichael, M., Hossain, F. (eds.): Satellite Rainfall Applications for Surface Hydrology. Springer Nature (2010)

GEO, GEO Task US-09-01a: Critical Earth observation priorities; Precipitation data characteristics and user types. L. Friedl (task lead), GEO, 60 pp. (2014). Available at http://sbageotask.larc.nasa.gov/GEO US0901a - Precipitation Data Report - Final.pdf

Goodison, B.E., Louie, P.Y.T., Yang, D.: WMO solid precipitation measurement intercomparison. Instruments and Observing Methods Rep. 67, WMO/TD-872, 212 pp. Available at https://library.wmo.int/records/item/28336-wmo-solid-precipitation-measurement-intercomparison (1998)

Grecu, M., Anagnostou, E.N., Adler, R.F.: Assessment of the use of lightning information in satellite infrared rainfall estimation. J. Hydrometeorol. **1**, 211–221 (2000). https://doi.org/10.1175/1525-7541(2000)001<0211:AOTUOL>2.0.CO;2

Greene, J.S., Klatt, M., Morrissey, M., Postawko, S.: The comprehensive Pacific Rainfall Database. J. Atmos. Ocean. Technol. **25**, 71–82 (2008). https://doi.org/10.1175/2007JTECHA904.1

Groisman, P.Y., Legates, D.R.: Documenting and detecting long-term precipitation trends: where are we and what should be done? Clim. Chang. **31**, 601–622 (1995). https://doi.org/10.1007/BF01095163

Großklaus, M., Uhlig, K., Hasse, L.: An optical disdrometer for use in high wind speeds. J. Atmos. Ocean. Technol. **15**(4), 1051–1059 (1998). https://doi.org/10.1175/1520-0426(1998)015<1051:AODFUI>2.0.CO;2

Guilloteau, C., Gosset, M., Vignolles, C., Alcoba, M., Tourre, Y.M., Lacaux, J.-P.: Impacts of satellite-based rainfall products on predicting spatial patterns of rift valley fever vectors. J. Hydrometeorol. **15**, 1624–1635 (2014). https://doi.org/10.1175/JHM-D-13-0134.1

Habib, E., Krajewski, W.F., Ciach, G.J.: Estimation of rainfall interstation correlation. J. Hydrometeorol. **2**(6), 621–629 (2001). https://doi.org/10.1175/1525-7541(2001)002<0621:EORIC>2.0.CO;2

Habib, E., Lee, G., Kim, D., Ciach, G.J.: Ground-based direct measurement. In: Testik, F., Gebremichael, M. (eds.) Rainfall: State of the Science. American Geophysical Union (2010). https://doi.org/10.1029/2010GM000953

Harrison, D., Kitchen, M., Scovell, R.W.: High-resolution precipitation estimates for hydrological uses. Water Manag. **162**(2), 125–135 (2009). https://doi.org/10.1680/wama.2009.162.2.125

Harrison, D., Norman, K., Pierce, C., Gaussiat, N.: Radar products for hydrological applications in the UK. Water Manag. **165**(2), 89–103 (2012). https://doi.org/10.1680/wama.2012.165.2.89

Haynes, J.M., L'Ecuyer, T.S., Stephens, G.L., Miller, S.D., Mitrescu, C., Wood, N.B., Tanelli, S.: Rainfall retrieval over the ocean with spaceborne W-band radar. J. Geophys. Res. Atmos. **114**, D00A22 (2009). https://doi.org/10.1029/2008JD009973

Henderson, D.S., Kummerow, C.D., Marks, D.A., Berg, W.: A regime-based evaluation of TRMM oceanic precipitation biases. J. Atmos. Ocean. Technol. **34**, 2613–2635 (2017). https://doi.org/10.1175/JTECH-D-16-0244.1

Herold, N., Alexander, L.V., Donat, M.G., Contractor, S., Becker, A.: How much does it rain over land? Geophys. Res. Lett. **43**(1), 341–348 (2016). https://doi.org/10.1002/2015GL066615

Hitschfeld, W., Bordan, J.: Errors inherent in the radar measurement of rainfall at attenuating wavelengths. J. Atmos. Sci. **11**(1), 58–67 (1954). https://doi.org/10.1175/1520-0469(1954)011<0058:EIITRM>2.0.CO;2

Hong, Y., Hsu, K.-L., Sorooshian, S., Gao, X.: Precipitation estimation from remotely sensed imagery using an artificial neural network cloud classification system. Bull. Am. Meteorol. Soc. **43**(12), 1834–1853 (2004). https://doi.org/10.1175/JAM2173.1

Hossain, F., Srinivasan, M., Peterson, C., Andral, A., Beighley, E., Anderson, E., Amini, R., Birkett, C., Bjerklie, D., Blain, C.A., Cherchali, S., David, C.H., Doorn, B., Escurra, J., Fu,

L.-L., Frans, C., Fulton, J., Gangopadhay, S., Ghosh, S., Gleason, C., Gosset, M., Hausman, J., Jacobs, G., Jones, J., Kaheil, Y., Laignel, B., Le Moigne, P., Li, L., Lefèvre, F., Mason, R., Mehta, A., Mukherjee, A., Nguy-Robertson, A., Ricci, S., Paris, A., Pavelsky, T., Picot, N., Schumann, G., Shrestha, S., Le Traon, P.-Y., Trehubenko, E.: Engaging the user community for advancing societal applications of the Surface Water Ocean Topography mission. Bull. Am. Meteorol. Soc. **98**(11), ES285–ES290 (2017). https://doi.org/10.1175/BAMS-D-17-0161.1

Hou, A.Y., Kakar, R.K., Neeck, S.A., Azarbarzin, A., Kummerow, C.D., Kojima, M., Oki, R., Nakamura, K., Iguchi, T.: The global precipitation measurement mission. Bull. Am. Meteorol. Soc. **95**, 701–722 (2014). https://doi.org/10.1175/BAMS-D-13-00164.1

Houze Jr., R.A., Rasmussen, K.L., Zuluaga, M.D., Brodzik, S.R.: The variable nature of convection in the tropics and subtropics: a legacy of 16 years of the Tropical Rainfall Measuring Mission satellite. Rev. Geophys. **53**(3), 994–1021 (2015). https://doi.org/10.1002/2015RG000488

Houze, R.A., McMurdie, L.A., Petersen, W.A., Schwaller, M.R., Baccus, W., Lundquist, J.D., Mass, C.F., Nijssen, B., Rutledge, S.A., Hudak, D.R., Tanelli, S., Mace, G.G., Poellot, M.R., Lettenmaier, D.P., Zagrodnik, J.P., Rowe, A.K., DeHart, J.C., Madaus, L.E., Barnes, H.C., Chandrasekar, V.: The Olympic Mountains Experiment (OLYMPEX). Bull. Am. Meteorol. Soc. **98**(10), 2167–2188 (2017). https://doi.org/10.1175/BAMS-D-16-0182.1

Hsu, K., Gao, X., Sorooshian, S., Gupta, H.V.: Precipitation estimation from remotely sensed information using artificial neural networks. J. Appl. Meteorol. Climatol. **36**, 1176–1190 (1997). https://doi.org/10.1175/1520-0450(1997)036<1176:PEFRSI>2.0.CO;2

Huffman, G.J., Adler, R.F., Arkin, P., Chang, A., Ferraro, R., Gruber, A., Janowiak, J., McNab, A., Rudolf, B., Schneider, U.: The global precipitation climatology project (GPCP) combined precipitation dataset. Bull. Am. Meteorol. Soc. **78**(1), 5–20 (1997). https://doi.org/10.1175/1520-0477(1997)0780005:TGPCPG2.0.CO;2

Huffman, G.J., Bolvin, D.T., Braithwaite, D., Hsu, K.L., Joyce, R.J., Kidd, C., Nelkin, E.J., Sorooshian, S., Stocker, E.F., Tan, J. and Wolff, D.B.: Integrated multi-satellite retrievals for the global precipitation measurement (GPM) mission (IMERG). In Satellite precipitation measurement: Volume 1, 343–353 (2020). Cham: Springer International Publishing. https://doi.org/10.1007/978-3-030-24568-9_19

Huffman, G.J., Bolvin, D.T., Nelkin, E.J., Wolff, D.B., Adler, R.F., Gu, G., Hong, Y., Bowman, K.P., Stocker, E.F.: The TRMM multisatellite precipitation analysis (TMPA): quasi-global, multiyear, combined-sensor precipitation estimates at fine scales. J. Hydrometeorol. **8**(1), 38–55 (2007). https://doi.org/10.1175/JHM560.1

Huffman, G.J., Adler, R.F., Behrangi, A., Bolvin, D.T., Nelkin, E.J., Gu, G., Ehsani, M.R.: The new version 3.2 Global Precipitation Climatology Project (GPCP) monthly and daily precipitation products. J. Clim. **36**(21), 7635–7655 (2023). https://doi.org/10.1175/JCLI-D-23-0123.1

Huuskonen, A., Saltikoff, E., Holleman, I.: The operational weather radar network in Europe. Bull. Am. Meteorol. Soc. **95**(6), 897–907 (2014). https://doi.org/10.1175/BAMS-D-12-00216.1

Jayanthi, H., Husak, G.J., Funk, C., Magadzire, T., Adoum, A., Verdin, J.P.: A probabilistic approach to assess agricultural drought risk to maize in southern Africa and millet in western Sahel using satellite estimated rainfall. Int. J. Disaster Risk Reduct. **10**, 490–502 (2014). https://doi.org/10.1016/j.ijdrr.2014.04.002

Jensen, M.P., Petersen, W.A., Bansemer, A., Bharadwaj, N., Carey, L.D., Cecil, D.J., Collis, S.M., Del Genio, A.D., Dolan, B., Gerlach, J., Giangrande, S.E., Heymsfield, A., Heymsfield, G., Kollias, P., Lang, T.J., Nesbitt, S.W., Neumann, A., Poellot, M., Rutledge, S.A., Schwaller, M., Tokay, A., Williams, C.R., Wolff, D.B., Xie, S., Zipser, E.J.: The Midlatitude Continental Convective Clouds Experiment (MC3E). Bull. Am. Meteorol. Soc. **97**(9), 1667–1686 (2016). https://doi.org/10.1175/BAMS-D-14-00228.1

Janowiak, J.E., Joyce, R.J., Yarosh, Y.: A real-time global half-hourly pixel-resolution infrared dataset and its applications. Bull. Am. Meteorol. Soc. **82**, 205–218 (2001). https://doi.org/10.1175/1520-0477(2001)0820205:ARTGHH2.3.CO;2

Joss, J., Waldvogel, A.: Raindrop size distribution and sampling size errors. J. Atmos. Sci. **26**, 566–569 (1969). https://doi.org/10.1175/1520-0469(1969)026<0566:RSDASS>2.0.CO;2

Joss, J., Waldvogel, A.: Raindrop size distributions and Doppler velocities. In: Preprints 14th Radar Meteorology Conf, pp. 153–156. American Meteorological Society, Tucson (1970)

Joyce, R.J., Janowiak, J.E., Arkin, P.A., Xie, P.: CMORPH: a method that produces global precipitation estimates from passive microwave and infrared data at high spatial and temporal resolutions. J. Hydrometeorol. **5**(3), 487–503 (2004). https://doi.org/10.1175/1525-7541(2004)005<0487:CAMTPG>2.0.CO;2

Joyce, R.J., Xie, P., Yarosh, Y., Janowiak, J.E., Arkin, P.A.: CMORPH: A "morphing" approach for high resolution precipitation product generation. In Satellite rainfall applications for surface hydrology 23-37.: Dordrecht: Springer. Netherlands. (2009). https://doi.org/10.1007/978-90-481-2915-7_2

Jutla, A., Akanda, A.S., Huq, A., Faruque, A.S.G., Colwell, R., Islam, S.: A water marker monitored by satellites to predict seasonal endemic cholera. Remote Sens. Lett. **4**(8), 822–831 (2013). https://doi.org/10.1007/978-3-642-40041-4_20

Kidd, C., Kniveton, D.R., Todd, M.C., Bellerby, T.J.: Satellite rainfall estimation using a combined passive microwave and infrared algorithm. J. Hydrometeorol. **4**, 1088–1104 (2003). https://doi.org/10.1175/1525-7541(2003)004<1088:SREUCP>2.0.CO;2

Kidd, C., Levizzani, V., Laviola, S.: Quantitative precipitation estimation from Earth observation satellites. In: Testik, F., Gebremichael, M. (eds.) Rainfall: State of the Science. American Geophysical Union (2010). https://doi.org/10.1029/2009GM000920

Kidd, C., Bauer, P., Turk, F.J., Huffman, G.J., Joyce, R., Hsu, K.-L., Braithwaite, D.: Intercomparison of high-resolution precipitation products over Northwest Europe. J. Hydrometeorol. **13**, 67–83 (2012). https://doi.org/10.1175/JHM-D-11-042

Kidd, C., Matsui, T., Chern, J., Mohr, K., Kummerow, C., Randel, D.: Global precipitation estimates from cross-track passive microwave observations using a physically based retrieval scheme. J. Hydrometeorol. **17**, 383–400 (2016). https://doi.org/10.1175/JHM-D-15-0051.1

Kidd, C., Levizzani, V.: Status of satellite precipitation retrievals. Hydrol. Earth Syst. Sci. **15**, 1109–1116 (2011). https://doi.org/10.5194/hess-15-1109-2011

Kidd, C., Tan, J., Kirstetter, P.-E., Petersen, W.A.: Validation of the Version 05 Level 2 precipitation products from the GPM Core Observatory and constellation satellite sensors. Q. J. R. Meteorol. Soc. **144**(S1), 313–328 (2018). https://doi.org/10.1002/qj.3175

Kidd, C., Huffman, G., Maggioni, V., Chambon, P., Oki, R.: The global satellite precipitation constellation: current status and future requirements. Bull. Am. Meteorol. Soc. **102**(10), E1844–E1861 (2021). https://doi.org/10.1175/BAMS-D-20-0299.1

Kidd, C., Matsui, T., Blackwell, W., Braun, S., Leslie, R., Griffith, Z.: Precipitation estimation from the NASA TROPICS mission: initial retrievals and validation. Remote Sens. **14**(13), 2992 (2022). https://doi.org/10.3390/rs14132992

Kirschbaum, D.B., Stanley, T.: Satellite-based assessment of rainfall-triggered landslide hazard for situational awareness. Earth's Future. **6**(3), 505–523 (2018). https://doi.org/10.1002/2017EF000715

Kirschbaum, D.B., Huffman, G.J., Adler, R.F., Braun, S., Garrett, K., Jones, E., McNally, A., Skofronick-Jackson, G., Stocker, E., Wu, H., Zaitchik, B.F.: NASA's remotely-sensed precipitation: a reservoir for applications users. Bull. Am. Meteorol. Soc. **98**, 1169–1184 (2017). https://doi.org/10.1175/BAMS-D-15-00296.1

Kirstetter, P.-E., Petersen, W., Kummerow, C.D., Wolff, D.B.: Integrated multi-satellite evaluation for the global precipitation measurement: impact of precipitation types on spaceborne precipitation estimation. In: Levizzani, V., Kidd, C., Kirschbaum, D.B., Kummerow, C.D., Nakamura, K., Turk, F.J. (eds.) Satellite Precipitation Measurement Advances in Global Change Research, vol. 69, pp. 583–608. Springer Nature (2020). https://doi.org/10.1007/978-3-030-35798-6_2

Kitchen, M., Illingworth, A.J.: From observations to forecasts - part 13: the UK weather radar network—past, present and future. Weather. **66**(11), 291–297 (2011). https://doi.org/10.1002/wea.861

Klepp, C.: The oceanic shipboard precipitation measurement network for surface validation—OceanRAIN. Atmos. Res. **163**, 74–90 (2015). https://doi.org/10.1016/j.atmosres.2014.12.014

Klepp, C., Michel, S., Protat, A., Burdanowitz, J., Albern, N., Kähnert, M., Dahl, A., Louf, V., Bakan, S., Buehler, S.A.: OceanRAIN, a new in-situ shipboard global ocean surface-reference dataset of all water cycle components. Sci. Data. **5**, 180122 (2018). https://doi.org/10.1038/sdata.2018.122

Knapp, K.R., Wilkins, S.R.: Gridded Satellite (GridSat) GOES and CONUS data. Earth Syst. Sci. Data. **10**, 1417–1425 (2018). https://doi.org/10.5194/essd-10-1417-2018

Krajewski, A., Banasik, K., Sikorska, A.E.: Stormflow and suspended sediment routing through a small detention pond with uncertain discharge rating curves. Hydrol. Res. **50**(4), 1177–1188 (2019). https://doi.org/10.2166/nh.2018.131

Kruger, A., Krajewski, W.F.: Two-dimensional video disdrometer: a description. J. Atmos. Ocean. Technol. **19**(5), 602–617 (2002). https://doi.org/10.1175/1520-0426(2002)019<0602:TDVDAD>2.0.CO;2

Kubota, T., Shige, S., Hashizume, H., Aonashi, K., Takahashi, N., Seto, S., Hirose, M., Takayabu, Y.N., Nakagawa, K., Iwanami, K., Ushio, T., Kachi, M., Okamoto, K.: Global precipitation map using satelliteborne microwave radiometers by the GSMaP Project: production and validation. IEEE Trans. Geosci. Remote Sens. **45**, 2259–2275 (2007). https://doi.org/10.1109/TGRS.2007.895337

Kubota, T., Aonashi, A., Ushio, T., Shige, S., Takayabu, Y.N., Kachi, M., Arai, Y., Tashima, T., Masaki, T., Kawamoto, N., Mega, T., Yamamoto, M.K., Hamada, A., Yamaji, M., Liu, G., Oki, R.: Global Satellite Mapping of Precipitation (GSMaP) products in the GPM era. In: Levizzani, V., Kidd, C., Kirschbaum, D.B., Kummerow, C.D., Nakamura, K., Turk, F.J. (eds.) Satellite Precipitation Measurement Advances in Global Change Research, vol. 67, pp. 355–374. Springer Nature (2020). https://doi.org/10.1007/978-3-030-24568-9_20

Kuligowski, R.J.: A self-calibrating GOES rainfall algorithm for short term rainfall estimates. J. Hydrometeorol. **3**(2), 112–130 (2002). https://doi.org/10.1175/1525-7541(2002)003<0112:ASCRTG>2.0.CO;2

Kummerow, C., Giglio, L.: A passive microwave technique for estimating rainfall and vertical structure information from space. Part I: algorithm description. J. Appl. Meteorol. Climatol. **33**, 3–18 (1994). https://doi.org/10.1175/1520-0450(1994)0332.0.CO;2

Kummerow, C., Barnes, W., Kozu, T., Shiue, J., Simpson, J.: The Tropical Rainfall Measuring Mission (TRMM) sensor package. J. Atmos. Ocean. Technol. **15**(3), 809–817 (1998). https://doi.org/10.1175/1520-0426(1998)015<0809:TTRMMT>2.0.CO;2

Kummerow, C., Hong, Y., Olson, W.S., Yang, S., Adler, R.F., McCollum, J., Ferraro, R., Petty, G., Shin, D.-B., Wilheit, T.T.: The evolution of the Goddard Profiling algorithm (GPROF) for rainfall estimation from passive microwave sensors. J. Appl. Meteorol. Climatol. **40**(11), 1801–1820 (2001). https://doi.org/10.1175/1520-0450(2001)040<1801:TEOTGP>2.0.CO;2

Kummerow, C.D., Ringerud, S., Crook, J., Randel, D., Berg, W.: An observationally generated a priori database for microwave rainfall retrievals. J. Atmos. Ocean. Technol. **28**, 113–130 (2011). https://doi.org/10.1175/2010JTECHA1468.1

Kummerow, C.D., Randel, D.L., Kulie, M., Wang, N.-Y., Ferraro, R., Munchak, S.J., Petkovic, V.: The evolution of the Goddard Profiling algorithm to a fully parametric scheme. J. Appl. Meteorol. Climatol. **32**(12), 2265–2280 (2015). https://doi.org/10.1175/JTECH-D-15-0039.1

Kummerow, C.D., Simpson, J., Thiele, O., Barnes, W., Chang, A.T.C., Stocker, E., Adler, R.F., Hou, A., Kakar, R., Wentz, F., Ashcroft, P., Kozu, T., Hong, Y., Okamoto, K., Iguchi, T., Kuroiwa, H., Im, E., Haddad, Z., Huffman, G., Ferrier, B., Olson, W.S., Zipser, E., Smith, E.A., Wilheit, T.T., North, G., Krishnamurti, T., Nakamura, K.: The status of the tropical rainfall measuring Mission (TRMM) after two years in orbit. J. Appl. Meteorol. **39**, 1965–1982 (2000). https://doi.org/10.1175/1520-0450(2001)0401965:TSOTTR2.0.CO;2

L'Ecuyer, T.S., Stephens, G.L.: An estimation-based precipitation retrieval algorithm for attenuating radars. J. Appl. Meteorol. **41**(3), 272–285 (2002). https://doi.org/10.1175/1520-0450(2002)0410272:AEBPRA2.0.CO;2

Lambrigtsen, B., Kangaslahti, P., Montes, O., Niamsuwan, N., Posselt, D., Roman, J., Schreier, M., Tanner, A., Wu, L., Yanovsky, I.: A geostationary microwave sounder: design, implementation and performance. IEEE J. Sel. Top. Appl. Earth Obs. Remote Sens. **15**, 623–640 (2022). https://doi.org/10.1109/JSTARS.2021.3132238

Lauderdale, J.M., Caminade, C., Heath, A.E., Jones, A.E., MacLeod, D.A., Gouda, K.C., Murty, U.S., Goswami, P., Mutheneni, S.R., Morse, A.: Towards seasonal forecasting of malaria in India. Malar. J. **13**, 310 (2014). https://doi.org/10.1186/1475-2875-13-310

Laviola, S., Levizzani, V.: The 183-WSL fast rain rate retrieval algorithm. Part I: retrieval design. Atmos. Res. **99**, 443–461 (2011). https://doi.org/10.1016/j.atmosres.2010.11.013

Lee, G., Jung, S.-H., Lee, J.-H., Cho, Y.-H., Ahn, K.-D., Heo, B.-H., Lee, C.-K.: Tri-agency radar networks in Korea: where are we heading? In: Moore, R.J., Cole, S.J., Illingworth, A.J. (eds.) Weather Radar and Hydrology. Proceedings of the International Symposium on Weather Radar and Hydrology, Exeter, vol. 351, pp. 15–19, 18–21 April 2011 (2012).

Lensky, I.M., Rosenfeld, D.: Clouds-Aerosols-Precipitation Satellite Analysis Tool (CAPSAT). Atmos. Chem. Phys. **8**(22), 6739–6753 (2008). https://doi.org/10.5194/acp-8-6739-2008

Levizzani, V., Kidd, C., Aonashi, K., Bennartz, R., Ferraro, R.R., Huffman, G.J., Roca, R., Turk, F.J., Wang, N.-Y.: The activities of the international precipitation working group. Q. J. R. Meteorol. Soc. **144**(S1), 3–15 (2018). https://doi.org/10.1002/qj.3214

Liguori, S., Rico-Ramirez, M.A., Schellart, A.N.A., Saul, A.J.: Using probabilistic radar rainfall nowcasts and NWP forecasts for flow prediction in urban catchments. Atmos. Res. **103**, 80–95 (2012). https://doi.org/10.1016/j.atmosres.2011.05.004

Lindsey, D.T., Heidinger, A.K., Sullivan, P.C., McCorkel, J., Schmit, T.J., Tomlinson, M., Vandermeulen, R., Frost, G.J., Kondragunta, S., Rudlosky, S.: GeoXO: NOAA's future geostationary satellite system. Bull. Am. Meteorol. Soc. **105**(3), E660–E679 (2025). https://doi.org/10.1175/BAMS-D-23-0048.1

Löffler-Mang, M., Joss, J.: An optical disdrometer for measuring size and velocity of hydrometeors. J. Atmos. Ocean. Technol. **17**(2), 130–139 (2000). https://doi.org/10.1175/1520-0426(2000)017<0130:AODFMS>2.0.CO;2

Ma, B.B., Nystuen, J.A., Lien, R.-C.: Prediction of underwater sound levels from rain and wind. J. Acoust. Soc. Am. **117**(6), 3555–3565 (2005). https://doi.org/10.1121/1.1910283

Machault, V., Vignolles, C., Borchi, F., Vounatsou, P., Pages, F., Briolant, S., Lacaux, J.-P., Rogier, C.: The use of remotely sensed environmental data in the study of malaria. Geospat. Health. **5**, 151–168 (2011). https://doi.org/10.4081/gh.2011.167

MacMillan, K., Monaghan, A.J., Apangu, T., Griffith, K.S., Mead, P.S., Acayo, S., Acidri, R., Moore, S.M., Mpanga, J.T., Enscore, R.E., Gage, K.L., Eisen, R.J.: Climate predictors of the spatial distribution of human plague cases in the West Nile region of Uganda. Am. J. Trop. Med. Hyg. **86**, 514–523 (2012). https://doi.org/10.4269/ajtmh.2012.11-0569

Maggioni, V., Massari, C.: On the performance of satellite precipitation products in riverine flood modeling: a review. J. Hydrol. **558**, 214–224 (2018). https://doi.org/10.1016/j.jhydrol.2018.01.039

Maggioni, V., Vergara, H.J., Anagnostou, E.N., Gourley, J.J., Hong, Y., Stampoulis, D.: Investigating the applicability of error correction ensembles of satellite rainfall products in river flow simulations. J. Hydrometeorol. **14**, 1194–1211 (2013). https://doi.org/10.1175/JHM-D-12-074.1

Maidment, R., Black, E., Greatrex, H., Young, M.: TAMSAT. In: Levizzani, V., Kidd, C., Kirschbaum, D.B., Kummerow, C.D., Nakamura, K., Turk, F.J. (eds.) Satellite Precipitation Measurement Advances in Global Change Research, vol. 67, pp. 393–407. Springer Nature (2020). https://doi.org/10.1007/978-3-030-24568-9_22

Makihara, Y., Uekiyo, N., Tabata, A., Abe, Y.: Accuracy of radar-AMeDAS precipitation. IEICE Trans. Commun. **79**(6), 751–762 (1996)

Mandapaka, P.V., Germann, U., Panziera, L., Hering, A.: Can Lagrangian extrapolation of radar fields be used for precipitation nowcasting over complex alpine orography? Weather Forecast. **27**(1), 28–49 (2012). https://doi.org/10.1175/WAF-D-11-00050.1

Marzano, F.S., Palmacci, M., Cimini, D., Giuliani, G., Turk, F.J.: Multivariate statistical integration of satellite infrared and microwave radiometric measurements for rainfall retrieval at the geostationary scales. IEEE Trans. Geosci. Remote Sens. **42**, 1018–1032 (2004). https://doi.org/10.1109/TGRS.2003.820312

Meneghini, R., Iguchi, T. Seto, S., Yoshida, N., Awaka, J., Kubota, T.: GPM/DPR Level 2 Algorithm Theoretical Basis Document (ATBD). Available at https://gpm.nasa.gov/resources/documents/gpmdpr-level-2-algorithm-theoretical-basis-document-atbd. Last accessed 9 Jan 2025 (2021)

Menne, M.J., Durre, I., Vose, R.S., Gleason, B.E., Houston, T.G.: An overview of the Global Historical Climatology Network-daily database. J. Atmos. Ocean. Technol. **29**, 897–910 (2012). https://doi.org/10.1175/JTECH-D-11-00103.1

Menzel, W.P., Purdom, J.F.W.: Introducing GOES-I: the first of a new generation of Geostationary Operational Environmental Satellites. Bull. Am. Meteorol. Soc. **75**(5), 757–782 (1994). https://doi.org/10.1175/1520-0477(1994)075<0757:IGITFO>2.0.CO;2

Midekisa, A., Senay, G., Henebry, G.M., Semuniguse, P., Wimberly, M.C.: Remote sensing-based time series models for malaria early warning in the highlands of Ethiopia. Malar. J. **11**, 165 (2012). https://doi.org/10.1186/1475-2875-11-165

Milani, L., Kidd, C.: The state of precipitation measurements at mid-to-high latitudes. Atmos. **14**(11), 1677 (2023). https://doi.org/10.3390/atmos14111677

Miller, S.W., Arkin, P.A., Joyce, R.J.: A combined microwave/infrared rain rate algorithm. Int. J. Remote Sens. **22**, 3285–3307 (2001). https://doi.org/10.1080/01431160152609155

Mimikou, M.A., Baltas, E.A.: Flood forecasting based on radar rainfall measurements. J. Water Resour. Plan. Manag. **122**(3), 151 (1996). https://doi.org/10.1061/(ASCE)0733-9496(1996)122:3(151)

Mitchell, T.D., Jones, P.D.: An improved method of constructing a database of monthly climate observations and associated high-resolution grids. Int. J. Climatol. **25**(6), 693–712 (2005). https://doi.org/10.1002/joc.1181

Mo, T.: A study of the Microwave Sounding Unit on the NOAA-12 satellite. IEEE Trans. Geosci. Remote Sens. **33**(5), 1141–1152 (1995). https://doi.org/10.1109/36.469478

Molod, A., Takacs, L., Suarez, M., Bacmeister, J.: Development of the GEOS-5 atmospheric general circulation model: evolution from MERRA to MERRA2. Geosci. Model Dev. **8**, 1339–1356 (2015). https://doi.org/10.5194/gmd-8-1339-2015

Monaghan, A.J., MacMillan, K., Moore, S.M., Mead, P.S., Hayden, M.H., Eisen, R.J.: A regional climatography of West Nile, Uganda, to support human plague modeling. J. Appl. Meteorol. Climatol. **51**, 1201–1221 (2012). https://doi.org/10.1175/JAMC-D-11-0195.1

Moore, S.M., Monaghan, A., Griffith, K.S., Apangu, T., Mead, P.S., Eisen, R.J.: Improvement of disease prediction and modeling through the use of meteorological ensembles: human plague in Uganda. PLoS One. **7**, e44431 (2012). https://doi.org/10.1371/journal.pone.0044431

Moore, S.M., Azman, A.S., Zaitchik, B.F., Mintz, E.D., Brunkard, J., Legros, D., Hill, A., McKay, H., Luquero, F.J., Olson, D., Lessler, J.: El Niño and the shifting geography of cholera in Africa. PNAS. **114**(17), 4436–4441 (2017). https://doi.org/10.1073/pnas.1617218114

Mugnai, A., Cooper, H.J., Smith, E.A., Tripoli, G.J.: Simulation of microwave brightness temperatures of an evolving hailstorm at SSM/I frequencies. Bull. Am. Meteorol. Soc. **71**(1), 2–13 (1990). https://doi.org/10.1175/1520-0477(1990)071<0002:SOMBTO>2.0.CO;2

Mugnai, A., Casella, D., Cattani, E., Dietrich, S., Laviola, S., Levizzani, V., Panegrossi, G., Petracca, M., Sanò, P., Di Paola, F., Biron, D., De Leonibus, L., Melfi, D., Rosci, P., Vocino, A., Zauli, F., Pagliara, P., Puca, S., Rinollo, A., Milani, L., Porcù, F., Gattari, F.: Precipitation products from the hydrology SAF. Nat. Hazards Earth Syst. Sci. **13**, 1959–1981 (2013). https://doi.org/10.5194/nhess-13-1959-2013

Nakakita, E., Yoshikai, T., Kim, S.: Application of error-ensemble prediction method to a short-term rainfall prediction model considering orographic rainfall. IAHS Publ. **351**, 311–316 (2012)

New, M., Todd, M., Hulme, M., Jones, P.: Precipitation measurements and trends in the twentieth century. Int. J. Climatol. **21**, 1889–1922 (2001). https://doi.org/10.1002/joc.680

Nitu, R., Wong, K.: CIMO survey on national summaries of methods and instruments for solid precipitation measurement at automatic weather stations. WMO Instruments and Observing Methods Rep. 102, WMO/TD-1544, 57 pp. Available at https://library.wmo.int/records/item/50474-cimo-survey-on-national-summaries-of-methods-and-instruments-for-solid-precipitation-measurement-at-automatic-weather-stations#.Y4YbjRTMLe8 (2010a)

Nitu, R., Wong, K.: Measurement or solid precipitation at automatic weather stations: challenges and opportunities. TECO-2010-WMO Technical Conf. on Meteorological and Environmental Instruments and Methods of Observation, Helsinki, Finland. Available at https://wmoomm.sharepoint.com/:w:/s/wmocpdb/Eb3QVGkF-PdBuJKIXC94fmoB2h46YhorgwHB0mKn5BJ16Q?rtime=dotkQGoj3Ug (2010b)

Olson, W.S.: Physical retrieval of rainfall rates over the ocean by multispectral microwave radiometry: application to tropical cyclones. J. Geophys. Res.: Atmos. **94**, 2267–2280 (1989). https://doi.org/10.1029/JD094iD02p02267

Overeem, A., Leijnse, H., Uijlenhoet, R.: Retrieval algorithm for rainfall mapping from microwave links in a cellular communication network. Atmos. Meas. Tech. **9**(5), 2425–2444 (2016). https://doi.org/10.5194/amt-9-2425-201628

Pall, P., Allen, M.R., Stone, D.A.: Testing the Clausius–Clapeyron constraint on changes in extreme precipitation under CO2 warming. Clim. Dyn. **28**, 351–363 (2006). https://doi.org/10.1007/s00382-006-0180-2

Panziera, L., Germann, U., Gabella, M., Mandapaka, P.: NORA—Nowcasting of Orographic Rainfall by means of Analogues. Q. J. R. Meteorol. Soc. **137**, 2106–2123 (2011). https://doi.org/10.1002/qj.878

Pegion, K., Becker, E.J., Kirtman, B.P.: Understanding predictability of daily southeast U.S. precipitation using explainable machine learning. Artif. Intell. Earth Syst. **1**(4) (2022). https://doi.org/10.1175/AIES-D-22-0011.1

Peters, G., Fischer, B., Andersson, T.: Rain observations with a vertically looking micro rain radar (MRR). Boreal Environ. Res. **7**, 353–362 (2002)

Petersen, W.A., Kirstetter, P.-E., Wang, J., Wolff, D.B., Tokay, A.: The GPM Ground Validation Program. In: Levizzani, V., Kidd, C., Kirschbaum, D.B., Kummerow, C.D., Nakamura, K., Turk, F.J. (eds.) Satellite Precipitation Measurement Advances in Global Change Research, vol. 69, pp. 471–502. Springer Nature (2020). https://doi.org/10.1007/978-3-030-35798-6_2

Petty, G.W.: Frequencies and characteristics of global oceanic precipitation from shipboard present-weather reports. Bull. Am. Meteorol. Soc. **82**, 1377–1396 (1995). https://doi.org/10.1175/1520-0477(1995)076<1593:FACOGO>2.0.CO;2

Petty, G.W., Tran, H.K.: Seventy-year trends in ship-reported oceanic precipitation frequency. Geophys. Res. Lett. **50**(14), e2023GL104270 (2023). https://doi.org/10.1029/2023GL104270

Pierce, C., Seed, A., Ballard, S., Simonin, D., Li, Z.: Nowcasting. In: Bech, J., Chau, J.L. (eds.) Doppler Radar Observations - Weather Radar, Wind Profiler, Ionospheric Radar, and Other Advanced Applications, pp. 97–142. IntechOpen (2012). https://doi.org/10.5772/39054

Piran, M.J., Wang, X., Kim, H.J., Kwon, H.H.: Precipitation nowcasting using transformer-based generative models and transfer learning for improved disaster preparedness. Int. J. Appl. Earth Obs. Geoinf. **132**, 103962 (2024). https://doi.org/10.1016/j.jag.2024.103962

Rasmussen, R., Baker, B., Kochendorfer, J., Meyers, T., Landolt, S., Fischer, A.P., Black, J., Thériault, J.M., Kucera, P., Gochis, D., Smith, C., Nitu, R., Hall, M., Ikeda, K., Gutmann, E.: How well are we measuring snow: the NOAA/FAA/NCAR winter precipitation test bed. Bull. Am. Meteorol. Soc. **93**, 811–829 (2012). https://doi.org/10.1175/BAMS-D-11-00052.1

Randel, D.L., Kummerow, C.D., Ringerud, S.: The Goddard profiling (GPROF) precipitation retrieval algorithm. In: Satellite precipitation measurement, V. Levizzani, C. Kidd., D. B. Kirschbaum, C. D. Kummerow, K. Nakamura, F. J. Turk, Eds., Springer Nature, Cham. Advances in Global Change Research. **67**, 141–152 (2020). https://doi.org/10.1007/978-3-030-24568-9_8

Rodgers, E., Siddalingaiah, H.: The utilization of Nimbus-7 SMMR measurements to delineate rainfall over land. J. Appl. Meteorol. Climatol. **22**, 1753–1763 (1983). https://doi.org/10.1175/1520-0450(1983)022<1753:TUONSM>2.0.CO;2

Roebeling, R.A., Holleman, I.: SEVIRI rainfall retrieval and validation using weather radar observations. J. Geophys. Res.: Atmos. **114**, D21202 (2009). https://doi.org/10.1029/2009JD012102

Rossow, W.B., Schiffer, R.A.: ISCCP data products. Bull. Am. Meteorol. Soc. **72**(1), 2–20 (1991). https://doi.org/10.1175/1520-0477(1991)072<0002:ICDP>2.0.CO;2

Saltikoff, E., Haase, G., Delobbe, L., Gaussiat, N., Martet, M., Idziorek, D., Leijnse, H., Novák, P., Lukach, M., Stephan, K.: OPERA the Radar Project. Atmosphere. **10**(6), 320 (2019). https://doi.org/10.3390/atmos10060320

Sanò, P., Panegrossi, G., Casella, D., Di Paola, F., Milani, L., Mugnai, A., Petracca, M., Dietrich, S.: The Passive microwave Neural network Precipitation Retrieval (PNPR) algorithm for AMSU/MHS observations: description and application to European case studies. Atmos. Meas. Tech. **8**, 837–857 (2015). https://doi.org/10.5194/amt-8-837-2015

Savage, R.C., Weinman, J.A.: Preliminary calculations of the upwelling radiance from rain clouds at 37.0 and 19.35 GHz. Bull. Am. Meteorol. Soc. **56**, 1272–1274 (1975) Available at https://www.jstor.org/stable/26216455

Schmetz, J., Pili, P., Tjemkes, S., Just, D., Kerkmann, J., Rota, S., Ratier, A.: An introduction to Meteosat Second Generation (MSG). Bull. Am. Meteorol. Soc. **83**(7), 977–992 (2002). https://doi.org/10.1175/1520-0477(2002)083<0977:AITMSG>2.3.CO;2

Senay, G., Verdin, J.: Characterization of yield reduction in Ethiopia using a GIS-based crop water balance model. Can. J. Remote. Sens. **29**, 687–692 (2003). https://doi.org/10.5589/m03-039

Sevruk, B., Klemm, S.: Types of standard precipitation gauges. In: Proc. WMI/IAHS/ETH Workshop on Precipitation Measurement, St. Moritz, Switzerland, WMO Tech. Doc. 32, 589 pp (1989a)

Sevruk, B., Klemm, S.: Catalogue of national standard precipitation gauges. Instruments and Observing Methods Rep. 39, WMO/TD-313, 50 pp. Available at https://library.wmo.int/records/item/41708-catalogue-of-national-standard-precipitation-gauges (1989b)

Shepherd, J.M., Burian, S., Liu, C., Bernardes, S.: Satellite precipitation metrics to study the energy-water-food nexus within the backdrop of an urbanized globe. IEEE Earthzine. Available at https://earthzine.org/satellite-precipitation-metrics-to-study-the-energy-water-food-nexus-within-the-backdrop-of-an-urbanized-globe/ (2016)

Sims, E.M., Liu, G.: A parameterization of the probability of snow–rain transition. J. Hydrometeorol. **16**(4), 1466–1477 (2015). https://doi.org/10.1175/JHM-D-14-0211.1

Skofronick-Jackson, G., Hudak, D., Petersen, W., Nesbitt, S.W., Chandrasekar, V., Durden, S., Gleicher, K.J., Huang, G.-J., Joe, P., Kollias, P., Reed, K.A., Schwaller, M.R., Stewart, R., Tanelli, S., Tokay, A., Wang, J.R., Wolde, M.: Global Precipitation Measurement Cold Season Precipitation Experiment (GCPEX): for measurement's sake, let it snow. Bull. Am. Meteorol. Soc. **96**(10), 1719–1741 (2015). https://doi.org/10.1175/BAMS-D-13-00262.1

Skofronick-Jackson, G., Petersen, W.A., Berg, W., Kidd, C., Stocker, E.F., Kirschbaum, D.B., Kakar, R., Braun, S.A., Huffman, G.J., Iguchi, T., Kirstetter, P.E., Kummerow, C., Meneghini, R., Oki, R., Olson, W.S., Takayabu, Y.N., Furukawa, K., Wilheit, T.: The Global Precipitation Measurement (GPM) Mission for science and society. Bull. Am. Meteorol. Soc. **98**(8), 1679–1695 (2017). https://doi.org/10.1175/BAMS-D-15-00306.1

Smith, E.A., Mugnai, A., Cooper, H.J., Tripoli, G.J., Xiang, X.: Foundations for statistical-physical precipitation retrieval from passive microwave satellite measurements. Part I: brightness-temperature properties of a time-dependent cloud-radiation model. J. Appl. Meteorol. Climatol. **31**, 506–531 (1992). https://doi.org/10.1175/1520-0450(1992)031<0506:FFSPPR>2.0.CO;2

Soderholm, J., Louf, V., Brook, J., Protat, A., Warren, R.: Australian operational weather radar level 2 dataset [Dataset]. National Computing Infrastructure (2022). https://doi.org/10.25914/JJWZ-0F13

Sorooshian, S., Hsu, K.-L., Gao, X., Gupta, H.V., Imam, B., Braithwaite, D.: Evaluation of PERSIANN system satellite-based estimates of tropical rainfall. Bull. Am. Meteorol. Soc. **81**, 2035–2046 (2000). https://doi.org/10.1175/1520-0477(2000)081<2035:EOPSSE>2.3.CO;2

Spencer, R.W.: A satellite passive 37-GHz scattering-based method for measuring oceanic rain rates. J. Appl. Meteorol. Climatol. **25**(6), 754–766 (1986). https://doi.org/10.1175/1520-0450(1986)025<0754:ASPGSB>2.0.CO;2

Spencer, R.W.: Global oceanic precipitation from the MSU during 1979–91 and comparison to other climatologies. J. Clim. **6**(7), 1301–1326 (1993). https://doi.org/10.1175/1520-0442(1993)006<1301:GOPFTM>2.0.CO;2

Stephens, G.L., Vane, D.G., Boain, R.J., Mace, G.G., Sassen, K., Wang, Z., Illingworth, A.J., O'Connor, E.J., Rossow, W.B., Durden, S.L., Miller, S.D., Austin, R.T., Benedetti, A., Mitrescu, C., the CloudSat Science Team: The CloudSat mission and the A-Train - a new dimension of space-based observations of clouds and precipitation. Bull. Am. Meteorol. Soc. **83**(12), 1771–1790 (2002). https://doi.org/10.1175/BAMS-83-12-1771

Stephens, G.L., Vane, D.G., Tanelli, S., Im, E., Durden, S., Rokey, M., Reinke, D., Partain, P., Mace, G.G., Austin, R., L'Ecuyer, T., Haynes, J., Lebsock, M., Suzuki, K., Waliser, D., Wu, D., Kay, J., Gettelman, A., Wang, Z., Marchand, R.: CloudSat mission: performance and early science after the first year of operation. J. Geophys. Res.: Atmos. **113**, D00A18 (2008). https://doi.org/10.1029/2008JD009982

Stephens, G., Winker, D., Pelon, J., Trepte, C., Vane, D., Yuhas, C., L'Ecuyer, T., Lebsock, M.: CloudSat and CALIPSO within the A-Train: ten years of actively observing the Earth system. Bull. Am. Meteorol. Soc. **99**, 569–581 (2018). https://doi.org/10.1175/BAMS-D-16-0324.1

Strangeways, I.: Measuring the Natural Environment, 2nd edn. Cambridge University Press (2003)

Strangeways, I.: A history of rain gauges. Weather. **65**, 133–138 (2010). https://doi.org/10.1002/wea.548

Sun, Q., Miao, C., Duan, Q., Ashouri, H., Sorooshian, S., Hsu, K.-L.: A review of global precipitation data sets: data sources, estimation, and intercomparisons. Rev. Geophys. **56**(1), 79–107 (2017). https://doi.org/10.1002/2017RG000574

Tapiador, F.J., Kidd, C., Levizzani, V., Marzano, F.S.: A neural networks-based fusion technique to estimate half-hourly rainfall estimates at 0.1° resolution from satellite passive microwave and infrared data. J. Appl. Meteorol. Climatol. **43**, 576–594 (2004). https://doi.org/10.1175/1520-0450(2004)043<0576:ANNFTT>2.0.CO;2

Thornes, J., Bloss, W., Bouzarovski, S., Cai, X., Chapman, L., Clark, J., Dessai, S., Du, S., van der Horst, D., Kendall, M., Kidd, C., Randalls, S.: Communicating the value of atmospheric services. Meteorol. Appl. **17**, 243–250 (2010). https://doi.org/10.1002/met.200

Todd, M.C., Kidd, C., Kniveton, D., Bellerby, T.J.: A combined satellite infrared and passive microwave technique for estimation of small-scale rainfall. J. Atmos. Ocean. Technol. **18**(5), 742–755 (2001). https://doi.org/10.1175/1520-0469(2001)058<0742:ACSIAP>2.0.CO;2

Tokay, A., Kruger, A., Krajewski, W.F.: Comparison of drop size distribution measurements by impact and optical disdrometers. J. Appl. Meteorol. Climatol. **40**(11), 2083–2097 (2001). https://doi.org/10.1175/1520-0450(2001)040<2083:CODSDM>2.0.CO;2

Tourre, Y.M., Lacaux, J.-P., Vignolles, C., Lafaye, M.: Climate impacts on environmental risks evaluated from space: a conceptual approach to the case of Rift Valley Fever in Senegal. Glob. Health Action. **2**, 2053 (2009). https://doi.org/10.3402/gha.v2i0.2053

Trenberth, K.E., Dai, A., Rasmussen, R.M., Parsons, D.B.: The changing character of precipitation. Bull. Am. Meteorol. Soc. **84**(9), 1205–1218 (2003). https://doi.org/10.1175/BAMS-84-9-1205

Turk, F.J., Rohaly, G., Hawkins, J., Smith, E.A., Marzano, F.S., Mugnai, A., Levizzani, V.: Meteorological applications of precipitation estimation from combined SSM/I, TRMM and geostationary satellite data. In: Pampaloni, P., Paloscia, S. (eds.) Microwave Radiometry and Remote Sensing of the Earth's Surface and Atmosphere, pp. 353–363. VSP Int. Sci. Publisher, Utrecht (2000)

Valle, D., Zaitchik, B., Feingold, B., Spangler, K., Pan, W.: Abundance of water bodies is critical to guide mosquito larval control interventions and predict risk of mosquito-borne diseases. Parasit. Vectors. **6**, 179 (2013). https://doi.org/10.1186/1756-3305-6-179

Verdin, J., Klaver, R.: Grid-cell-based crop water accounting for the famine early warning system. Hydrol. Process. **16**(8), 1617–1630 (2002). https://doi.org/10.1002/hyp.1025

Vicente, G.A., Scofield, R.A., Menzel, W.P.: The operational GOES infrared rainfall estimation technique. Bull. Am. Meteorol. Soc. **79**, 1883–1898 (1998). https://doi.org/10.1175/1520-0477(1998)079<1883:TOGIRE>2.0.CO;2

Vuerich, E., Monesi, C., Lanza, L.G., Stagi, L., Lanzinger, E.: WMO field intercomparison of rainfall intensity gauges. WMO Instruments and Observing Methods Rep. 99, WMO/TD-1504, 290 pp. Available at https://library.wmo.int/records/item/50453-wmo-field-intercomparison-of-rainfall-intensity-gauges#.Y4Yb-RTMLe8 (2009)

Wallace, J.M., Hobbs, P.V.: Atmospheric Science: an Introductory Survey, 2nd edn. Academic (2006)

Wang, X., Parrish, D., Kleist, D., Whitaker, J.: GSI 3DVar-based ensemble-variational hybrid data assimilation for NCEP Global Forecast System: single-resolution experiments. Mon. Weather Rev. **141**, 4098–4117 (2013). https://doi.org/10.1175/MWR-D-12-00141.1

Whiton, R.C., Smith, P.L., Bigler, S.G., Wilk, K.E., Harbuck, A.C.: History of operational use of weather radar by U.S. Weather Services. Part I: the pre-NEXRAD era. Weather Forecast. **13**(2), 219–243 (1998a). https://doi.org/10.1175/1520-0434(1998)013<0219:HOOUOW>2.0.CO;2

Whiton, R.C., Smith, P.L., Bigler, S.G., Wilk, K.E., Harbuck, A.C.: History of operational use of weather radar by U.S. Weather Services. Part II: development of operational Doppler weather radars. Weather Forecast. **13**(2), 244–252 (1998b). https://doi.org/10.1175/1520-0434(1998)013<0244:HOOUOW>2.0.CO;2

Wilheit, T. J.: The Electrically Scanning Microwave Radiometer (ESMR) experiment. Nimbus 5 Users Guide, NASA, Goddard Space Flight Center, pp. 55–105 (1972)

Wilheit, T.T., Chang, A.T.C., Rao, M.S.V., Rodgers, E.B., Theon, J.S.: A satellite technique for quantitatively mapping rainfall rates over the oceans. J. Appl. Meteorol. Climatol. **16**(5), 551–560 (1977). https://doi.org/10.1175/1520-0450(1977)016<0551:ASTFQM>2.0.CO;2

Wilheit, T., Adler, R., Avery, S., Barrett, E., Bauer, P., Berg, W., Chang, A., Ferriday, J., Grody, N., Goodman, S., Kidd, C., Kniveton, D., Kummerow, C., Mugnai, A., Olson, W., Petty, G., Shibata, A., Smith, E.: Algorithms for the retrieval of rainfall from passive microwave measurements. Remote Sens. Rev. **11**, 163–194 (1994). https://doi.org/10.1080/02757259409532264

Winterrath, T., Rosenow, W., Weigl, E.: On the DWD quantitative precipitation analysis and nowcasting system for real-time application in German flood risk management. IAHS Publ. **351**, 323–329 (2012)

WMO: Guide to meteorological instruments and methods of observation, 7th edn. World Meteorological Organization WMO-8, 681 pp. Available at https://community.wmo.int/en/activity-areas/imop/wmo-no_8 (2008)

Wood, N.B., L'Ecuyer, T.S., Bliven, F.L., Stephens, G.L.: Characterization of disdrometer uncertainties and impacts on estimates of snowfall rate and radar reflectivity. Atmos. Meas. Tech. **6**, 3635–3648 (2013). https://doi.org/10.5194/amt-6-3635-2013

Wood, N.B., L'Ecuyer, T.S., Heymsfield, A.J., Stephens, G.L., Hudak, D.R., Rodriguez, P.: Estimating snow microphysical properties using collocated multisensor observations. J. Geophys. Res. Atmos. **119**, 8941–8961 (2014). https://doi.org/10.1002/2013JD021303

Wu, Q., Wang, Y.: Comparison of oceanic multisatellite precipitation data from Tropical Rainfall Measurement Mission and Global Precipitation Measurement mission datasets with rain gauge data from ocean buoys. J. Atmos. Ocean. Technol. **36**(5), 903–920 (2019). https://doi.org/10.1175/JTECH-D-18-0152.1

Wu, H., Adler, R.F., Hong, Y., Tian, Y., Policelli, F.: Evaluation of global flood detection using satellite-based rainfall and a hydrologic model. J. Hydrometeorol. **13**, 1268–1284 (2012). https://doi.org/10.1175/JHM-D-11-087.1

Wu, H., Adler, R.F., Tian, Y., Huffman, G.J., Li, H., Wang, J.J.: Real-time global flood estimation using satellite-based precipitation and a coupled land surface and routing model. Water Resour. Res. **50**(3), 2693–2717 (2014). https://doi.org/10.1002/2013WR014710

Xie, P., Arkin, P.A.: Global precipitation: A 17-year monthly analysis based on gauge observations, satellite estimates, and numerical model outputs. Bull. Am. Meteorol. Soc. **78** (11), 2539-2558 (1997). (1997). https://doi.org/10.1175/1520-0477(1997)0782539:GPAYMA2.0.CO;2

Xie, P., Chen, M., Shi, W.: CPC unified gauge-based analysis of global daily precipitation. In: Proc. 24th Conf. on Hydrology, Atlanta, Amer. Meteor. Soc, 2.3A. Available online at https://ams.confex.com/ams/90annual/techprogram/paper_163676.htm (2010)

Xue, Z., Gebremichael, M., Ahmad, R., Weldu, M.L., Bagtzoglou, A.C.: Impact of temperature and precipitation on propagation of intestinal schistosomiasis in an irrigated region in Ethiopia: suitability of satellite datasets. Trop. Med. Int. Health. **16**, 1104–1111 (2011). https://doi.org/10.1111/j.1365-3156.2011.02820.x

Yang, M., Wang, G., Wu, S., Block, P., Lazin, R., Alexander, S., Lala, J., Haider, M.R., Dokou, Z., Atsbeha, E.A., Koukoula, M., Shen, X., Peña, M., Nikolopoulos, E., Bagtzoglou, A., Anagnostou, E.: Seasonal prediction of crop yields in Ethiopia using an analog approach. Agric. For. Meteorol. **331**, 109347 (2023). https://doi.org/10.1016/j.agrformet.2023.109347

Yu, W., Doutriaux, M., Sèze, G., Le Treut, H., Desbois, M.: A methodology study of the validation of clouds in GCMs using ISCCP satellite observations. Clim. Dyn. **12**, 389–401 (1996). https://doi.org/10.1007/BF00211685

Zaitchik, B.F., Feingold, B., Valle, D., Pan, W.: Integrating earth observations to support malaria risk monitoring in the Amazon. IEEE Earthzine. Available at https://earthzine.org/integrating-earth-observations-to-support-malaria-risk-monitoring-in-the-amazon/ (2014)

Zhang, Y., Reed, S., Kitzmiller, D.: Effects of retrospective gauge-based readjustment of multisensor precipitation estimates on hydrologic simulations. J. Hydrometeorol. **12**, 429–443 (2011). https://doi.org/10.1175/2010JHM1200.1

Zhang, G., Li, X., Perrie, W., Zhang, B., Wang, L.: Rain effects on the hurricane observations over the ocean by C-band Synthetic Aperture Radar. J. Geophys. Res. Oceans. **121**(1), 14–26 (2016). https://doi.org/10.1002/2015JC011044

Zinszer, K., Kigozi, R., Charland, K., Dorsey, G., Brewer, T.F., Brownstein, J.S., Kamya, M.R., Buckeridge, D.L.: Forecasting malaria in a highly endemic country using environmental and clinical predictors. Malar. J. **14**, 245 (2015). https://doi.org/10.1186/s12936-015-0758-4

The manufacturer's authorised representative in the EU is Springer Nature Customer Service Centre GmbH, Europaplatz 3, 69115 Heidelberg, Germany. If you have any concerns regarding our products, please contact ProductSafety@springernature.com

Printed and bound by CPI Group (UK) Ltd, Croydon, CR0 4YY
30/09/2026
02225567-0016